Introduction to Abstract Mathematics

Second Edition

John F. Lucas
University of Wisconsin at Oshkosh

Ardsley House, PUBLISHERS, INC.
New York

Address orders and editorial
correspondence to:
Ardsley House, Publishers, Inc.
320 Central Park West
New York, NY 10025

ISBN: 0-912675-73-X

Printed in the United States of America

10 9 8 7 6 5 4 3

To my Daughters,

      Christine and Kathryn

whom it was my joy to have also as my students of mathematics and to

      Shirley

with love and admiration

Preface

This is a book about mathematics and mathematical thinking. It is intended for the serious learner who is interested in studying some deductive strategies in the context of a variety of elementary mathematical situations. No background beyond single-variable calculus is presumed.

The first edition, published in 1986, was written for a transition course between the standard first-semester calculus and the more theory-oriented, upper-level courses, such as abstract or linear algebra, analysis, or topology. The main topical structure, humanistic emphasis, and problem-solving philosophy of the first edition still remain. These features are even more germane now in view of current thinking of mathematicians with respect to calculus reform, problem solving, and discrete mathematics. Thus, this book continues to serve as a useful transition mechanism between introductory and more advanced mathematics, while reinforcing certain ideas in calculus through problem solving and emphasizing mathematical reasoning and proof in the context of topics from discrete mathematics. The principal topics treated herein are those that characterize elementary foundations of mathematics — mathematical logic, proof strategies, informal set theory and cardinality, binary relations, and functions. This second edition includes several changes designed to enhance and deepen the treatment of some topics as well as to provide the learner with a respectable introductory course in discrete foundations. New topics in the second edition include an extension of mathematical induction to strong induction and other variations of induction, a deeper treatment of cardinality with some applications of the Schröder-Bernstein theorem, and a comprehensive new section on graph theory including directed and undirected graphs, modeling binary relations using digraphs, Euler circuits, and application of proof techniques to ideas of elementary graph theory. Additionally, diagonalization proofs of the countability of the rational numbers and the uncountability of the real numbers appear in Section 3.2. Another significant change is the inclusion, at the back of the text, of answers and hints to selected assigned problems. Finally, an instructor's manual containing solutions to problems not solved in the answers/hints section, together with a set of sample tests, has been written to accompany this text.

The set of topics in the second edition comprises about two-thirds of those topics identified by the MAA Panel on Discrete Mathematics at the annual joint meeting of AMS/MAA (1985) as an ideal framework for a two-semester university course in discrete mathematics. It must be noted, however, that this book is focused on mathematical reasoning, heuristic problem solving and deductive proof strategies; it is not aimed at algorithmic thinking. Therefore, it is not a compendium of loosely-related topics in discrete mathematics, as many texts in that area are, but rather a cohesive treatment of foundational mathematics presented in a medium of selected topics from discrete mathematics. Thus, it fits best into the mathematics curriculum at the second-semester freshman or first-semester sophomore level, after the student has had at least one course in calculus. The author was a participant in the Calculus for a New Century Colloquium in Washington, D.C. (1987), and current thinking in calculus reform seems to be that transcendental functions be included in the first course and numerical/graphical computation be emphasized in the second course. If a modern core curriculum in mathematics is to include a two- or three-semester sequence in calculus with a two-semester sequence in discrete mathematics, then this text, used for the first course in the discrete-mathematics sequence would serve multiple purposes of reinforcing basic ideas of calculus, providing a bridge to abstract mathematics, and simultaneously introducing some key concepts of discrete mathematics. An alternate approach might be to provide students with an algorithm-oriented first course in discrete mathematics followed by this course, which would serve to deepen their understanding of discrete topics while preparing them for future work in mathematics. In this manner, computer-science majors would be well-served by the first course, and mathematics majors would be well-served by both courses. In any event, this book is designed particularly to provide solid foundational work for pre-service and in-service teachers of mathematics as well as for those preparing to engage in graduate work in mathematics or related fields.

At this point, some commentary on the philosophy of this book and its problem sets is in order. Doing mathematics amounts to solving problems. Problems are the medium through which mathematical information ebbs and flows; they are the most important part of any textbook on mathematics. In this text, the problems have been very carefully selected to fit the "cognitive scheme" of each related section. Many of these problems are not exercises intended to drill a particular concept or technique. Rather, they are bona fide problems that require some thought and reflection on the part of the student. As such, they form an integral part of the course, and the instructor should plan to allow time in each class period for some give-and-take discussion of the problems. Students should be encouraged to attempt all the problems. It should be noted, however, that subsequent sections do not generally depend on specific ideas or concepts arising from earlier problems. (In the handful of situations where that does occur, it is referenced as such in the text.) A missed problem, therefore, will generally have the consequence of lessened practice and reflection on a thinking skill, rather than lack of reinforcement in or preparation for a particular mathematical concept.

The eminent mathematician George Polya separated the problems of mathematics into two major categories — problems to *find* and problems to *prove*. This book aims squarely at the latter, yet incorporates the heuristics of general problem solving to accomplish the goal. Through copious examples, the reader is implicitly encouraged to use diagrams, schematics, and representational models to assist thinking, to reason by analogy to similar situations, to develop a multiplicity of perspectives in viewing the facets of a mathematical idea, to work backwards — creating subgoals leading from desired status to given status, to distinguish between pattern search and mathematical induction as well as between conjecture and verified generalization, to search for counterexamples, to exploit symmetry whenever possible, and to learn to ask those most important key questions: "Why?" and "What if . . . ?"

Every field of endeavor has at least two internal dimensions — information and information management. In mathematics, the informational aspect includes definitions, axioms, theorems, basic ideas, and algorithmic techniques. These things characterize *what* we learn. Equally important, if not more so, however, are those reasoning skills that allow us to shape and mold mathematical information so as to make it work for us, either in application or in theory extension. Such things characterize *how* we manage the information we have learned. To access those information-management strategies that involve deduction, we can study mathematics under a microscope in a rigorous, axiomatic, and formal manner, hoping to distill the essential abstract qualities of that reasoning or we can study the reasoning itself and apply it to a variety of mathematical problems for reinforcement. The approach taken here is in the latter vein; it is semiformal, somewhat macroscopic, and intuitive. The book is written for students to read; the style is frequently conversational; both the content and style of instruction concentrate on *thinking mathematically*. The reader is encouraged to reflect on his or her own thinking in the context of a framework of reasoning generally accepted by mathematicians; to compare, contrast, sift, and winnow those patterns of reasoning; and to learn to carry on a mathematical conversation with himself or herself by asking strategy-oriented internalized questions. By studying this text, one should expect to develop skill in planning, formulating, communicating, executing, analyzing, and clarifying one's own arguments and proofs, as well as understanding the deductive thinking of others.

The preparation of this Second Edition has been assisted by various people who deserve my gratitude. My students, especially those who are excited by the beauty and discovery of mathematics, are a constant source of inspiration to me. Several of my colleagues, Norman Frisch, Robert Prielipp, and C. Patrick Collier, taught from this book over the past several years and made helpful suggestions and comments. It was Bob Prielipp's keen perception that caught various misprints and errors not discovered in the First Edition. Special thanks also are due to my friend and colleague, Hosien S. Moghadam, who edited the new section on graph theory and provided many excellent suggestions. My daughter, Christine, a high-school mathematics teacher, tested some of the logic and proof strategies with her high-school students, and provided me with insights from a precollege perspective. I am also grateful to Professor Martin Bunder of The University of Wollongong (New South Wales) for taking the time to make a number of helpful suggestions. Thoughtful comments from mathematics users throughout the United States and abroad were much appreciated. I am particularly thankful that Martin Zuckerman provided careful editing for this Second Edition and friendly suggestions to its completion. Assistance with typing and word processing was provided by our department secretary, Mary Gutsmeidl.

Finally, I thank my wife and best friend, Shirley, for her gifts of love and vision. She is my greatest source of inspiration.

Contents

1

Elementary Logic

1.1   Sentences and Symbols

Symbolic logic has been described as the analytic theory of the art of reasoning. In this course, our major concern will be the *process of reasoning* in mathematics. Therefore, we begin by studying the language by which mathematical actions can be described—logic. Our immediate goals will be to systematize and codify the language structure, and eventually the principles of valid reasoning; our long-range goals will be to understand the reasoning, strategies, and structure of mathematical proof and to be able to construct our own proofs of mathematical statements. Our study of logic will not take place in a vacuum; we shall use it quite frequently throughout the course to clarify difficult points and to judge the validity of mathematical arguments.

Linguistic structure consists of basic units that are combined by various rules of grammar to form more complex structures. For example, letters can be combined to form words; words to form phrases, clauses, sentences; sentences to form paragraphs; paragraphs to form theories, and the like. There are certain sets of rules that govern the formation and interpretation of language.

- *Syntactics* are rules to determine if a sequence of words belongs to the language.

- *Semantics* are rules to establish the meaning of a statement.

- *Pragmatics* are rules to discern the relationship between the language and the user.

For example, consider the three statements below and note the different perspectives.

Syntactics: The book is good.   (Grammar: The sentence makes sense.)

Semantics: The book is good.   (Meaning: The sentence is true.)

Pragmatics: The book is good.   (Opinion: This is what *I* think.)

In our discussion of logic we shall be interested in declarative sentences that can be combined under certain rules to form other declarative sentences. Our primary concern will be in the *form* of sentences, not in their *content* (meaning or opinion). Thus, we shall strip away the semantics and pragmatics and examine the syntactical structure of a language whose units are sentences. This study is called *sentential logic* (also sentential calculus, logic of propositions, and symbolic logic of sentences). Since we are not concerned with meanings (only forms), we have no need for words, and we will introduce symbols to represent sentences. As an initial convention, we will use capital letters $P, Q, R, \ldots$ as symbols for simple sentences. For a while, this will probably appear rather sterile and not useful (playing around with symbols), but there is a very specific reason for this. By concentrating only on symbols and rules involving sentence forms, we have the opportunity to see objectively how our language and reasoning works without having bias introduced through numerous meanings and interpretations that vary across individuals. If we can come to agreement now on structural characteristics, then later, when we are dealing with English sentences and mathematical arguments, we can always translate into our structural language (logic). In this way, if difficulties arise, we can see whether the source of difficulty is due to our reasoning or our interpretation.

In various treatments of logic, sentences are also referred to as *propositions* or *formulas*. We shall use the term *sentence* to mean any simple declarative statement. From simple (also called *atomic*) sentences, we will build compound (also called *molecular*) sentences by using parentheses and connectives. There are five basic connectives in symbolic logic. They are:

not	symbolized by $\sim$
or	symbolized by $\vee$
and	symbolized by $\wedge$
if..., then...	symbolized by $\rightarrow$
if and only if	symbolized by $\leftrightarrow$

Using one or more of these five connectives and sets of parentheses (brackets, braces, etc.), we can generate compound sentences from the building blocks (simple sentences). For example, if a simple English sentence is

Joe is here.

then we can form a compound from it using the connective "not" by

not (Joe is here.)

which, in standard English grammar, we state

Joe is not here.

To symbolize this, suppose the letter "*P*" is used to represent the statement "Joe is here." Then "~ *P*" would represent the statement "Joe is not here."

It turns out that most of our declarative discourse breaks down to sentences made up of combinations involving these five connectives (not, or, and, if ... then, if and only if). As another example, consider the mathematical sentence

$$x < 3 \quad \text{or} \quad x = 3$$

This can be viewed as two simple sentences ($x < 3$; $x = 3$) connected by the word "or." If *P* represents the first sentence and *Q* represents the second sentence, we symbolize the compound by

$$P \lor Q$$

Each of the compound sentences that arises by using one of the five connectives has a particular name. The names, definitions, and symbols follow.

1. *Negation:* If *P* is a sentence, then the sentence "*not-P*" is called the *negation* of *P*, symbolized by

   $$\sim P$$

2. *Disjunction:* If *P*, *Q* are sentences, then the sentence "*P* or *Q*" is called the *disjunction* of *P*, *Q*, symbolized by

   $$P \lor Q$$

3. *Conjunction:* If *P*, *Q* are sentences, then the sentence "*P* and *Q*" is called the *conjunction* of *P*, *Q*, symbolized by

   $$P \land Q$$

4. *Conditional:* If *P*, *Q* are sentences, then the sentence "If *P*, then *Q*" is called the *conditional* of *P*, *Q*, symbolized by

   $$P \to Q$$

   In a conditional sentence, the simple sentence in the "if" part is called the *antecedent* (or hypothesis, or condition), and the simple sentence in the "then" part is called the *consequent* (or conclusion).

5. *Biconditional:* If *P*, *Q* are sentences, then the sentence "*P* if and only if *Q*" is called the *biconditional* of *P*, *Q*, symbolized by

   $$P \leftrightarrow Q$$

The following table summarizes these definitions.

Simple	*Compound*	*Symbol*	*Name*
P	not *P*	$\sim P$	negation
P, *Q*	*P* or *Q*	$P \lor Q$	disjunction
P, *Q*	*P* and *Q*	$P \land Q$	conjunction
P, *Q*	If *P*, then *Q*	$P \to Q$	conditional
P, *Q*	*P* if and only if *Q*	$P \leftrightarrow Q$	biconditional

Example 1   English Compounds

Negation:	It is not cloudy outside.
Disjunction:	This sentence is atomic or it is molecular.
Conjunction:	Jill is angry and Sue is regretful.
Conditional:	If this is December, then Christmas will come soon.
Biconditional:	I will go home if and only if it starts snowing.

Example 2   Mathematical Compounds

Negation:	$2 \neq 3$
Disjunction:	$x < 0$ or $x > 5$
Conjunction:	$-3 < x < 3$ $(-3 < x$ and $x < 3)$
Conditional:	If $x + 3 = 7$, then $x = 4$.
Biconditional:	f is continuous at a if and only if $\lim_{x \to a} f(x) = f(a)$.

In mathematics, most definitions can be interpreted as biconditionals.

Parentheses ( ), *brackets* [ ], and *braces* { } are used in logic to enclose compound sentences so that they play the same role as that of simple sentences in relation to other sentences and connectives that may be joined to them. Every sentence, regardless of how complex it may appear, has a major connective (i.e., the entire sentence may be classified as exactly one of the five types). It is customary to not place parentheses around negations. For example, in the sentence

$$[(P \to Q) \wedge (\sim R \vee S)] \to T$$

P, Q, R, S, and T represent simple sentences, but the grouping $(P \to Q)$ —a conditional itself—acts like a simple sentence in relation to the grouping $(\sim R \vee S)$, which is a disjunction having one of its components a negation. The bracketed [ ] sentence is a conjunction, but it acts as an antecedent in the main sentence, which is a conditional. The major connective of the entire sentence is $(\to)$.

Sometimes it is helpful to place small numbers above connective symbols so that we can establish an order of importance. Later, this will also be helpful in deciding the truth value of complicated sentences. For the sentence above, this ordering might be

$$\overset{\text{①}}{[(P} \overset{\text{④}}{\to} \overset{\text{②}}{Q) \wedge} (\overset{\text{③}}{\sim R} \overset{\text{⑤}}{\vee S)] \to T}$$

In such an ordering of connectives, the last numbered connective is the sentence's major connective. Connectives can be ordered in terms of "strength," in which cases parentheses can sometimes be eliminated. The biconditional ($\leftrightarrow$) is regarded as the strongest of all connectives, the conditional ($\to$) is stronger than the remaining three, the disjunction ($\vee$) and conjunction ($\wedge$) are about the same strength, and the negation ($\sim$) is weakest. It is best to use parentheses, however, for the sake of clarity.

weakest

$\leftrightarrow$

$\to$

$\wedge, \vee$

$\sim$

Strongest

As far as translation of the above sentence is concerned, consider the following:

P: x is an integer

Q: $x^2 - 8x + 15 = 0$

R: $x = 3$

S: $x > 4$

T: $x = 5$

Using these mathematical statements in place of the sentence symbols, the sentence

$$[(P \to Q) \wedge (\sim R \vee S)] \to T$$

would read:

If (when x is an integer, then $x^2 - 8x + 15 = 0$) and

($x \neq 3$ or $x > 4$)

then $x = 5$.

There are certain words in English that play similar roles in English grammar. For example, *but* plays a role similar to *and*. Therefore, in our logic translation we shall translate *but* using the symbol $\wedge$. Since English is quite complicated, it would take much time and effort to discuss every possible grammatical structure and symbolize it in logic. Thus, for the sake of simplicity and in view of our mathematical objectives, we will consider the five obvious compounds in which the words *not, or, and, if then*, and *if and only if* actually appear as well as some situations that occasionally arise in mathematical discussion. The latter, together with their symbolic translation, follow.

English Sentence	Symbolic Translation
P or *Q* or both	$P \lor Q$
P or *Q* but not both	$(P \lor Q) \land \sim (P \land Q)$ also $\sim (P \leftrightarrow Q)$
P is not true	$\sim P$
P implies *Q*	$P \rightarrow Q$
if *P*, then *Q*	$P \rightarrow Q$
P is a sufficient condition for *Q*	$P \rightarrow Q$
Q if *P*	$P \rightarrow Q$
P is a necessary condition for *Q*	$Q \rightarrow P$
Q only if *P*	$Q \rightarrow P$
P is implied by *Q*	$Q \rightarrow P$
P if and only if *Q*	$P \leftrightarrow Q$

P is a necessary and
   sufficient condition
     for *Q*
	$P \leftrightarrow Q$

P is equivalent to *Q*	$P \leftrightarrow Q$
P but not *Q*	$P \land \sim Q$ also $\sim (P \rightarrow Q)$
P unless *Q*	$\sim Q \leftrightarrow P$
P when (whenever) *Q*	$Q \rightarrow P$

Assigned Problems

1. Symbolize the following sentences, indicating what simple sentence each letter symbol represents.

   (a) If it is after five, then the meeting is over.

   (b) Either my watch is wrong or we are going to be late.

   (c) If the plant cell does not have chlorophyll, then it cannot make food.

   (d) Sandstone is produced by layers of sand hardening and limestone is produced by shells of small animals.

   (e) If the tribe was nomadic, then it did not build permanent shelters.

2. Symbolize the following sentences, using the given symbols for the simple sentences.

Let P = "Pat arrives early."

Q = "Quinn stays late."

R = "Robert is happy." $(P \cap Q) \to R$

(a) If Pat arrives early and Quinn stays late, then Robert is happy.

(b) If either Pat arrives early or Quinn stays late, then Robert is happy. $(P \cup Q) \to R$

(c) If Pat arrives early and Quinn does not stay late, then Robert is not happy. $(P \cap \sim Q) \to \sim R$

(d) If Robert is happy, then Quinn stays late. $(R \to Q)$

(e) Robert is happy, Pat arrives early, and Quinn stays late. $(R \wedge P \wedge Q)$

(f) If Robert is not happy, then Pat does not arrive early. $\sim R \to \sim P$

(g) (Quinn does not stay late or Pat arrives early) if and only if Robert is happy. $(\sim Q \cup P) \longleftrightarrow R$

(h) Robert is happy and either Pat arrives early or Quinn stays late. $R \wedge (P \cup Q)$

(i) Quinn stays late and if Pat arrives early, then Robert is happy. $Q \cap (P \to R)$

(j) It is not the case that both Pat arrives early and Quinn stays late. $\sim (P \cap Q)$

(k) If Pat does not arrive early and Quinn stays late, then it is not the case that Robert is not happy. $\sim P \cap Q \to R$

(l) Pat arrives early if and only if Quinn stays late, but it is not the case that if Quinn does not stay late then Robert is happy. $P \longleftrightarrow Q \wedge \sim (\sim Q \to R)$

3. Complete the translation of the following compound sentences into logical symbols by replacing the words for the connectives by the proper symbol.

(a) If P, then Q. $P \to Q$

(b) Either P or Q but not both P and Q. $(P \cup Q) \wedge \sim (P \cap Q)$

(c) If either P or Q, then not R. $(P \cup Q) \cap \sim R$

(d) Either not P or not Q. $\sim P \cup \sim Q$

(e) Either P and Q or R and S. $(P \cap Q) \cup (R \cap S)$

(f) It is not the case that both P and Q. $\sim (P \cap Q)$

(g) It is not the case that either P or Q. $\sim (P \cup Q)$

(h) If not P, then not Q and R. $\sim P \to (\sim Q \cap R)$

(i) It is not the case that if P, then Q.

$$\sim (P \to Q)$$

$\sim (P \cap \sim P)$

(j) It is not the case that both P and not P.

(k) P and either Q or R. $P \cap (Q \cup R)$

(l) Either P and Q or R. $(P \cap Q) \cup R$

(m) P and if Q, then not R. $P \cap (Q \to \sim R)$

(n) R if and only if not Q.

$R \leftrightarrow \sim Q$

4. Match each of the words on the left with the examples or definitions in the list at the right.

(a) disjunction

(b) negation

(c) conditional sentence

(d) compound sentence

(e) antecedent

(f) consequent

(g) conjunction

(h) simple sentence

(1) $P \to Q$

(2) $\sim (P \wedge Q)$

(3) $P \vee Q$

(4) Q in the sentence $P \to Q$

(5) $\sim P$

(6) P in the sentence $P \to Q$

(7) $P \wedge Q$

(8) $\sim P \vee \sim Q$

(9) any sentence with a connective

(10) any sentence without a connective

5. Put parentheses in each sentence to generate the kind of compound sentence indicated at the left.

(a) conjunction $P \vee Q \wedge R$ (g) conditional $\sim P \to \sim R$

(b) negation $\sim P \wedge Q$ (h) disjunction $P \vee Q \wedge R$

(c) conjunction $\sim P \wedge Q$ (i) negation $\sim P \to Q$

(d) conditional $P \wedge Q \to R$ (j) conjunction $P \wedge Q \to R$

(e) negation $\sim P \wedge \sim R$ (k) biconditional $P \leftrightarrow Q \to \sim R$

(f) disjunction $P \to Q \vee R$

6. Number each connective (reading generally left to right) in order of importance in each of the following compound sentences. State the major connective in each case (i.e., identify the type of sentence).

(a) $\sim \{[(P \leftrightarrow Q) \leftrightarrow (\sim R \wedge S)] \vee \sim T\}$

(b) $[\sim (R \vee S) \wedge \sim (P \vee Q)] \rightarrow (P \rightarrow \sim Q)$

(c) $[(P \rightarrow Q) \wedge (Q \rightarrow P)] \leftrightarrow (P \leftrightarrow Q)$

(d) $(P \rightarrow Q) \wedge [(Q \rightarrow R) \vee (R \rightarrow S)]$

(e) $\{[P \vee \sim (R \wedge S)] \rightarrow (Q \vee T)\} \vee [\sim (P \wedge \sim S)]$

1.2 Truth Value

Symbolic logic deals with sentences. We can reduce compound sentences to simple components, and we can build up compound sentences from simple components using any combination of five fundamental connectives ($\sim$, $\vee$, $\wedge$, $\rightarrow$, $\leftrightarrow$) to produce sentences we call negation, disjunction, conjunction, conditional, and biconditional, respectively. When we analyze or synthesize sentences, we are operating on sentences. We could view the situation as though we have some huge system consisting of a set of elements (the elements being sentences) and various operations (ways of producing sentences from one or two given sentences) on the set.

Next we shall assign to each sentence in our system exactly one of two values—true (symbolized by "T"), or false (symbolized by "F"). This assignment can be regarded as a function that associates every sentence in our system with exactly one value in the two-element set $\{T, F\}$. Every sentence in our system has a truth value; no sentence in our system has more than one truth value. We call such a function a truth-value assignment. Since our system deals with only two values, it is sometimes referred to as "two-valued" logic.

Translating the above remarks to the system of English declarative sentences, what we are saying is that our discussion will be restricted to only those sentences that we can judge "true" or "false." Following are some examples.

1. Five is an odd number.

2. The square root of 13 is rational.

3. There are infinitely many grains of sand on earth.

4. The first name of (the Democratic) President Roosevelt was "Franklin."

In English discourse, not all sentences can be judged true or false; for example,

1. Come here.   (imperative)

2. Is the universe finite?   (interrogative)

3. If only I had seen that car!   (exclamatory)

4. He is President of the United States.   (declarative)

5. Mahatma Gandhi was probably a pantheist.   (declarative)

Note that the declarative sentence (4) cannot be judged true or false as it stands until some replacement is made for the pronoun "he." Also, statements of uncertainty (e.g., sentence (5)) are not included in our formulation of logic.

It is interesting to note that not all logics (systems of reasoning) are two-valued. A four-valued logic might include true, false, both true and false, and neither true nor false. For example, sentence (4) might be regarded as neither true nor false. Certain paradoxes appear to be both true and false simultaneously. Consider, for example, the famous "Russell paradox," named after the English philosopher and logician Bertrand Russell (1872–1970):

The Russell Paradox (stated in terms of sets)

Let $A = \{$all sets that do not contain themselves as elements$\}$.

Consider the following two sentences:

If $A \in A$, then $A \notin A$.

If $A \notin A$, then $A \in A$.

This paradox once raised quite a stir in the mathematical world because it introduced an apparent inconsistency in the foundations of set theory. We "skirt" the paradox by asserting that there is no set like A, or by asserting that A is not a well-defined set (see Chapter 3).

Sometimes the way we reason in daily life is not strictly dichotomous—there are often many "shades of gray." For example, there may be many defendants in a courtroom, each partially responsible for environmental pollution of a certain area, and the judgment might take the form of 10% negligence, 30% negligence, 50% negligence, etc. However, for most mathematical situations, two-valued logic is the most effective system of representation, and it is the one we shall use here.

In this section, our objective will be to come to agreement on what truth value each type of compound sentence should have for the various possible truth value combinations of its simple components. Then, using these agreements, we shall see how to judge *any* sentence (no matter how complex) if we have a truth-value assignment for each of its parts.

Example 1   Consider the logic sentence

$$P \to Q$$

If P is assigned *true* and Q is assigned *false*, what truth value assignment should the sentence $P \to Q$ have? To guide us in answering this question, we might go to our own language for an example. Suppose we say

If Karen is hungry, then Karen has not eaten for four hours.

Additionally, suppose we *know* that it is true that Karen is hungry, but we also know that she had eaten only two hours ago. What truth value assignment (T or F) would you agree the whole conditional sentence should have? Can we all come to the same agreement on that truth value? If so, then we have the beginnings of a "logic"—a system of reasoning and deciding with agreed-upon rules.

A convenient mechanism for organizing our truth value decisions is the *truth table*. We use a truth table for reference as a decision procedure to tell us the truth value of a compound sentence for any assignment of the truth values of its component parts. Following are truth tables for the five fundamental compound sentences.

Negation Table

P	$\sim P$
T	F
F	T

This says the negation of a sentence has the "opposite" truth value as the original sentence.

For example, consider the following two sentences:

S_1 $\sqrt{5}$ is rational. (F)

S_2 $\sqrt{5}$ is not rational. (T)

Disjunction Table

P	Q	$P \vee Q$
T	T	T
T	F	T
F	T	T
F	F	F

The disjunction table asserts that a disjunction of two sentences is true if and only if *at least one* of the sentences is true. This also means the only situation in which a disjunction is false occurs when *both* constituent sentences are false. For example, consider the following sentences.

S_1 Grass is green or Einstein was a physicist. (T, T; T)

S_2 $2 + 3 < 7$ or $5 = 3$. (T, F; T)

S_3 Farmers dislike rain or $\varnothing \subseteq \varnothing$. (F, T; T)

S_4 29 is even or 2^3 is odd. (F, F; F)

Conjunction Table

P	Q	$P \wedge Q$
T	T	T
T	F	F
F	T	F
F	F	F

The conjunction table asserts that a conjunction of two sentences is true if and only if *both* parts are true. This can also be stated as "A conjunction of two sentences is false if and only if *at least one* of the sentences is false." Note the duality between the disjunction and conjunction tables. The following sentences illustrate truth value combinations for conjunctions.

S_1 π is irrational and the square root function is increasing.   (T, T; T)

S_2 Chicago is a large city but* Oshkosh is larger.   (T, F; F)

S_3 $|-5| < |-3|$ and the Civil War ended in 1865.   (F; T; F)

S_4 9 is prime and 47 is composite.   (F, F; F)

Next we examine the truth values for conditional sentences. These are particularly important for at least two reasons: (1) most theorems in mathematics are stated in conditional form and (2) truth values for conditionals tend to confuse some people, and as a consequence, subtle errors in logical reasoning may arise. Let us study some examples first, then construct the truth table. Suppose a salesman makes the following statement to his family:

"If I get the raise, I shall take the family on vacation."

This sentence is a conditional having the form $P \rightarrow Q$. Now we shall make various combinations of truth assignment to P and Q, and try to decide what truth value the entire sentence should have with respect to each combination. The final judgment shall rest upon whether the salesman was "telling the truth" or "lying" when he made the assertion. Suppose he gets the raise (P true), and he takes the family on vacation (Q true). This shouldn't be a very difficult decision —it appears he was telling the truth ($P \rightarrow Q$ true). Now what if he gets the raise (P true), but he does *not* take the family on vacation (Q false). Again the decision should not be too difficult—he did not tell the truth ($P \rightarrow Q$ false). Finally, suppose he is unfortunate and he does not get the raise (P false). How are we to decide? Can we call him a liar whether or not he takes the family on vacation? The cases where P is false (he does not get the raise) and Q is true (he takes the family on vacation) or Q is false (he does not take the family on

---

*Translate *but* as *and*.

vacation) are sometimes difficult to decide. But in this example, we must remember the form of the salesman's assertion—he said "*If* I get the raise..."; he did not say what would happen if he did *not* get the raise. Therefore, it seems reasonable to decide that the original sentence is true (and give him the benefit of the doubt) as opposed to false when the antecedent is false. We summarize these observations in the truth table for conditionals.

Conditional Table

P	Q	$P \rightarrow Q$
T	T	T
T	F	F
F	T	T
F	F	T

In other words, a conditional sentence is true if and only if the consequent is true or the antecedent is false. The only case in which a conditional is false occurs when the antecedent is true and the consequent false.

The last of the basic compounds—the biconditional ($P \leftrightarrow Q$)—is easy to decide if you think of it as a conjunction of two conditionals [$(P \rightarrow Q) \wedge (Q \rightarrow P)$]. This will be verified later when we consider logically equivalent sentences.

Biconditional Table

P	Q	$P \leftrightarrow Q$
T	T	T
T	F	F
F	T	F
F	F	T

For example, consider the following sentences:

S_1 This is Monday. $\leftrightarrow$ It is the second day of the week. (T, T; T)

S_2 This is Monday. $\leftrightarrow$ It is the fourth day of the week. (T, F; F)

S_3 This is Tuesday. $\leftrightarrow$ It is the third day of the week. (F, T; F)

S_4 This is Tuesday. $\leftrightarrow$ It is the fourth day of the week. (F, F; T)

From the table we see that a biconditional of two sentences is true if and only if both sentences have the same truth value (both true or both false), and it is false if they have different truth values.

Having decided truth values for the various combinations of truth for each of the five basic sentence forms, we can now easily extend our observations to making decisions about more complicated sentences built up from the basic forms. For example, what should the truth value of the sentence below be if W is true, A is false, and J is false?

$$[(W \land A) \to (\sim J \lor W)] \to [\sim (J \land A)]$$

This sentence has *three* sentence variables—W, A, and J. Since each sentence has exactly two truth value possibilities (true or false) then the three have $2 \cdot 2 \cdot 2$ or eight different truth value combinations.

There is a convenient, fast way to determine all truth value combinations for any number of sentence variables. Called *lexicographical ordering*, it goes like this. Suppose there are n different sentence variables. Then there will be 2^n different truth value combinations of these variables. (Why?) Now take the first sentence, place it at the top of the first column in the truth table, and underneath it place all T for the first half of the possibilities ($1/2$ of 2^n, or 2^{n-1}), and F for the second half. Next take the second variable and alternate T's and F's in the following manner—T's for the first fourth of the possibilities (2^{n-2}), F's for the next fourth, T's for the third fourth, and F's for the last fourth. Continue this way for each of the sentence variables—alternating T's and F's for half the alternation cycle in the previous table—until finally you reach the last sentence variable. Here the alternation will be one-by-one; that is, T-F-T-F, and so on to the end of the table under that last variable. For the three-variable example above, we would have as the first three tables

W	A	J	$\ldots$
T	T	T	
T	T	F	
T	F	T	
T	F	F	$\ldots$
F	T	T	
F	T	F	
F	F	T	
F	F	F	

This method of lexicographical ordering very quickly picks up all eight truth value combinations (rows in the table). The case where W is true and A and J are both false would occur in the fourth row of the table.

What we have done so far is to describe a quick, organized method for determining all truth value combinations of the sentence variables. Next we need to complete the entire truth table to reflect all the various truth values for the original complicated sentence corresponding to any of the eight truth value combinations. Our last column in the truth table would thus yield all possible truth values for the sentence in any possible situation for its component parts. We can finish the truth table in two different ways: (1) create a series of subtables—one for each compound part of the original sentence taken in order of importance, e.g., $(W \wedge A)$, $\sim J$, $\sim J \vee W$, $(W \wedge A) \rightarrow (\sim J \vee W)$, $J \wedge A$, $\sim (J \wedge A)$, and finally putting it all together:

$$[(W \wedge A) \rightarrow (\sim J \vee W)] \rightarrow [\sim (J \wedge A)]$$

or we can (2) number each connective in order of importance, and create a table under each connective, the last table being that for the entire sentence. This, in effect, condenses the method outlined in (1). For the sake of simplicity, we shall use method (2), as illustrated in the following table.

			①	④	②	③	⑦	⑥	⑤
W	A	J	$[(W \wedge A)$	$\rightarrow$	$(\sim J$	$\vee W)]$	$\rightarrow$	$[\sim$	$(J \wedge A)]$
T	T	T	T	T	F	T	F	F	T
T	T	F	T	T	T	T	T	T	F
T	F	T	F	T	F	T	T	T	F
T	F	F	F	T	T	T	T	T	F
F	T	T	F	T	F	F	F	F	T
F	T	F	F	T	T	T	T	T	F
F	F	T	F	T	F	F	T	T	F
F	F	F	F	T	T	T	T	T	F

The encircled column is the table of values for the entire sentence (which was a conditional). We see that in the fourth row, when W is true, A false, and J false, the original sentence is true for that particular combination.

Truth tables are quite helpful in logic and mathematics. In addition to laying out all the possible truth values for a sentence given any combination of truth values for its parts, truth tables are useful for establishing logical theorems and equivalences, and for judging the validity of logical argument. Further, they have application in certain counting problems using Venn diagrams and sets, and they serve to clarify combinations and outcomes in any situation where there are two possible outcomes per event (e.g., circuit on/circuit off, pipe flow open/pipe flow closed, positive charge/negative charge, etc.).

Assigned Problems

1. Decide the truth of each of the following sentences, using your knowledge of mathematics and logic. Assume that x represents any real number and f any function.

   (a) For every real number x, $x^2 \geq 0$.

   (b) If $x = 5$, then $x \leq 5$.

   (c) If $f(x) = x^7$, then $f'(x) = 7x^6$.

   (d) If $x = 0$ or $x = 1$, then $x^2 = x$.

   (e) If $x^2 = x$, then $x = 0$ or $x = 1$.

   (f) A series either converges or it diverges.

(g) If a function is continuous, then it is differentiable.

(h) $(|x| < 3) \leftrightarrow (-3 < x < 3)$

(i) $\displaystyle\lim_{n \to \infty} \frac{1}{n} = 2$, where n is a positive integer $(n = 1, 2, 3, \ldots)$.

2. Decide the truth value of each of the following, using your knowledge of mathematics and logic.

(a) $\displaystyle\lim_{n \to \infty} \frac{1}{n} = 0$ or e is rational.

(b) $\sqrt{a^2} = |a|$ and $7 \neq 5$.

(c) π is rational or π is real.

(d) Integers $\cup$ Rationals $=$ Reals or $\int \sin x \, dx = \cos x + C$.

(e) It is not the case that (if $x = 3$, then $5x = 15$).

(f) If a convex polygon has no diagonals, it is a triangle.

(g) If S is a square, then S is a rectangle.

(h) l_1 is parallel to $l_2 \leftrightarrow l_1 \cap l_2 = \emptyset$ (assume that l_1 and l_2 lie in the same plane).

(i) π is real $\leftrightarrow \pi$ is irrational.

3. Construct a truth table for the following sentence:
$$(\sim P \vee Q) \to (P \to Q)$$
Do you notice any patterns? How can you interpret this?

4. Construct a truth table for the following sentence:
$$(P \to Q) \wedge (P \wedge \sim Q)$$
Any patterns? Your interpretation?

5. Construct a truth table for the sentence
$$[(P \wedge Q) \vee \sim R] \to (R \vee \sim Q)$$
How is this sentence different from those in Problems 3 and 4?

6. Construct truth tables for each of the following.

(a) $(P \to Q) \wedge \sim R$

(b) $(P \to Q) \leftrightarrow (\sim P \vee Q)$

(c) $[P \wedge (P \to Q)] \to Q$

(d) $[Q \wedge (P \to Q)] \to Q$

(e) $\{[(A \to B) \wedge (C \to D)] \wedge (A \vee C)\} \to (B \vee D)$

(f) $S \to (T \vee \sim R)$

(g) $\sim [(A \wedge B) \to A]$

7. Suppose

N = "New York is larger (in population) than Chicago." $\quad T$

W = "New York is north of Washington, D.C." $\quad T$

C = "Chicago is larger (in population) than New York." $\quad F$

Indicate which of the following sentences are true, and which are false.

(a) $N \vee C \quad\quad T \vee F$

(b) $N \leftrightarrow (\sim W \vee C)$

(c) $(W \vee \sim C) \to N$

(d) $(W \leftrightarrow \sim N) \leftrightarrow (N \leftrightarrow C)$

(e) $(W \to N) \to [(N \to \sim C) \to (\sim C \to W)]$

8. Let "$a = 0$" and "$a = b$" be true sentences and let "$b = c$" and "$b = d$" be false sentences. Find the truth values of each of the following:

(a) If $a = 0$ and $a = b$, then $b \neq c$. $\quad T$

(b) If $a \neq 0$ and $b = d$, then $b = c$. $\quad F \cap F$

(c) If $a \neq b$ or $b \neq c$, then $b = d$. $\quad\quad F \to F \quad T$

(d) If $a = 0$, then $a \neq b$ and $b \neq d$.

(e) If $a \neq 0$ or $b \neq c$, then $a \neq b$.

1.3   *Tautology and Equivalence*

There are certain sentences that are *always true* regardless of the truth value combinations of constituent parts. Such sentences are called *tautologies*. These sentences exhibit truth because of *form* and not because of meaning. As an example, one of the simplest of all tautologies is the statement

$P \vee \sim P$

Regardless of whether P is true or P is false, the disjunction $P \vee \sim P$ is always true. This particular sentence is known in logic as the *law of the excluded middle*. There are many other examples of tautologies, and they play an extremely important role in the development of logic and mathematical reasoning. We shall list in this section those tautologies that are most important and well known; in fact, we shall provide word names for each of them to assist in memorizing them.

Since a tautology is a sentence that is true for all possible truth value combinations of its variables, we can determine whether a given sentence is a tautology by constructing a truth table and looking for all "T" under the major connective.

Another sentence that occurs in logic is the negation of a tautology, or *contradiction*. A contradiction is any sentence that is *false* for all possible truth value combinations of its parts; it is false because of *form*, not meaning. The simplest example of a contradiction is the form

$$P \wedge \sim P$$

which says, in effect, a statement and its negation cannot hold simultaneously. The truth table for a contradiction exhibits all "F" under its main connective. One could feasibly construct a list of contradictions by negating every tautology in the list of tautologies, but contradictions do not play as important a role in the development of logic as tautologies, so we shall not. The idea of contradiction will become important later when we examine the strategy of indirect proof (sometimes called "proof by contradiction").

Sentences that are neither tautologies nor contradictions—whose truth tables exhibit some T and some F entries—are called *indeterminate*.

Example 1   (a) $(P \rightarrow Q) \leftrightarrow (\sim P \vee Q)$ is a *tautology*.

(b) $[(P \vee Q) \wedge \sim P] \wedge (\sim Q)$ is a *contradiction*.

(c) $(P \vee Q) \rightarrow P$ is *indeterminate*.

Some of the most important tautologies are given in the following list. It will be helpful to read carefully over the list and see if the forms "make sense." That is, regardless of what sentence replacements you might make, do you develop a sense of feeling that the forms always lead to truth? Also, it will be helpful to become familiar with these forms for later reference.

List of Tautologies

T1	$P \vee \sim P$	(excluded middle)
T2	$\sim (P \wedge \sim P)$	(noncontradiction)
T3	$[P \wedge (P \rightarrow Q)] \rightarrow Q$	(modus ponens)
T4	$[(P \rightarrow Q) \wedge (Q \rightarrow R)] \rightarrow (P \rightarrow R)$	(syllogism)
T5	$\sim (\sim P) \leftrightarrow P$	(double negation)
T6	$(P \rightarrow Q) \leftrightarrow (\sim Q \rightarrow \sim P)$	(contraposition)
T7	$[(P \rightarrow Q) \wedge (\sim Q)] \rightarrow \sim P$	(modus tollens)

List of Tautologies   (continued)

T8	$[(P \lor Q) \land \sim P] \to Q$	(disjunctive syllogism)
T9	$(P \to Q) \leftrightarrow (\sim P \lor Q)$	(conditional disjunction)
T10	$[\sim P \to (Q \land \sim Q)] \to P$	(reductio ad absurdum)
T11	$\{[(P \to R) \land (Q \to S)] \land (P \lor Q)\} \to (R \lor S)$	(dilemma)
T12	$[P \to (Q \to R)] \leftrightarrow [(P \land Q) \to R]$	(exportation)
T13	$(P \leftrightarrow Q) \leftrightarrow [(P \to Q) \land (Q \to P)]$	(biconditional)
T14	$(P \land Q) \to P$	(simplification)
T15	$P \to (P \lor Q)$	(simplification)
T16	$(P \land P) \leftrightarrow P$	(idempotence)
T17	$(P \lor P) \leftrightarrow P$	(idempotence)
T18	$\sim (P \lor Q) \leftrightarrow (\sim P \land \sim Q)$	(De Morgan's Law)
T19	$\sim (P \land Q) \leftrightarrow (\sim P \lor \sim Q)$	(De Morgan's Law)
T20	$\sim (P \to Q) \leftrightarrow (P \land \sim Q)$	(negation of conditional)
T21	$\sim (P \leftrightarrow Q) \leftrightarrow (\sim P \leftrightarrow Q)$	(negation of biconditional)
T22	$(P \lor Q) \leftrightarrow (Q \lor P)$	(commutativity $\lor$)
T23	$(P \land Q) \leftrightarrow (Q \land P)$	(commutativity $\land$)
T24	$(P \leftrightarrow Q) \leftrightarrow (Q \leftrightarrow P)$	(commutativity $\leftrightarrow$)
T25	$[(P \lor Q) \lor R] \leftrightarrow [P \lor (Q \lor R)]$	(associativity $\lor$)
T26	$[(P \land Q) \land R] \leftrightarrow [P \land (Q \land R)]$	(associativity $\land$)
T27	$[(P \leftrightarrow Q) \leftrightarrow R] \leftrightarrow [P \leftrightarrow (Q \leftrightarrow R)]$	(associativity $\leftrightarrow$)
T28	$[P \land (Q \lor R)] \leftrightarrow [(P \land Q) \lor (P \land R)]$	(distributivity $\land$ over $\lor$)
T29	$[P \lor (Q \land R)] \leftrightarrow [(P \lor Q) \land (P \lor R)]$	(distributivity $\lor$ over $\land$)
T30	$[P \to (Q \lor R)] \leftrightarrow [(P \to Q) \lor (P \to R)]$	(distributivity $\to$ over $\lor$)
T31	$[P \to (Q \land R)] \leftrightarrow [(P \to Q) \land (P \to R)]$	(distributivity $\to$ over $\land$)
T32	$[(P \lor Q) \to R] \leftrightarrow [(P \to R) \land (Q \to R)]$	(disjunction/conditional)
T33	$[(P \land Q) \to R] \leftrightarrow [(P \to R) \lor (Q \to R)]$	(conjunction/conditional)
T34	$(P \to Q) \to [(R \land P) \to (R \land Q)]$	(factorization)
T35	$(P \to Q) \to [(R \lor P) \to (R \lor Q)]$	(summation)

You may have noticed that a number of the above 35 tautologies were stated in biconditional ($\leftrightarrow$) form. Such tautologies are called *equivalences* (or *logical equivalences*). That is to say, whenever a statement in the form of a biconditional is a tautology (true in every possible case), then the statement on the left of the biconditional symbol *means the same* as the statement on the right. We also say, under such conditions, that the two statements are *logically equivalent*. For example, consider tautology T9,

$$(P \rightarrow Q) \leftrightarrow (\sim P \vee Q)$$

which says that "any conditional $P \rightarrow Q$ is logically equivalent to the disjunction $\sim P \vee Q$." If you substitute actual sentences for P and Q, you will see why they have the "same meaning." Suppose

$P =$ It is raining.

$Q =$ It is cloudy.

Then the statement

If it is raining, then it is cloudy.

is logically equivalent to (means the same as)

It is not raining or it is cloudy.

Similarly, tautology T6 says that any conditional is logically equivalent to its contrapositive; T13 states that any biconditional is logically equivalent to a conjunction of the two conditionals involved (both directions $\rightarrow$ and $\leftarrow$); T18 says to negate a disjunction all you have to do is negate both parts, change it into a conjunction and you will have an equivalent sentence, and so on. In this manner, each tautology that is a biconditional can be interpreted as a logical equivalence of two sentences. If two sentences are logically equivalent, it is reasonable to assume that we can replace either by the other in any sentence combination without affecting the meaning or the truth value of the latter. We can use this observation to generate new tautologies by modifying known ones and retaining equivalent forms.

Example 2   Consider T9: $(P \rightarrow Q) \leftrightarrow (\sim P \vee Q)$.

Since this is a tautology, both sides of the biconditional have the same truth value. Therefore, we could negate both sides and still have a tautology:

$$\sim (P \rightarrow Q) \leftrightarrow \sim (\sim P \vee Q)$$

Now the right side is a negation of a disjunction, and tautology T18 gives an equivalent form to the negation of a disjunction (negate both parts and change disjunction to conjunction). Thus we have

$$\sim (\sim P \vee Q) \leftrightarrow \sim (\sim P) \wedge \sim Q$$

But $\sim (\sim P)$ is equivalent to P (T5 double negation). Therefore, we have

$$\sim (\sim P \vee Q) \leftrightarrow P \wedge \sim Q$$

and since $\sim (P \rightarrow Q)$ is logically equivalent to $\sim (\sim P \vee Q)$, which in turn is logically equivalent to $P \wedge \sim Q$, we have

$$\sim (P \rightarrow Q) \leftrightarrow (P \wedge \sim Q)$$

as a tautology. Note that this tautology does appear in our list (T20) as one of the fundamental tautologies.

Example 3   Start with T12 (exportation):

$$[P \rightarrow (Q \rightarrow R)] \leftrightarrow [(P \wedge Q) \rightarrow R]$$

The right side is equivalent to $(Q \wedge P) \rightarrow R$ by commutativity of conjunction (T23). This, in effect, interchanges the roles of P and Q on the right side. Thus, by interchanging P and Q on the left side, and using the exportation tautology T12, we obtain a related tautology:

$$[Q \rightarrow (P \rightarrow R)] \leftrightarrow [(Q \wedge P) \rightarrow R]$$

or

$$[Q \rightarrow (P \rightarrow R)] \leftrightarrow [(P \wedge Q) \rightarrow R]$$

The last tautology does not appear in our list. So if we wanted, we could add it to our list, and use it in the future as an accepted tautology, since it was derived from known tautologies. An alternate mode of verifying that indeed it is a tautology is to construct a truth table and obtain all T's under the main connective ($\leftrightarrow$).

We have seen that if a biconditional sentence is a tautology, then the two sentences connected via the biconditional are logically equivalent. There are other sentences in our list of tautologies that are simply *conditionals* and not biconditionals. When a conditional sentence turns out to be a tautology, we call it *logical implication*, and we say the antecedent *implies* the consequent, or the consequent *is implied by* the antecedent. For example, T15 asserts that a statement logically implies the disjunction of itself with any other statement; the law of factorization (T34) says that a conditional logically implies the result of conjoining of its components with any arbitrary statement; T8 (disjunctive syllogism) says if we have a disjunction together with (and) the negation of one of its parts, then that implies the other part; T14 (and T23) state that a conjunction logically implies either of its parts, and so on. We can test a conditional to see if it is a *logical implication* by forming a truth table and thereby seeing if the conditional is a tautology.

Assigned Problems

1. Judge, by constructing truth tables, whether each of the following is a tautology, contradiction, or indeterminate.

   (a) $(A \rightarrow B) \wedge (A \wedge \sim B)$

   (b) $[(A \rightarrow B) \wedge A] \rightarrow B$

   (c) $(A \rightarrow B) \wedge [(C \wedge B) \wedge \sim (C \wedge A)]$

(d) $[\{[(A \to B) \wedge (C \to D)] \wedge (A \vee C)\} \wedge \sim B] \to D$

(e) $A \wedge [\sim (A \vee B)]$

2. Using the tautologies that deal with negating sentences (T5, T18, T19, T20, and T21), give negations of each of the following sentences. That is, find equivalent sentences that have the negation symbol ($\sim$) only on single sentence variables.

(a) $\sim P \vee Q$ (f) $A \to (B \vee C)$

(b) $A \wedge \sim B$ (g) $(A \vee B) \to C$

(c) $\sim (A \wedge \sim B)$ (h) $[(A \to B) \wedge (B \to C)] \to (A \to C)$

(d) $\sim A \to B$ (i) $[(A \to B) \wedge A] \to B$

(e) $A \to (B \wedge C)$ (j) $(A \wedge \sim B) \vee (\sim C \vee \sim D)$

3. Negate, in English, each of the following sentences in the same way you did the negations in Problem 2. (That is, nothing more than a simple sentence should be negated in the logically equivalent form.)

(a) Art is smart and Willy is silly.

(b) If Jane is plain, then Willy is not silly.

(c) It is not the case that if Art is smart, then Jane is not plain.

(d) Either Willy is silly or Jane is plain, but Art is not smart.

(e) Art is smart only if Jane is plain and Willy is silly.

4. The following theorem is important in elementary mathematics:

> If $xy = 0$, then $x = 0$ or $y = 0$.   (x, y are real numbers)

Usually, a proof will begin as follows:

> "Assume $xy = 0$ and $x \neq 0$; we will show that $y = 0$."

In other words, the prover is saying that he intends to replace the given statement by

> If $xy = 0$ and $x \neq 0$, then $y = 0$.

Can he validly do this? Why?

5. Verify, using truth tables, that the following tautologies on our list really are tautologies.

     T6   (contraposition)
     T10   (reductio ad absurdum)
     T13   (biconditional)
     T19   (De Morgan's Law)
     T32   (disjunction/conditional)

6. Negate (give logical equivalents with only simple sentences negated) the following mathematical assertions.

   (a) If $x \le y$, then $f(x) \le f(y)$.

   (b) If f is continuous, then f is differentiable.

   (c) If f is integrable, then f is continuous or f is differentiable.

   (d) The series $\sum_{n=1}^{\infty} a_n$ diverges if and only if $\lim_{n \to \infty} a_n \ne 0$.

   (e) p is a prime or p^2 is rational.

   (f) Q is a rectangle and Q is convex.

   (g) It is not the case that irrational numbers cannot be represented as quotients of integers.

7. There is a theorem in calculus that states

$$f'(x) = f(x) \text{ if and only if } f(x) = ke^x \text{ (where } k \text{ is some constant}$$
   and f is a function mapping real numbers to real numbers).

   To prove this, one might see in the textbook: $P \leftrightarrow Q \leftrightarrow \boxed{P \to Q \wedge (Q \to P)}$

   (a) ($\to$) assume f is a function and $f'(x) = f(x)$; we shall show $f(x) = ke^x$ for some constant k, and later...

   (b) ($\leftarrow$) assume $f(x) = ke^x$; we now show $f'(x) = f(x)$.

   To which tautology in our list is the theorem-prover appealing? Explain.

8. It happens to be true that the basic connectives $\sim$ and $\vee$ are "primitive" in logic. This means sentences with the other three connectives can always be reduced to sentences involving only the two connectives $\sim$ and $\vee$. For example, tautology T9 shows directly the equivalent of a conditional in terms of $\sim$, $\vee$ symbols: $\sim (P \to Q) \leftrightarrow \text{Nil} P \wedge \sim Q$

$$(P \to Q) \leftrightarrow (\sim P \vee Q)$$

   $(\sim P \vee Q) \vee True$

   What about conjunctions $(P \wedge Q)$ and biconditionals $(P \leftrightarrow Q)$? Using a series of substitutions through equivalent forms (using any of the tautologies of our list), show how the conjunction $(P \wedge Q)$ and the biconditional $(P \leftrightarrow Q)$ can be reduced to equivalent sentences involving only the symbols $\sim$ and $\vee$.

9. This problem involves reasoning much like that required in Problem 8, only somewhat more complex. There is a single symbol $|$ (called the Scheffer stroke, named after Professor Henry M. Scheffer of Harvard University), which is defined as follows:

   $P|Q$ means the same as $\sim P \vee \sim Q$

It turns out that every statement of logic could be written in terms of Scheffer strokes alone. Take each of the five fundamental statement forms ($\sim P$, $P \vee Q$, $P \wedge Q$, $P \rightarrow Q$, and $P \leftrightarrow Q$) and show how (using tautologies from our list, the definition of Scheffer stroke, and substitution of equivalent forms) each can be reduced to an equivalent statement involving only Scheffer strokes.

10. Negate, in English, the following sentence:

> If he receives a good commission, then he will make a profit
> if he invests wisely, but he will not make a profit
> if he does not invest wisely.

1.4   Conditional Forms

The *conditional* sentence is of particular importance in mathematics because many statements and theorems take the form *if* (antecedent, hypothesis, condition), *then* (consequent, conclusion). In this brief section we shall examine the nature of the conditional form and its variants. Suppose we take the conditional form

Conditional: $P \rightarrow Q$

and construct other conditionals related to it by interchanging or negating the hypothesis and the conclusion. In this manner we shall form new conditionals called the *converse*, *inverse*, and *contrapositive*.

- *Converse of $P \rightarrow Q$:* $Q \rightarrow P$

    The *converse* of a conditional is formed by interchanging the hypothesis (antecedent) and the conclusion (consequent).

- *Inverse of $P \rightarrow Q$:* $\sim P \rightarrow \sim Q$

    The *inverse* of a conditional is formed by negating both the hypothesis and the conclusion.

- *Contrapositive of $P \rightarrow Q$:* $\sim Q \rightarrow \sim P$

    The *contrapositive* of a conditional is formed by interchanging *and* negating both the hypothesis and conclusion.

We know from our study of tautologies and truth tables that any conditional is logically equivalent to its contrapositive. Recall the tautology

T6 $(P \rightarrow Q) \leftrightarrow (\sim Q \rightarrow \sim P)$ (contraposition)

This means that a contrapositive is really saying the same thing as the conditional from which it is derived and that it could be substituted for the original conditional in any sentence or argument. The following question arises: "Is a conditional logically equivalent to its converse, or its inverse?" We can test this easily by constructing truth tables and looking to see if the truth values match for each particular combination of the variables. But first let us develop a sense of feeling from English statements. Consider the conditional

> If it is midnight, then it is dark outside.

Now consider the converse of that conditional:

> If it is dark outside, then it is midnight.

Assuming they do not live within the Arctic circle, most persons would agree with the conditional, but not many (hopefully) would agree with the converse. That is, it can be dark outside and not necessarily be midnight (antecedent true, consequent false; therefore converse statement false). Therefore, it appears that the given conditional and its converse are *not* saying the same thing; that is, they are *not* logically equivalent.

Example 1   *A Mathematical Example*

Conditional:

> If $x = y$, then $x^2 = y^2$. $(x, y \in R)$

Converse:

> If $x^2 = y^2$, then $x = y$. $(x, y \in R)$

Here, the conditional is true (since squaring real numbers is a functional relationship) but the converse is false via the counterexample: let $x = -3$, $y = 3$; then

> $(-3)^2 = 3^2$, but $-3 \neq 3$

Suppose we examine the inverse and the contrapositive:

Inverse:

> If $x \neq y$, then $x^2 \neq y^2$. $(x, y \in R)$

Contrapositive:

> If $x^2 \neq y^2$, then $x \neq y$. $(x, y \in R)$

Note that the inverse is false (take $x = -3$, $y = 3$) and the contrapositive is true. This could have been predicted since a conditional is logically equivalent (has the same truth value) as its contrapositive, and the inverse is really the contrapositive of the converse.

Now go back and consider the four variants of the English conditional posed earlier:

Conditional: If it is midnight, then it is dark outside.

Converse: If it is dark outside, then it is midnight.

Inverse: If it is not midnight, then it is not dark outside.

Contrapositive: If it is not dark outside, then it is not midnight.

Is it clear that the conditional and contrapositive are saying the same thing, that the conditional and inverse are not equivalent, and that the converse and inverse do have the same meaning?

To verify the equivalences and nonequivalences above, we construct a truth table exhibiting all four conditional forms:

P	Q	$\sim P$	$\sim Q$	Conditional $P \to Q$	Converse $Q \to P$	Inverse $\sim P \to \sim Q$	Contrapositive $\sim Q \to \sim P$
T	T	F	F	T	T	T	T
T	F	F	T	F	T	T	F
F	T	T	F	T	F	F	T
F	F	T	T	T	T	T	T

logically equivalent

logically equivalent

We summarize these observations with the emphatic statements:

---

1. A conditional form is logically equivalent to its contrapositive.
2. A conditional form is *not* logically equivalent to its converse.

Example 2   One of the properties of functions we shall study later is the *uniqueness property*. This property says, in words, that if two domain elements are equal, then their images under the function must also be equal. In mathematical symbols, for each $x, y \in \text{dom } f$,

If $x = y$, then $f(x) = f(y)$.

A negation of the uniqueness property would assert that there is a pair of domain elements x and y where $x = y$ but $f(x) \neq f(y)$; that is, the preimages would be the same, but the images would be different. In pictures, this would look like

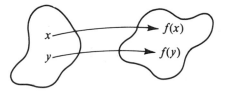

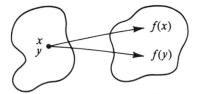

(a) Uniqueness property holding   (b) Uniqueness property failing

Now the *contrapositive* of the uniqueness property would be

If $f(x) \neq f(y)$, then $x \neq y$.

This says that if the images are unequal, then the preimages must be unequal. What would the *converse* say? There is a certain property of some functions that would hold if the converse were true. What property is it?

Example 3   The operation of addition of natural numbers can be thought of as a function associating the set of ordered pairs of natural numbers with the set of natural numbers. For example,

$$+(2,3) = 5$$
$$+(7,7) = 14$$
$$+(1,9) = 10$$

Since this is a function, the uniqueness property holds (see Example 2) and

If $(a, b) = (c, d)$, then $+(a, b) = +(c, d)$.

In standard forms of communicating, this last conditional says (using the definition of ordered pair equality; see Section 4.1)

If $a = c$ and $b = d$, then $a + b = c + d$.

Is the converse also true? That is, is the statement

If $a + b = c + d$, then $a = c$ and $b = d$.

true for all natural number replacements of a, b, c, and d? It is easy to find a counterexample, such as $a = 1$, $b = 4$, $c = 2$, and $d = 3$:

$$1 + 4 = 2 + 3 \text{ but } 1 \neq 2 \text{ or } 4 \neq 3.$$

Note the form of negation for the converse conditional—assert the antecedent, and deny the consequent (which involves negating a conjunction).

Thus, in the last two examples, we see once again the nonequivalence of a conditional and its converse.

Assigned Problems   1. State the converse, inverse, and contrapositive of each of the following conditionals (and simplify so that negations appear only on simple sentences).

(a) $(A \vee B) \rightarrow (A \wedge B)$

(b) $\sim A \rightarrow [\sim B \vee (C \rightarrow D)]$

(c) If John is Mary's brother, then Mary is John's sister.

(d) If it rains, then the grass grows.

2. State the converse, inverse, and contrapositive for each of the conditionals below, and judge the truth of the original conditional as well as its other variants (using your knowledge of mathematics). If a conditional is false, provide a counterexample to prove it is false. Assume that x and y are real numbers, and that f and g are real-valued functions.

(a) If $x = y$, then $x^3 = y^3$.

(b) If $\sqrt{x} = \sqrt{y}$, then $x = y$. $(x, y \geq 0)$

(c) If $|x| = |y|$, then $x = y$.

(d) If $x = y$, then $[\![x]\!] = [\![y]\!]$.   ($[\![x]\!]$ means "the greatest integer in x")

(e) If $f(x) = g(x)$, then $f'(x) = g'(x)$.

(f) If $f(x) = g(x)$, then $\int f(x)\,dx = \int g(x)\,dx$.

(g) If a function is a polynomial function, then the function is continuous.

(h) If $a - b = c - d$, then $a = c$ and $b = d$. $(a, b, c, d \in \text{Integers})$

(i) If $e^x = e^y$, then $x = y$.

(j) If $\sin x = \sin y$, then $x = y$.

(k) If the series $\sum\limits_{n=1}^{\infty} a_n$ converges, then $\lim\limits_{n \to \infty} a_n = 0$.

(l) If an increasing sequence is bounded above, then the sequence converges.

3. Sometimes a statement is quite difficult to prove in its original form, but a restatement via some logically equivalent form makes it quite a bit easier. Consider the statement

If $2^n - 1$ is prime, then n is prime.   (n a positive integer)

How would you prove this?

4. Can you write a false conditional where its converse is also false? Why or why not?

5. (a) Prove the statement

If a is an odd integer, then a^2 is an odd integer.

(b) Now suppose you were asked to prove the statement

If a^2 is an even integer, then a is an even integer.

Any comments?

1.5   *Quantifiers*

Phrases of the form "for all x" or "there exists at least one x such that..." are called *quantifiers*. They tell us about the generality or existence of the objects in the statement they precede. The study of quantified sentences belongs to a branch of logic called *quantificational logic* (or quantificational calculus, or predicate logic). We shall only scratch the surface of quantificational logic in this section, enough to become familiar with the symbolism, meaning, use, and negation of quantified sentences.

In most mathematical discussions, we generally have in mind some "domain of discourse" or "universe" about which we are talking and to which the variables we are using relate. For example, if we are discussing algebra, we generally use the variables x, y, and so on to refer to real numbers; in calculus, the symbols f, g, and so on usually refer to functions; in geometry, l_1, l_2 might represent lines. When we use variables that refer to objects in some set U (universe), we call such variables *object variables*. Then the statements we make about them will be symbolized by $P(x)$, $Q(y)$, $R(x, y)$, and such. The latter are called *predicates*. In this book, we assume the universe U is nonempty.

Example 1   (a) Suppose universe U is all real numbers, x is an (object) variable on U (this means that x takes replacements from the set of real numbers), and P represents the predicate "is less than 50." Then $P(x)$ would be the statement

x is less than 50.

In this case, P is called a *one-place predicate* because it is used with only one object variable (x).

(b) Using U as all real numbers and two object variables x and y on U, let $Q(x, y)$ be the predicate statement

x is less than y.

Here Q is called a *two-place predicate*.

(c) The symbol $P(x_1, x_2, \ldots x_n)$ is called an *n-place predicate* where P represents some meaningful statement relating the n object variables $x_1, x_2, \ldots, x_n$.

We shall introduce the following two symbols for quantifiers:

Universal Quantifier

$$\forall \text{ means } \begin{cases} \text{for all} & \text{everything} \\ \text{for every} & \text{every} \\ \text{for each} & \text{each} \\ \text{all} \end{cases}$$

Existential Quantifier

$$\exists \text{ means} \begin{cases} \text{there exists} \\ \text{there is at least one} \\ \text{there is} \end{cases} \begin{array}{l} \text{for one (at least one)} \\ \text{something} \\ \text{for some} \end{array}$$

As a mnemonic, note that the *universal quantifier* is really an inverted capital A (representing *all*), and the *existential quantifier* is a backwards capital E (representing *exists*).

We shall now link up quantifiers with predicates and examine the translation process. Suppose U is some universe of objects we wish to discuss, x is an (object) variable taking its replacements from U, and P is some property about x (we won't specify it here). Then some basic symbolic statements involving quantifiers and predicates would translate as follows:

Symbolic Statement	*Translation*
$\forall_x P(x)$	Everything (in U) has property P.
$\forall_x(\sim P(x))$	Nothing (in U) has property P. (For every object in U, property P does not hold.)
$\exists_x P(x)$	Something (in U) has property P. (There exists at least one object in U having property P.)
$\exists_x(\sim P(x))$	Something (in U) does not have property P. (There exists at least one object in U that does not have property P.)

Example 2   Let the universe U be the set of all positive integers.
Let $A(x)$ mean "x is a prime number," and
  $B(x)$ mean "x is an odd number."

There are four basic forms of sentences whose meanings and translations are quite important in mathematics. They are stated, symbolized, and diagrammed using the sets shown on page 31.

Form 1

All *A*'s are *B*'s.
All primes are odd.
$\forall_x(A(x) \to B(x))$
[for all positive integers *x*, if *x* is a prime, then *x* is odd].

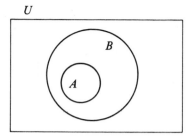

Form 2

No *A*'s are *B*'s.
No primes are odd.
$\forall_x(A(x) \to \sim B(x))$
[for all positive integers *x*, if *x* is a prime, then *x* is *not* odd].

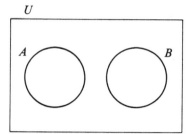

Form 3

Some *A*'s are *B*'s.
Some primes are odd.
$\exists_x(A(x) \wedge B(x))$
[there is at least one positive integer that is prime and odd].

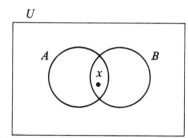

Form 4

Some *A*'s are not *B*'s:
Some primes are not odd.
$\exists_x(A(x) \wedge \sim B(x))$
[there is at least one positive integer that is prime and *not* odd].

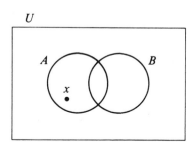

The reader is cautioned to look back over Example 2 carefully. Some common errors in logical reasoning arise by misinterpretation of these kinds of statements. For example, to some persons it may seem reasonable to translate the statement

Some primes are odd.

by

$$\exists_x (A(x) \rightarrow B(x))$$

But this logical error can be discovered if we write the conditional in the logically equivalent form $\sim A(x) \vee B(x)$ to obtain

$$\exists_x (\sim A(x) \vee B(x))$$

or

There exists at least one integer x such that x is not prime or x is odd.

The latter sentence clearly does not mean the same as "Some primes are odd."

Another error in logical reasoning occurs when some persons translate the statement "all primes are odd" into the form

$$\forall_x (A(x) \wedge B(x))$$

This latter statement really says "every integer is both prime and odd," but what we *mean* to say is *if* we have a prime, then it must be odd. Thus it is very important to clarify our meanings and be cautious when we work with quantified statements.

Frequently in mathematics we may wish to restrict the universe of a quantified object variable, and we indicate that restriction as a subscripted condition directly after the quantifier. For example,

$$\exists_{b > 0} \quad \text{means} \quad \text{"there exists a (number) } b \text{ that is greater than 0"}$$

or

$$\forall_{x \in \text{dom} f} \quad \text{means} \quad \text{"for every element } x \text{ in the domain of } f \text{"}$$

This is not standard notation in logic, but we use it in mathematics for speed and simplicity when the context of such usage is clear.

The *scope of a quantifier* is the sentence to which the quantifier applies. This is usually clarified by using parentheses, but in complex expressions involving many quantifiers it is important to observe the scope of each given quantifier.

Moving quantifiers inside or outside parentheses changes their scope and usually results in changed interpretations of the sentences.

Example 3   (a) $(\forall_x P(x)) \wedge Q(y)$
        (Scope of $\forall_x$ is just the predicate $P(x)$.)

(b) $\exists_x [P(x) \rightarrow \forall_y Q(x, y)]$
        (Scope of $\exists_x$ is the entire conditional; scope of $\forall_y$ is the two-place predicate $Q(x, y)$.)

(c) $\forall_x [\exists_z (P(x) \wedge \forall_y Q(y, z)) \rightarrow \exists_y R(x, y)]$
        (Scope of $\forall_x$ is the entire conditional; scope of $\exists_z$ is the antecedent of the conditional; scope of $\forall_y$ is the two-place predicate $Q(y, z)$; scope of $\exists_y$ is the two-place predicate $R(x, y)$.)

There are many instances in mathematics where several quantifiers appear together, such as "for every x and for every y," "for every x there exists a y such that," "there exists a y for every x," and so on. It is important to be familiar with the meaning of such expressions. For example, there is no difference in meaning if we say "for every x and for every y..." or if we say "for every y and for every x...," but there can be a significant difference in meaning if we mix quantifiers such as "for every x there exists a y such that..." and "there exists a y such that for every x...."

Example 4   In the system of integers, there are two properties of the operation of addition that deal with the concepts of *additive identity element* and *additive inverses*. They are correctly stated as follows.

Additive Identity Element
$$\exists_y \forall_x (x + y = x)$$

Additive Inverse Property
$$\forall_x \exists_y (x + y = 0)$$

The first property says that there is a particular element y (at least one) such that for every integer x, when you add this element y to x you obtain x itself as the result (y "preserves the identity" of x under addition). As you know, the particular element y is what we know to be "0." The second property says to take any integer x and (for that integer) there will exist an integer y such that when you add the two together, you obtain 0 (the additive identity element). As you know, given any integer x, the integer you add to it to produce 0 is "$-x$." Note clearly the difference in meaning when the universal and existential quantifiers are mixed.

$\exists_y \forall_x$ means   "one particular element works for all in the system"

$\forall_x \exists_y$ means   "given any element in the system, there is an element that works for *it*"

To reinforce the various interpretations of sentences involving two quantifiers, study the next example.

Example 5   Suppose x and y are variables on the system of integers and $D(x, y)$ means "x divides y." The statements on the left are symbolic statements involving double quantifiers; the statements on the right are colloquial interpretations.

$\forall_x \forall_y D(x, y)$	Every integer divides every integer.
$\exists_x \exists_y D(x, y)$	There is an integer that divides some integer.
$\forall_y \forall_x D(x, y)$	Every integer is divided by every integer.
$\exists_y \exists_x D(x, y)$	Some integer is divided by an integer.
$\forall_x \exists_y D(x, y)$	Every integer divides some integer (for every integer there is one that the first divides).
$\exists_x \forall_y D(x, y)$	There is an integer that divides every integer.
$\forall_y \exists_x D(x, y)$	Every integer is divided by some integer.
$\exists_y \forall_x D(x, y)$	There is an integer that every integer divides.

We have seen how to translate sentences with one or two quantifiers and with one or more single or two-place predicates. We turn now to the question of how to *negate* quantified sentences and the issue of *equivalence* of such forms. Let us start with an example.

Example 6   Consider the following three universes:

$$U_1 = \{3, 5, 17, 100\}$$
$$U_2 = \{1, -4, 6\}$$
$$U_3 = \{-3, 0, 1, 1.5\}$$

Let x be a variable on each universe, in turn, and let $P(x)$ be the statement "x is greater than 2."

(a) For universe U_1, the following statements are true:

$\forall_x P(x)$	All numbers are greater than 2.
$\sim [\exists_x (\sim P_x)]$	There is no number that is not greater than 2.
$\exists_x P(x)$	There is at least one number that is greater than 2.
$\sim [\forall_x (\sim P(x))]$	It is not the case that all numbers are not greater than 2.

(b) For universe U_2, the following statements are true:

$\exists_x P(x)$

$\sim [\forall_x (\sim P(x))]$

$\exists_x (\sim P(x))$ There is a number that is not greater than 2.

$\sim [\forall_x P(x)]$ It is not the case that all numbers are greater than 2.

(c) For universe U_3, the following statements are true:

$\forall_x (\sim P(x))$ Every number is not greater than 2.

$\sim [\exists_x P(x)]$ No number is greater than 2.

$\exists_x (\sim P(x))$

$\sim [\forall_x P(x)]$

Note that among these examples, there are pairs of statements that seem to be saying the same thing. To assert "all numbers are greater than 2" really says the same as "there is no number that is not greater than 2." That is, it would appear that the two sentences,

$$\forall_x P(x), \quad \sim [\exists_x (\sim P(x))]$$

are logically equivalent. Similarly, the assertion "there is at least one number that is greater than 2" says the same thing as "it is not the case that all numbers are not greater than 2." Here, the two sentences

$$\exists_x P(x), \quad \sim [\forall_x (\sim P(x))]$$

appear to be logically equivalent.

From these remarks it should be apparent that certain statements involving quantifiers can be made in several different ways and mean the same thing. This suggests the notion of logical equivalence in quantificational logic. Intuitively, there are four basic equivalences that deal with *negating* quantified statements.

1. *All true* means the same as *none false*.

$$\forall_x P(x) \leftrightarrow \sim [\exists_x (\sim P(x))]$$

2. *All false* means the same as *none true*.

$$\forall_x (\sim P(x)) \leftrightarrow \sim [\exists_x P(x)]$$

3. *Not all true* means the same as *at least one false*.

$$\sim \forall_x P(x) \leftrightarrow \exists_x (\sim P(x))$$

4. *Not all false* means the same as *at least one true*.

$$\sim [\forall_x(\sim P(x))] \leftrightarrow \exists_x P(x)$$

We summarize these observations in the following rule:

> To negate a quantified expression (one quantifier), change to the other quantifier and negate the sentence in its scope.

This rule extends to more complex sentences involving multiple quantifiers. Observe the original form (conjunction, conditional, etc.) of the sentence to be negated and wherever a multiply quantified expression is to be negated, simply reverse each quantifier and negate the sentence in its scope, working from outside toward inside and left to right. One must apply this rule carefully, however, observing what is to be negated and what is not. For example, to negate the sentence

$$\forall_x[P(x) \to Q(x)]$$

we have as the negation

$$\exists_x[P(x) \wedge \sim Q(x)]$$

Also, the negation of

$$\exists_x[\forall_y P(y) \to Q(x, y)]$$

would be

$$\forall_x[\forall_y P(y) \wedge \sim Q(x, y)]$$

(Note: The inside quantifier is unchanged. Why?)

But to negate the more complex sentence

$$\forall_x\exists_y[\forall_z(P(x) \to R(y, z)) \wedge (P(y) \vee \exists_z R(x, z))]$$

we would have

$$\exists_x\forall_y[\exists_z(P(x) \wedge \sim R(y, z)) \vee (\sim P(y) \wedge \forall_z \sim R(x, z))]$$

Note that the more complex sentence we started with was a conjunction with two quantifiers on it. We start by reversing the quantifiers and negating the conjunction. To negate the conjunction, we change it to a disjunction and negate both parts. The first part is a conditional with a universal quantifier on it, etc.

Unfortunately, there is no compact decision procedure (like truth tables) to demonstrate equivalence of sentences involving quantifiers. There are, however, some fundamental tautologies in quantificational logic. These are stated below with a brief explanation of each.

Q1 $\forall_x P(x) \leftrightarrow \sim [\exists_x (\sim P(x))]$

Q2 $\exists_x P(x) \leftrightarrow \sim [\forall_x (\sim P(x))]$

These first two tautologies were discussed in examples above relating to negating quantified statements.

Q3 $\forall_x P(x) \rightarrow \exists_x P(x)$

This says that if a statement is true for every object in a universe, then it must be true for at least one object in the universe. For example, if every even integer is divisible by 2, then there is an even number that is divisible by 2.

Q4 $\forall_x P(x) \rightarrow P(y)$

This is called *universal specification*. It says that if a statement is true for every object in a universe, then for any *specific object* from that universe, the statement is true. If we have a statement such as "for each positive integer $n \ldots$" this allows us to make the statement for each positive integer taken individually. For example, if every even number is divisible by 2, then 4 is divisible by 2.

Q5 $P(y) \rightarrow \exists_x P(x)$

If a statement is true for a *specific object* in a universe, then we can assert the existential statement shown. For example, if 6 is a multiple of 3, then there is at least one integer that is a multiple of 3.

Q6 $[\forall_x (P(x) \wedge Q(x))] \leftrightarrow [\forall_x P(x) \wedge \forall_x Q(x)]$

This tautology says that the $\forall$ symbol distributes over conjunctions. That is, everything is large and green if and only if everything is large and everything is green. Also, "every finite vector space has a basis and dimension" is equivalent to saying "every finite vector space has a basis and every finite vector space has a dimension."

Q7 $[\exists_x (P(x) \vee Q(x))] \leftrightarrow [\exists_x P(x) \vee \exists_x Q(x)]$

This says in effect that the symbol $\exists$ distributes over disjunctions ($\vee$). For example, to say "there exists a number that is prime or rational" is equivalent to saying "there is a number that is prime or there is a number that is rational."

Note that $\forall$ does *not* distribute over disjunctions. Saying "everything is small or blue" is *not* the same as saying "everything is small or everything is blue." While this is not an equivalence, there is a related implication, which we discuss next.

Q8 $[\forall_x P(x) \lor \forall_x Q(x)] \to \forall_x(P(x) \lor Q(x))$

If a disjunction has both parts universally quantified, then we could "factor" out the universe quantifier, but not vice-versa. For example, from the statement "every positive number is greater than zero or every positive number has no negative factors" we can deduce the statement "every positive number is greater than zero or has no negative factors."

Q9 $[\exists_x(P(x) \land Q(x))] \to (\exists_x P(x) \land \exists_x Q(x))$

This tells us that the existential quantifier may be pushed inside a conjunction. The converse is not true. For example, if there exists a number that is prime and even, then there exists a number that is prime and there exists a number that is even. Conversely, however, there may exist a number having one property and a number having another property but this does *not* imply that there must be a number having both properties.

Q10 $\forall_x[P(x) \to Q(x)] \to [\forall_x P(x) \to \forall_x Q(x)]$

This last tautology says that the universal quantifier can be pushed inside a conditional (placed on both antecedent and consequent). Note that again the converse is not necessarily true. For example, if we believe the statement "All men are mortal" then we would also believe the statement "If everything is a man, then everything is mortal."

One final concept in mathematical and logical quantification is that of *unique existence*. Often in mathematics we wish to convey not only that *there exists* some object having a given property, but that it is the *only one*. As a cue, we see such word combinations as *one and only one* or *exactly one*. There is a symbol that is used to describe this situation—the existential quantifier directly followed by an exclamation point ($\exists!$). For example, the statement

$$\exists!_x[P(x) \land E(x)]$$

might mean

there exists only one x that is prime and even, or

there is exactly one number that is both prime and even, or

there is one and only one number that is prime and even.

Assigned Problems

1. Let $P(x)$ mean "x is prime," $E(x)$ mean "x is even," $O(x)$ mean "x is odd," and $D(x, y)$ mean "x divides y." Translate each of the following into English.

(a) $P(7)$

(b) $E(2) \wedge P(2)$

(c) $\forall_x[D(2, x) \to E(x)]$

(d) $\exists_x[E(x) \wedge D(x, 6)]$

(e) $\forall_x[\sim E(x) \to \sim D(2, x)]$

(f) $\forall_x[E(x) \to \forall_y(D(x, y) \to E(y))]$

(g) $\forall_x[P(x) \to \exists_y(E(y) \wedge D(x, y))]$

(h) $\forall_x[O(x) \to \forall_y(P(y) \to \sim D(x, y))]$

(i) $[\exists_x(E(x) \wedge P(x))] \wedge (\sim \exists_x[((E(x) \wedge P(x)) \wedge (\exists_y((x \neq y) \wedge E(y) \wedge P(y))])$

2. Translate each of the following, using the indicated predicates and terms.

(a) All judges are lawyers. $(J(x), L(x))$

(b) Some lawyers are shysters. $S(x)$

(c) No judge is a shyster.

(d) Some judges are old but vigorous. $(O(x), V(x))$

(e) Judge Jones is neither old nor vigorous. $J(j)$

(f) Not all lawyers are judges.

(g) Some lawyers who are politicians are congressmen. $(P(x), C(x))$

(h) No Congressman is not vigorous.

(i) All Congressmen who are old are lawyers.

(j) Some women are both lawyers and congressmen. $W(x)$

(k) No woman is both a politician and a housewife. $H(x)$

(l) There are some women lawyers who are housewives.

(m) All women who are lawyers admire some judge. $A(x, y)$

(n) There are both lawyers and shysters who admire Judge Jones.

(o) Some lawyers admire only judges.

(p) Some lawyers admire women.

(q) Some shysters admire no lawyer.

(r) Judge Jones does not admire any shyster.

(s) Only judges admire judges.

(t) All judges admire only judges.

3. Negate each of the following quantified sentences.

(a) $\exists_x(P(x) \rightarrow Q(x))$

(b) $\forall_x \exists_y \forall_z(xy = z)$

(c) $\exists_x \exists_y \forall_z(x + y = z)$

(d) $\forall_x \exists_y[(P(x) \rightarrow Q(y)) \wedge \exists_z(R(x, y, z)]$

(e) $\forall_x \exists_n\left[(0 \le n) \wedge \left(x < \dfrac{n}{10}\right)\right]$

4. The definition of limit of a function f at a point a is given in calculus as

$$\left(\lim_{x \to a} f(x) = L\right) \text{ if and only if}$$

$$(\forall_{\varepsilon > 0})(\exists_{\delta > 0})(\forall_{x \in \text{dom} f})[\text{if } 0 < |x - a| < \delta, \text{ then } |f(x) - L| < \varepsilon]$$

To say, then, that a function f fails to have a limit at some point a would be to negate the right side of the biconditional definition. Negate that quantified statement, and interpret what the negation means to you. (You might use a diagram or visual to assist yourself.)

5. A function f is said to be *even* if and only if

$$(\forall_{x \in \text{dom} f})f(-x) = f(x)$$

Negate this statement and give an example of a function that is not even.

6. A function f is said to be *odd* if and only if

$$(\forall_{x \in \text{dom} f})f(-x) = -f(x)$$

Negate this statement and give an example of a function that is not odd.

7. If a function is not even, must it be odd? If a function is not odd, must it be even? Explain your answers. Can a function be both odd and even? Can a function be neither odd nor even?

8. Negate the concept being defined in each case:

(a) A function f is *periodic* $\leftrightarrow (\exists_{p > 0})(\forall_{x \in \text{dom} f})(f(x + p) = f(x))$.

What does it mean to *not* be periodic?

(b) A function f is *decreasing* $\leftrightarrow (\forall_{x \in \text{dom} f})(\forall_{y \in \text{dom} f})$

(If $x \leq y$, then $f(x) \geq f(y)$.) What does it mean to be *nondecreasing*?

(c) A function f is one-to-one $\leftrightarrow (\forall_{x \in \text{dom} f})(\forall_{y \in \text{dom} f})$

(If $f(x) = f(y)$, then $x = y$.) What does it mean to be not one-to-one?

(d) A function f is *continuous at a point a* $\leftrightarrow$

$(\forall_{\varepsilon > 0})(\exists_{\delta > 0})(\forall_{x \in \text{dom} f})$

(If $0 < |x - a| < \delta$, then $|f(x) - f(a)| < \varepsilon$.) What does it mean to be discontinuous at a?

(e) A function f is *continuous on a set E* $\leftrightarrow$

$(\forall_{x \in E})(\forall_{\varepsilon > 0})(\exists_{\delta > 0})[\forall_y(\text{if } y \in E \text{ and } |x - y| < \delta, \text{ then } |f(x) - f(y)| < \varepsilon)]$

9. The definition of limit of a function f as x gets large positively is

$$\lim_{x \to \infty} f(x) = L \text{ if and only if}$$

$$(\forall_{\varepsilon > 0})(\exists_{N > 0})(\forall_{x \in \text{dom} f})[\text{if } x > N, \text{ then } |f(x) - L| < \varepsilon]$$

Sketch a graph that illustrates this situation. Next, negate the statement (i.e., indicate what it means for a limit not to exist as $x \to \infty$) and sketch a graph that illustrates that situation.

10. Abraham Lincoln once said,

You can fool all of the people some of the time and you can fool some of the people all of the time, but you cannot fool all of the people all of the time.

Using the symbols x and y for persons from some universe of people, t for times from some universe of times, and the 3-place predicate $F(x, y, t)$ as "x fools y at time t," write a symbolic translation of this quotation. (Translate the "you" as though it is impersonal—that is, assume Abraham Lincoln was speaking to any given person x and don't quantify that variable.)

2

Mathematical Proof

2.1 *Inference and Deduction; Direct Proof*

Mathematics has been called the *queen of the sciences*, and indeed it is an exact science. However, the *reasoning* processes through which mathematics is discovered, developed, and verified are more an art than a science. If we oversimplify mathematical problems, we might see, as the mathematician George Polya did, two large classes of problems—problems to find and problems to prove. This course is focused on *problems to prove*, and this chapter represents the core of the course. Here, we shall examine the art of mathematical reasoning as it applies in mathematical proofs. We shall study various forms of proof—direct, conditional, indirect, proof by cases, and proof by mathematical induction.

To begin to understand the nature of proof and why it is necessary, we need to ask ourselves what it is that we are proving, what constitutes a "proof," and what it accomplishes mathematically. We start our investigation in the domain where we speak about mathematics—logic. Most of this section shall be devoted to the structure of *logical proof*, where we shall see how the reasoning process itself works. Once we develop some facility with the reasoning process, we will turn our attention to the direct form of mathematical proof.

Inference and Deduction

We have seen how to separate sentences into logical parts and symbolize these parts. In this way we have learned about the logical *form* of sentences (e.g., the sentence $P \rightarrow Q$ is the same in logical form regardless of what the English sentences for P and Q may be). Sentential *connectives* determine the logical form of a sentence.

Knowing the logical forms and having the means of symbolization at our command, we proceed now to a most important aspect of logic, *logical inference and deduction*. Deductive inference is like a game: *rules of inference* control the moves (steps) we make (i.e., each step must be justified by a rule of inference). The game is played with logical *formulas* (symbolized sentences). We begin with

43

sets of formulas called *premises* (or *assumptions* or *hypotheses*). The object of the game is to start with the premises and use the rules of inference in such a way as to lead to other formulas that are called *conclusions*. To proceed logically from premises to conclusion is to perform a *deduction* (or *proof*, or *derivation*). The conclusion you reach is said to *follow logically from* (or be *derived from*, *deduced from*, or *implied by*) the premises *if* every move you make to reach that conclusion is justified by a rule of inference.

The whole idea of inference is: *From premises that are true we reach only conclusions that are true* (i.e., if the premises are true, then the conclusions that are drawn logically from them must be true).

A *proof*, then, is a finite sequence of steps proceeding from a set of premises (stated or known information) to a conclusion (a desired piece of information), where each step is justified by some agreed-upon rule of reasoning. Formally speaking, the last statement in a proof is a theorem that adds to the available body of knowledge in the system and thus qualifies to be used as a premise in subsequent proofs.

The rules, or patterns of reasoning we use to justify steps in a proof (or equivalently to make inferences), are called *rules of inference*. We shall introduce five rules of inference in our total discussion of logic proof; we begin by examining the first two.

(P) *The Rule of Premises:*   A premise may be introduced at any point in a proof.

Inference Pattern: (1) Formula introduced as premise   P

(MP) *Modus (Ponendo) Ponens:*   From statements $P \rightarrow Q$ and P, we may deduce Q.

Inference Pattern:
(1) $P \rightarrow Q$	P
(2) P	P
(3) $\therefore \quad Q$	MP (1)(2)

Example 1   *A Proof Using the First Two Rules of Inference*

Prove: $\sim S$

Given the premises:

$$\begin{cases} T \\ T \rightarrow \sim Q \\ \sim Q \rightarrow \sim S \end{cases}$$

Proof
(1)	T	P	Since these are given as premises, they are introduced in steps (1) and (2) and justified by the rule of premises.
(2)	$T \rightarrow \sim Q$	P	
(3)	$\sim Q$	MP(1)(2)	Step (3) is deduced from steps (1) and (2) by modus ponens.

(4)	$\sim Q \rightarrow \sim S$	P	This was a given premise; it is introduced here by the rule of premises.
(5)	$\sim S$	MP(3)(4)	Step (5) (the desired *conclusion*) is deduced from steps (3) and (4) by modus ponens.

The five steps shown above constitute a proof that the conclusion $\sim S$ follows logically from the three premises T, $T \rightarrow \sim Q$, and $\sim Q \rightarrow \sim S$. Each step is justified by some rule of inference (in this case, P or MP). The format for a formal proof in logic will include the following:

1. Each line is numbered.

2. Each line is justified by a rule of inference.

3. After the abbreviation for the rule, we put the numbers of the lines from which that line was derived.

We have seen above a proof that a conclusion does follow logically from a set of premises, even though we could not go from the premises to the conclusion in one step. By going through several steps, each permitted by a rule of inference, we have *formally derived* the conclusion from the premises. That is, we have a proof—it is embodied in the five steps above.

In contrast to a proof, an *argument* is simply a set of premises and a conclusion. For example,

$$\begin{cases} T \\ T \rightarrow \sim Q \\ \sim Q \rightarrow \sim S \end{cases}$$

$$\therefore \sim S$$

is an argument. Here we have learned a precise way to prove that an argument is *valid*. To say an argument consisting of premises $P_1, P_2, P_3, \ldots P_n$ and a conclusion C is *valid* is to say:

1. Conclusion C can be logically deduced from premises $P_1, P_2, P_3, \ldots P_n$ (i.e., there exists a proof in which $P_1, P_2, \ldots, P_n$ appear as premises and after a finite number of steps are taken—each justified by a rule of inference—the conclusion C appears in the last line), or

2. Conclusion C is tautologically implied by the conjunction of all premises $P_1, P_2, \ldots P_n$—i.e., $(P_1 \wedge P_2 \wedge P_3 \wedge \cdots \wedge P_n) \rightarrow C$ is a tautology.

Thus, the construction of a proof is one way (the major way) to show that an argument is valid. Before we proceed to the other three rules of inference, let us take a moment to reflect on and summarize what we have so far. Inference, or proof, is *the derivation of conclusions from sets of premises.* In order to perform deductions, we make use of certain agreed-upon rules, called *rules of inference.* These rules operate like the rules of any game. They permit us to make certain moves. Every move permitted by the rules is a step in inference, a sentence that can be derived if we are given certain other sentences. In formal derivations, we justify every step of inference taken by referring to the particular rule of inference that permits that step. We indicate this rule by putting the abbreviation for its name to the right of the step of inference. It is also necessary to indicate the numbers of the lines in the inference from which each step was derived.

The rules of logic and mathematical reasoning are not just any arbitrarily chosen rules. They are generally agreed-upon rules, and as such, they allow us to make only valid inferences (valid, provided we agree with the rules).

A valid inference is one that follows logically from the premises. This means that when the premises are true, the conclusion that follows must also be true. The reason for the rules of inference is to insure that when we are given sets of true sentences, the conclusions that can be derived logically from those sentences will also be true.

To proceed further in the study of mathematical reasoning it is essential that you become very familiar with the idea of valid inference and also with each particular rule of inference that permits a logical step to be taken. If you do not know these logical rules and the common tautologies well, you will be unable to plan a strategy for the construction of logical or mathematical proofs.

The following exercises will provide you with some practice in proof using the two rules of inference, P and MP, which we have studied so far.

Assigned Problems

1. For each of the following, derive a conclusion and then write a proof that shows that the conclusion logically follows from the given premises.

    (a) $\begin{cases} \sim A \to \sim B \\ \sim A \end{cases}$ (b) $\begin{cases} R \\ R \to (\sim T \lor Q) \end{cases}$ (c) $\begin{cases} \sim B \to (\sim D \land A) \\ \sim B \end{cases}$

2. Construct formal proofs (derivations) for each of the following:

    (a) Prove: $\sim T$

    Premises: $\begin{cases} R \to \sim T \\ S \to R \\ S \end{cases}$

    (b) Prove: G

    Premises: $\begin{cases} \sim H \to \sim J \\ \sim H \\ \sim J \to G \end{cases}$

We now extend our rules of inference to include three more. Listed below, with examples, are the *five* rules of inference that we shall use in this course. Note particularly that instantiation (INS) relates back to our list of tautologies. Therefore, using modus ponens (MP) or biconditional replacement (BR), *every tautology* has the effect of being a rule of inference. So it is important to study carefully not only the five rules of inference, but also the complete list of tautologies.

---

1. The Rule of Premises (abbreviated "P")

A premise may be introduced at any point in a proof.

2. Modus (Ponendo) Ponens (abbreviated "MP")

From statements $P \rightarrow Q$ and P, we may deduce Q.

3. Adjunction (abbreviated "Adj")

From statements P, Q individually, we may deduce $P \wedge Q$.

To illustrate the use of adjunction, consider the following argument and proof.

Example 2   **A Proof Using Adjunction**

Prove: D

Premises:
$$\begin{cases} A \\ A \rightarrow B \\ (A \wedge B) \rightarrow C \\ C \rightarrow D \end{cases}$$

Proof   (1) A P

(2) $A \rightarrow B$ P

(3) B MP (1)(2)

(At this point in the proof we see the remaining two premises $(A \wedge B) \rightarrow C$ and $C \rightarrow D$; we know we can get D if we have C (by MP), and we can get C if we have $A \wedge B$ (by MP). The problem is how to get $A \wedge B$. Note that the individual sentences A, B appear independently in steps (1) and (3), so by *adjunction*, we can introduce the conjunction $A \wedge B$ at this point and proceed according to plan.)

(4) $A \wedge B$ Adj (1)(3)

(5) $(A \wedge B) \rightarrow C$ P

(6) C MP (4)(5)

(7) $C \rightarrow D$ P

(8) D MP (6)(7)

Thus, adjunction is a way of introducing a conjunction of two statements into a proof, providing we have those two statements appearing individually elsewhere prior in the proof.

---

4. Biconditional Replacement (abbreviated "BR")

If P is logically equivalent to Q ($P \leftrightarrow Q$ is a tautology), then either may be substituted for the other to introduce a new sentence, or either may be replaced for the other to introduce a revision of a prior sentence.

Examples of form:

(1) $P \leftrightarrow Q$	(1) $P \leftrightarrow Q$	(1) $P \leftrightarrow Q$	(1) $P \leftrightarrow Q$
(2) P	(2) Q	(2) sentence involving P	(2) sentence involving Q
(3) $\therefore$ Q	(3) $\therefore$ P	(3) same sentence involving Q	(3) same sentence involving P

To illustrate the biconditional replacement rule of inference, consider the following argument and proof of its validity.

Example 3   ***A Proof Using Biconditional Replacement***

Prove: S

Premises:
$$\begin{cases} R \\ T \rightarrow Q \\ R \leftrightarrow T \\ P \\ (P \wedge Q) \rightarrow S \end{cases}$$

Proof (1) $R \leftrightarrow T$ P

(2) R P

(3) T BR (1)(2) $\begin{cases} R \text{ in step (2) is replaced by} \\ \text{the logical equivalent } T. \end{cases}$

(4) $T \rightarrow Q$ P

(5) Q MP (3)(4)

Note that this could have been done in a different way: Instead of replacing R by T, we could have replaced the T in the premise $T \rightarrow Q$ by R—a logical equivalent from step (1). Then, leaving the premise R alone, we would have R (premise), $R \rightarrow Q$ (variation of $T \rightarrow Q$ after R substituted for T), from which we could deduce Q by MP.

(6) P P

(7) $P \wedge Q$ Adj (5)(6)

(8) $(P \wedge Q) \rightarrow S$ P

(9) S MP (7)(8)

The biconditional replacement rule of inference plays a role in logic like "equals" does in mathematics. If we know that two expressions are equal, either can be substituted for the other in other expressions; similarly, if we know two sentences are logically equivalent (i.e., their biconditional is a tautology), then either can be substituted for the other in other sentences.

The fifth rule of inference, called *instantiation*, has the effect of converting every known tautology into an available premise to help us in a proof. It is very much like using theorems we have proved in mathematics to help us prove new theorems in mathematics.

> **5. Instantiation (abbreviated "Ins"; some description of the referred-to tautology)**
> *Any instance of a known tautology may be introduced at any point in a proof.*

To illustrate *instantiation*, observe the following proof.

Example 4   *A Proof Using Instantiation*

Prove: $A \rightarrow C$

Premises: $\begin{cases} A \rightarrow B \\ B \rightarrow C \end{cases}$

Proof   (1) $A \rightarrow B$ P

(2) $B \rightarrow C$ P

(At this point in the proof, we look for a tautology that has a form that would allow us to deduce (using MP or BR) the sentence $A \rightarrow C$ from the premises $A \rightarrow B$ and $B \rightarrow C$; such a tautology is the law of syllogism, except this tautology is stated in terms of P, Q, R

instead of A, B, C. If we recognize an appropriate replacement of the letters, we use it and proceed.)

(3) $[(A \rightarrow B) \wedge (B \rightarrow C)] \rightarrow (A \rightarrow C)$ Ins (law of syllogism)

(This says line (3) is an instance of a known tautology; here P is replaced by A, Q by B, and R by C.)

(4) $(A \rightarrow B) \wedge (B \rightarrow C)$ Adj (1)(2)

(In order to use modus ponens on step (3), we need the antecedent, which is the conjunction of steps (1) and (2).)

(5) $A \rightarrow C$ MP (3)(4)

Direct Proof

To help you develop a feeling for logical proof and the structure of mathematical reasoning, most of the proofs required in the assigned problems will be symbolic two-column proofs like the preceding examples. However, *mathematical proofs* are usually much more informal than that. In most cases, mathematical proofs are embodied in written paragraphs where only the most important steps (equations, specific inferences, etc.) are highlighted by being offset. Also, while we operate in our mathematical reasoning via the rules of logical inference, we usually don't specify them. This is traditional but rather unfortunate, because often it is very difficult for the reader to see how the prover "came up" with the ideas that led to the creation of the proof. Only in very complex mathematical proofs is any time spent on clarifying the underlying logic, and in most cases it is *assumed* that the reader agrees with the logical basis of the proof.

Thus, while our major goal in this course is to understand and construct mathematical proofs, it is absolutely necessary to have a good working knowledge of the basic logical maneuvers by which these mathematical proofs are created. Hence, we have examined the logical structure first.

It will be particularly helpful in both logical and mathematical proofs if before you start attempting to write out a proof, you first work out a plan "in your head." Ask yourself questions like: "What does this lead to?" "I need this; to get it I need to obtain that first," "Can I build a bridge from what I know to what I want?" "What does this term mean?" "Do I know its definition?" "How does one usually get that kind of information?" "Do I recognize the structure of this argument?" "Have I seen this before?" Once a plan is roughly established, the proof is simply a carefully worded execution of the plan. The importance of *planning* cannot be overstated in mathematics—it's where the real creativity exists.

To prove a statement in mathematics by *direct proof*, we start by assembling all our known information (definitions, axioms, theorems, results of problems, familiar techniques, etc.) that may relate to the situation at hand. This known information plays the role that premises did in logic, and we naturally feel free to use it whenever and wherever we want in the proof (recall the use of the premises and instantiation rules of inference). If a statement is more accessible in a logically equivalent form (e.g., using the contrapositive form of a conditional), we simply adjust or replace the statement with its logical equivalent (biconditional replacement). We make no fuss about putting statements together in conjunctive form (adjunction), and *modus ponens* reasoning appears virtually everywhere, however subtle. Actually, modus ponens is the heart of direct proof. We proceed from known information A through a series of implications $A \rightarrow B \rightarrow C \rightarrow D$ to desired information D. We believe that if our known information is true, then the above mechanisms permit us to derive only information that is true. Therefore, we do not specify the logical rules by which we reason, and we focus instead on the mathematical bits and pieces of information that we have selected to introduce into our proof.

Example 5   A Mathematical (Direct) Proof

Prove: For any natural number $n \geq 2$, $n^3 - n$ is divisible by 6.

  Our premises here include the given information that n is a natural number and $n \geq 2$, together with everything we know previously about natural numbers: factoring, divisibility, prime numbers, and so on. It might occur to us that a "more accessible" form of $n^3 - n$ is its algebraic equivalent in factored form

$$n^3 - n = n(n^2 - 1)$$
$$= n(n - 1)(n + 1)$$
$$= (n - 1)n(n + 1)$$

So we turn our attention to $(n - 1)n(n + 1)$ rather than $n^3 - n$. We might also notice that $n - 1$, n, and $n + 1$ describe three *consecutive* natural numbers. Now we ask ourselves, "What do we know about three *consecutive* natural numbers?" It is not too difficult to see that one of them must be a multiple of 3 (we may have to prove that in general to satisfy ourselves). Analogously, among any *two* consecutive natural numbers, one of them is even (divisible by 2). Therefore we conclude that $(n - 1)n(n + 1)$ must be divisible by both 2 and 3, hence divisible by 6.

  Notice in the foregoing discussion that no reference was made to logical rules of inference, yet they were being used implicitly "all over the place" in the construction of the proof.

  It should be clearly understood that there may be regular rules by which we reason, but *there are no set rules by which we put them together to create proofs.* This is where mathematics becomes an art—when we, as human beings, work out the strategy that takes us from previously known information to something unknown. There are probably as many ways of doing that as there are human beings since proof-making and problem-solving involve reasoning processes whose chemistry calls forth creativity, ingenuity, insight, and organization. Many people find constructing proofs to be a very difficult task. That is understandable when one considers what a proof really is. Additionally, it is not sufficient, in learning how to construct mathematical proofs, to watch someone else construct mathematical proofs, just as it is not sufficient, in learning how to ride a bicycle, to watch someone else ride a bicycle. You must get actively involved and construct proofs for yourself.

Example 6   **Another Mathematical (Direct) Proof, This One Involving a Discovery**

Find an explicit formula for the sum of the first n natural numbers.

$$\sum_{i=1}^{n} i = 1 + 2 + 3 + \cdots + n = ?$$

The reader may recall (or perhaps learn for the first time) a little technique that is useful for searching for patterns involving certain general sums. Write the sum out in one line, then reverse the order of terms and write that sum out in the next line. Add the two lines and observe the emerging pattern.

$$\sum_{i=1}^{n} i = 1 + 2 + 3 + \cdots + n$$

$$\sum_{i=1}^{n} i = n + (n-1) + (n-2) + \cdots + 1$$

$$2\sum_{i=1}^{n} i = (n+1) + (n+1) + \cdots + (n+1)$$

The right side of the last equation above has n terms (use the top line as "counters"), each being $(n+1)$, so it can be "condensed" to the number $n(n+1)$. Therefore, after dividing both sides of the equation by 2, we have

$$\sum_{i=1}^{n} i = \frac{n(n+1)}{2}$$

It might be helpful (and reassuring) to test this formula on some special cases, even though we have "proved" it holds in general. There are some tacit premises in the above proofs. Can you find them?

Assigned Problems   1. Construct direct, two-column logic proofs for each of the arguments given below. Use the standard format shown in the examples.

(a) Prove: $M \vee N$

Premises: $\begin{cases} \sim J \rightarrow (M \vee N) \\ (F \vee G) \rightarrow \sim J \\ F \vee G \end{cases}$

(b) Prove: $\sim N$

Premises: $\begin{cases} R \rightarrow \sim S \\ R \\ \sim S \rightarrow Q \\ N \rightarrow \sim Q \end{cases}$

2. Before constructing a logic proof for each of the following symbolic arguments, sketch a brief plan that indicates how you can proceed directly from the premises (through a series of intermediate steps) to the conclusion. Then construct the formal proof.

   (a) Prove: B

   Premises:
   $$\begin{cases} \sim G \to E \\ E \to K \\ \sim G \\ K \to \sim L \\ M \to L \\ \sim M \to B \end{cases}$$

   (b) Prove: $T \wedge S$

   Premises:
   $$\begin{cases} E \to S \\ \sim T \to \sim J \\ E \wedge J \end{cases}$$

3. Symbolize the argument given below (P, W, J, E) and then write a two-column proof of its validity.

   > If Peter gets a car for the summer, then he goes west.
   > If John gets a car for the summer, then he goes east.
   > Peter or John gets a car for the summer.
   > Peter does not go west.
   > ∴   John goes east.

4. Using truth tables or counterexamples, test the following arguments for validity (or invalidity).

   (a) Joseph is neurotic or irresponsible.
       Joseph is neurotic.
       ∴   Joseph is not irresponsible.

   (b) $(A \vee B) \to (C \vee D)$
       $A \wedge \sim D$
       ∴ $B \to C$

   (c) Deduce $M \leftrightarrow N$

   from premises
   $$\begin{cases} M \vee N \\ N \leftrightarrow [\sim M \to \sim P] \\ P \vee (\sim N \wedge Q) \end{cases}$$

5. Construct a mathematical proof of the following statement.

   > The cube of any odd integer is an odd integer.

   (Use the fact that every odd integer can be expressed in the form $2k + 1$, where k is some integer.)

6. Prove that the product of any four consecutive integers, increased by one, is a perfect square.

   $$\left(n^2 + 3n + 1\right)^2$$

7. There is a little trick for squaring two-digit numbers that end in 5. Increase the tens digit by one, multiply that by the tens digit unchanged, and attach the two digits 25 to this product. For example, $(25)^2$: Increase 2 to 3, multiply that by the unchanged 2 (get 6), attach 25 to get 625, the answer. Why does this work?

8. Establish an explicit formula for the general sum of odd numbers,

$$\sum_{i=1}^{n} (2i - 1) = 1 + 3 + 5 + \cdots + (2n - 1)$$

9. Use calculus to prove the following statement:

> Of all rectangles having fixed perimeter, the square has the largest area.

Take note of all the information and techniques you use in the proof.

10. A standard definition of the inequality relation ($<$) in the system of real numbers goes as follows:

> $a < b$ means there exists a positive number k such that $a + k = b$.

Using this definition and what you know about real numbers, properties of equalities, etc., prove the statement:

> If $a < b$ and c is positive, then $ac < bc$.

2.2   Conditional Proof

Many theorems in mathematics are stated in the form of conditionals $(P \rightarrow Q)$; that is, if a certain condition or conditions (P) are met, then we try to deduce a conclusion or conclusions (Q). An important question both logically and mathematically can be posed here: *What is it that we are trying to prove—the conditional sentence $P \rightarrow Q$ or the piece of information Q itself?* A brief reflection back at elementary theorems of algebra and geometry should help clarify the answer.

Theorem   1

(algebra)   If $2x + 5 = 13$, then $x = 4$.

Theorem   2

(geometry)   If in $\triangle ABC$, $\overline{AB} \cong \overline{BC}$, then $\angle A \cong \angle C$.

For example, is the conclusion $x = 4$ deducible by itself from the axioms and theorems of algebra? Similarly, is the conclusion $\angle A \cong \angle C$ deducible by itself from the axioms and theorems of geometry? It should be clear that when we have a statement to prove, and that statement is posed in conditional (if ... then) form, then we are trying to prove that the *entire conditional* is a theorem that is deducible from the known information at that point. Another (logically equivalent) way of looking at this situation is that we are trying to deduce conclusion Q from known information (premises) *together with* the added premise P.

The above observations hint at a method of proof originally validated in 1929 by the logician Alfred Tarski. He stated and proved a theorem of metalogic called the *deduction theorem*, which leads to a proof strategy we shall call *conditional proof*.

The rule of conditional proof is stated below.

> If a sentence C can be derived from a sentence R and a set of premises, then the conditional sentence $R \rightarrow C$ can be derived from the set of premises alone.

The associated proof strategy (conditional proof) can be outlined in several steps:

1. We have a set of premises $P_1, P_2, \ldots P_n$ (the premises P_i represent known information—definitions, axioms, theorems, results of problems, etc.).

2. We wish to deduce a statement in *conditional* form, $R \rightarrow C$ (i.e., we wish to derive as a conclusion the conditional $R \rightarrow C$ from the premises P_i).

3. We introduce the premise R as an *added* premise (recorded as P_a), and treat it as though it were known (or given) information.

4. Using the known information (premises $P_1, P_2, \ldots P_n$) together with the added information (premise R), we try to deduce the conclusion C.

5. If we are successful, we invoke the rule of conditional proof, and assert that we have derived the desired conditional $R \rightarrow C$ from the original set of premises alone.

Conditional proof is based on the exportation tautology
$$[(P \wedge Q) \rightarrow R] \leftrightarrow [P \rightarrow (Q \rightarrow R)]$$
and it is an extension of this tautology to include any number of premises:

$$\left\{ \left[\underbrace{(P_1 \wedge P_2 \wedge \cdots \wedge P_n) \wedge R}_{\text{known information + added premise}} \right] \underbrace{\rightarrow C}_{\text{implies } C} \right\} \leftrightarrow \left[\underbrace{(P_1 \wedge P_2 \wedge \cdots \wedge P_n)}_{\text{known information alone}} \rightarrow \underbrace{(R \rightarrow C)}_{\text{implies } R \rightarrow C} \right]$$

Following is a two-column proof of an argument in symbolic logic using the conditional proof strategy.

Example 1   Prove: $R \rightarrow \sim P$

Premises: $\begin{cases} P \rightarrow Q \\ R \rightarrow \sim Q \end{cases}$

Note that we wish to prove a statement in the form of a conditional. That, in itself, might suggest that we try conditional proof with R as the added premise. However, we may feel even more strongly about trying conditional proof after we look at the premises. Both premises are conditionals and at first glance it would appear there is no way to "start" the proof (there are alternate ways, but often they are not very apparent).

Proof

(1)	$R \rightarrow \sim Q$	P
(2)	R	P_a (the premise R is being added)
(3)	$\sim Q$	MP (1)(2)
(4)	$P \rightarrow Q$	P
(5)	$(P \rightarrow Q) \leftrightarrow (\sim Q \rightarrow \sim P)$	Ins (contraposition)
(6)	$\sim Q \rightarrow \sim P$	BR (4)(5)
(7)	$\sim P$	MP (3)(6)

(We have at this point deduced $\sim P$ from the original premises *and* the added premise R; the next step involves conditional proof and says we have deduced the conditional sentence $R \rightarrow \sim P$ from the original premises alone.)

(8)	$R \rightarrow \sim P$	Conditional proof (2)(7)

You may have noticed that the foregoing argument could have been handled by direct proof, using the following reasoning. We need a conditional $R \rightarrow \sim P$ and we have two conditionals $P \rightarrow Q$ and $R \rightarrow \sim Q$. If we link up (adjoin) the two premise conditionals in the right way, perhaps we can produce the conclusion. This suggests reasoning by syllogism: we have $R \rightarrow \sim Q$; the conditional $P \rightarrow Q$ is equivalent to $\sim Q \rightarrow \sim P$ (contraposition); thus we could produce $R \rightarrow \sim P$.

As one might expect, reasoning by conditional proof strategy plays a significant role in mathematics because of the frequency of conditional statements in mathematical discourse. Following are several examples of conditional proof in mathematical reasoning. As you study the examples and construct conditional proofs, it is important to keep in mind the nature of conditional proof—to prove $P \rightarrow Q$, you first assume P to be true ("add" it to your existing knowledge), and try to deduce Q from all your existing knowledge and that extra piece P. Once Q is deduced, you have proved $P \rightarrow Q$. You have *not* shown that Q (alone) is true, but rather that Q is true *if* P is true. Whether P by itself is true or Q by itself is true are different questions; your main concern here is the truth of $P \rightarrow Q$.

Example 2   Prove: If *m* is an even integer and *n* is an odd integer, then $m + n$ is an odd integer.

We start by *assuming* that *m* is an even integer and *n* is an odd integer (this is the premise added to our existing knowledge). Next we call on our existing knowledge to recall the definitions of even and odd integers.

m even   means   "$m = 2k$ for some integer k"

n odd   means   "$n = 2j + 1$ for some integer j"

(Note that we do not use the same integer *k* or *j* in both cases, since this would really be saying that *m* and *n* are consecutive integers—a special case of what we wish to prove.) Therefore, we add *m* and *n* and see what happens:

$$m + n = 2k + (2j + 1)$$
$$= (2k + 2j) + 1 \qquad \text{(associative property of +)}$$
$$= 2(k + j) + 1 \qquad \text{(distributive property)}$$

This is an odd integer since $k + j$ is an integer if *k* and *j* are integers; thus, $m + n$ has the form "two times some integer, plus 1," an odd integer by definition.

We have thus proved: if *m* is even and *n* is odd, then $m + n$ is odd.

Example 3   In the diagram, *A*, *B*, *C*, and *θ* represent angles, and l_1, l_2 represent lines. Prove the following statement:

If $m(\angle A) = m(\angle B)$, then $l_1 \cap l_2 = \varnothing$.

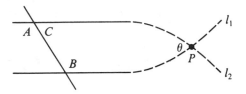

Suppose we start by restating the theorem in the form of its logically equivalent contrapositive:

If $l_1 \cap l_2 \neq \varnothing$, then $m(\angle A) \neq m(\angle B)$.

So, let us assume $l_1 \cap l_2 \neq \varnothing$ (added premise), and try to deduce $m(\angle A) \neq m(\angle B)$ using that together with what we know about plane geometry. Since $l_1 \cap l_2 \neq \varnothing$, they must intersect in some point *P*, creating an angle *θ*. Now

$$m(\angle A) + m(\angle C) = 180 \qquad \text{(supplementary angles)}$$

also

$$m(\angle B) + m(\angle C) + m(\theta) = 180 \qquad \text{(angles of a triangle)}$$

Therefore, after a little algebra, we see

$$m(\angle A) = m(\angle B) + m(\theta)$$

But θ is an angle of a triangle, so its measure must be positive. Therefore we conclude $m(\angle A) \neq m(\angle B)$. Thus we have shown

$$\text{If } l_1 \cap l_2 \neq \varnothing, \text{ then } m(\angle A) \neq m(\angle B).$$

or, equivalently,

$$\text{If } m(\angle A) = m(\angle B), \text{ then } l_1 \cap l_2 = \varnothing.$$

In words, if two interior alternate angles created by a transversal intersecting two distinct lines are of equal measure, then the lines must be parallel.

Example 4   Prove, using calculus, that if $a > 1$ and r and s are rational numbers such that $r < s$, then $a^r < a^s$.

We assume (a) $a > 1$, a some real number
(b) $r < s$, r, s rational numbers

We try to deduce the conclusion $a^r < a^s$. Since $r < s$ (added premise), then $s - r > 0$. Now consider the function

$$f(x) = x^{s-r}$$

$$f(1) = 1$$

$$f'(x) = (s - r)x^{s-r-1}$$

Note that since $s - r > 0$, and for all $x > 0$, x^{s-r-1} is positive, we have $f'(x)$ is positive for all positive x. This means that f is an increasing function. Therefore,

if $a > 1$ (our added premise)

we obtain $f(a) > f(1)$

or, equivalently, $a^{s-r} > 1$

Now multiply both sides of this inequality by a^r (a positive number) and we obtain

$$a^s > a^r$$

or, equivalently,

$$a^r < a^s$$

Thus we have deduced the conclusion $a^r < a^s$ *providing* the assumptions hold; namely, $a > 1$ and $r < s$ (a real, r, s rational).

Assigned Problems   1. Construct, using conditional proof strategy, two-column logic proofs of each of the following arguments.

(a) Prove: $D \to C$

    Premises: $\begin{cases} A \to (B \to C) \\ \sim D \vee A \\ B \end{cases}$

(b) Prove: $P \to (\sim Q \to R)$

    Premises: $\begin{cases} \sim P \vee M \\ M \to (Q \vee R) \end{cases}$

(c) Prove: $\sim (P \vee R) \to T$

    Premises: $\begin{cases} Q \to P \\ T \vee S \\ Q \vee \sim S \end{cases}$

2. Symbolize the argument below and write a conditional (logic) proof.

Prove: If many students like logic, then mathematics is not easy.

Premises:

(a) Either logic is difficult or not many students like logic.

(b) If mathematics is easy, then logic is not difficult.

3. Prove in conditional format (n is any integer):

(a) If n is odd, then n^2 is odd.

(b) If n^2 is odd, then n is odd.

After having proved both (a) and (b), what can you conclude?

4. Prove: If $2^n - 1$ is a prime number, then n is a prime number. (*Hint:* Can you restate this sentence in "more accessible" form?)

5. Mathematicians frequently prove

(a) sentences of form $P \to (Q \wedge R)$ by proving $P \to Q$ and $P \to R$,

(b) sentences of form $P \to (Q \to R)$ by proving $(P \wedge Q) \to R$,

(c) sentences of form $(P \to Q) \to (S \to R)$ by proving $[(P \to Q) \wedge S] \to R$, and

(d) sentences of form $P \to (Q \vee R)$ by proving $(P \wedge \sim Q) \to R$.

Are such restatements (or substitutions) justifiable? Why?

6. Show that if $0 < x < \pi/2$, then $\sin x + \cos x > 1$.

7. Show that if a and b are nonnegative real numbers and $a \le b$, then $\sqrt{a} \le \sqrt{b}$.

8. Verify that if n is an odd integer, then $n^3 - n$ must be divisible by 8.

9. Suppose that the product of two positive real numbers is a constant A. How is the first number related to the second if the sum of the first plus twice the second is a minimum?

(Note the logical form comparison of this problem's statement and the form in Problem 5(b) above; also, you might use calculus.)

10. Show that if r and s are roots of the quadratic equation $x^2 + px + q = 0$, then $r^3 + s^3 = -p^3 + 3pq$.

11. Answer the following two questions on symmetry in the Euclidean plane:

    (a) If a plane figure is symmetric with respect to a point, then is it also symmetric with respect to some line?

    (b) If a plane figure is symmetric with respect to a line, is it also symmetric to some point? (This is the converse of part a.)

2.3   *Indirect Proof*

Indirect proof is a powerful proof strategy that is also known as *proof by contradiction* or *reductio ad absurdum*. We shall discuss here the strategy of indirect proof from two perspectives—its logical basis and structure and its application in mathematical reasoning and proof.

Logical Basis of Indirect Proof

As a theorem of mathematical logic, the *rule of indirect proof* can be stated as follows:

> If a contradiction can be derived from a set of premises and the negation of sentence C, then C can be derived from the set of premises alone.

In terms of both logical and mathematical proof strategy this means, step-by-step:

1. We have a set of premises $P_1, P_2, P_3, \ldots P_n$ (the premises P_i represent known information—definitions, axioms, theorems, results of problems, etc.).

2. We wish to obtain a certain piece of information C (i.e., we wish to derive C as a conclusion from the premises P_i).

3. We start by introducing the negation of C ($\sim C$) as an *added* premise.

4. From $\sim C$ and the original premises (information), $P_1, P_2, \ldots P_n$, we derive a contradiction (some statement having logical form $Q \wedge \sim Q$).

5. Invoking the *rule of indirect proof*, we have essentially deduced the conclusion *C* from the original set of premises alone.

The logical basis of indirect proof is conditional proof and the *reductio ad absurdum* tautology, which is:

$$[(\sim P) \to (Q \wedge (\sim Q))] \to P$$

That is, if the negation of *P* leads to a contradiction, we can deduce *P*. Indirect proof is a form of conditional proof because in practice, we add the premise $\sim C$ (negation of what we want) to the original premises $P_1, P_2, \ldots P_n$ and from this chemistry of information deduce a contradiction $Q \wedge \sim Q$. The argument is symbolized as:

$$[(P_1 \wedge P_2 \wedge \cdots \wedge P_n) \wedge (\sim C)] \to (Q \wedge \sim Q)$$

By the strategy of conditional proof, this is logically equivalent to

$$(P_1 \wedge P_2 \wedge \cdots \wedge P_n) \to [(\sim C) \to (Q \wedge (\sim Q))]$$

That is, the conclusion $(\sim C) \to (Q \wedge (\sim Q))$ has been derived from the original premises alone. Now, by the *reductio ad absurdum* tautology, the latter statement leads to *C*. Therefore, *C* is derived from the original premises alone (using the law of syllogism as an intervening step).

Example 1   Indirect proof is applied here to a symbolic logic argument.
Prove: $\sim P$
Premises: $\begin{cases} \sim Q \vee R \\ P \to \sim R \\ Q \end{cases}$

Proof

(1)	$\sim Q \vee R$	P
(2)	Q	P
(3)	$(\sim Q \vee R) \wedge Q$	Adj (1)(2)
(4)	$[(\sim Q \vee R) \wedge Q] \to R$	Ins (disjunctive syllogism)
(5)	R	MP (3)(4)
(6)	$P \to \sim R$	P
(7)	P	P_a (added premise)
(8)	$\sim R$	MP (6)(7)
(9)	$R \wedge \sim R$	Adj (5)(8) (contradiction shows up here)
(10)	$\sim P$	Indirect proof (7)(9)

Notice that in the above proof, certain steps that would have appeared in a conditional proof were omitted. These are continued below. (They are not part of the indirect proof, but are given here to show the relationship between indirect and conditional proof.)

$$\vdots$$

(9) $R \wedge \sim R$ Adj (5)(8) (contradiction shows up here)
                                   Indirect proof

(10) $P \to (R \wedge \sim R)$ Conditional proof (7)(9)

(11) $\sim (R \wedge \sim R)$ Ins (law of noncontradiction)

(12) $[P \to (R \wedge \sim R)] \wedge [\sim (R \wedge \sim R)]$ Adj (10)(11)

(13) $\{[P \to (R \wedge \sim R)]$
          $\wedge [\sim (R \wedge \sim R)]\} \to \sim P$ Ins (Modus Tollens)

(14) $\sim P$ MP (12)(13)

Example 2   A different indirect proof for the argument appearing in Example 1 is schematized below. Prove: $\sim P$

Premises: $\begin{cases} \sim Q \vee R \\ P \to \sim R \\ Q \end{cases}$

Proof Schema:

$$\frac{\dfrac{P \to \sim R \qquad\qquad P}{\sim R \qquad\qquad\qquad \sim Q \vee R}}{\dfrac{(\sim Q \vee R) \wedge \sim R}{\dfrac{Q \qquad\qquad\qquad \sim Q}{\dfrac{Q \wedge \sim Q}{\sim P}}}} \quad \text{disjunctive syllogism}$$

Thus, to summarize the logic of indirect proof, we assume (as an added premise or piece of information) the negation of what we wish to conclude; using this added premise, we try to produce a contradiction; if we are successful, we conclude (by indirect proof) the negation of the added premise (which is the information we wanted to obtain).

No proof strategy works successfully in all cases, and it is a matter of experience and effort through which one becomes a good proof-maker. However, from a heuristic standpoint, there are certain cues to look for in attempting an indirect proof. Quite frequently in mathematics we encounter uniqueness arguments of the form: "*show there is only one . . . .*" It is often helpful to begin such proofs by assuming that "*there are at least two distinct . . . ,*" and trying to produce some contradiction (usually with a piece of known information). Also, indirect proof can be helpful when there seems to be no starting point for a proof. In such cases, we search for a "foothold" that gives us leverage or a "way into" the proof, and the added premise sometimes is precisely the information we want.

We turn now to examine how this important proof strategy plays a role in mathematical reasoning.

Mathematical Application of Indirect Proof

Indirect proof is not restricted to just symbolic logic and mathematics. It is an important part of the way we reason in daily life. To preface the subsequent discussion of the mathematical application of indirect proof, consider the following scenario involving argumentation. Keep in mind that indirect proof is analogous to a debating tactic known as "reductio ad absurdum," which means "reduction to the absurd." Using this strategy, the debator assumes the opponent's position is true, and then logically pursues that line of reasoning to a ridiculous conclusion (some contradiction).

> Imagine a dinner conversation between yourself as host and a guest. The guest claims that the poor of the world are poor because they are essentially lazy, and not because of the greed of others. A direct argument against this position might involve world economics, religion, issues of morality, war, famine, natural disaster, crime, and so on. However, suppose you begin by assuming your guest is absolutely correct. He would probably agree that we do not need laws to protect people against the greed of others, since if they are poor, it is through their own fault and the fortunes of life, with which everyone must learn to deal. Therefore, it would be completely reasonable for you to rob your guest of all his belongings and push him out the door to fend for himself. Do you think this line of reasoning would convince your guest?

If you think about it for a while, you will probably recognize many instances in your own reasoning and discourse where you have applied indirect proof, particularly in situations where you emphatically wished to drive a point home.

The following examples should help to illustrate the manner and effectiveness of indirect proof in mathematics.

Example 3   Prove: For any real number x, if $x \neq 0$, then $x^{-1} \neq 0$.

The sentence we wish to prove is in the form of a conditional $\forall_x (P(x) \to Q(x))$, so we start by assuming its negation, $\exists_x (P(x) \land \sim Q(x))$; that is, assume there exists a real number x such that

$$x \neq 0 \quad \text{and} \quad x^{-1} = 0$$

Now, by definition of multiplicative inverse, we know $x(x^{-1}) = 1$. But since $x^{-1} = 0$, and we know any product involving a factor of 0 is 0, from algebra, then

$$x(x^{-1}) = x(0) = 0$$

From this, we conclude $1 = 0$. But that contradicts known information that $1 \neq 0$ (the logical contradiction is $(1 = 0) \wedge \sim (1 = 0)$). Therefore, by indirect proof, we conclude that our original assumption $\exists_x (x \neq 0 \text{ and } x^{-1} = 0)$ was incorrect, and its negation must be true. Hence, $\forall_x$ (if $x \neq 0$, then $x^{-1} \neq 0$).

Example 4   Prove: If x is rational and y is irrational, then $x + y$ is irrational.

For simplicity, we will not include quantifiers in this argument. As in Example 3, the statement to prove is a conditional, only here it has the logical form $(P \wedge Q) \to R$. To negate this, we assume $(P \wedge Q) \wedge \sim R$, which translates to the assumption:

   (x is rational and y is irrational) and $x + y$ is rational.

If, as background knowledge, we know that the difference of two rational numbers is a rational number, then we can argue as follows:

   Consider the number

$$(x + y) - x$$

If $x + y$ is rational and x is rational, then this number must be rational. But $(x + y) - x = y$, and y is irrational. Hence, we have a contradiction:

   y is rational and y is irrational.

Therefore, our assumption is incorrect, and we have proved that if x is rational and y is irrational, then $x + y$ is irrational.

Example 5   A uniqueness proof. Show that there is only one identity element for addition in the system of integers.

Going back to definitions, we first recall what an identity element for addition in the integers means. It is a number (we symbolize it "0") having the property that

   for any integer n, $n + 0 = 0 + n = n$

That is, it preserves the "identity" of that number with which it is added. Now suppose we make the assumption that there are two distinct such identities for addition, and we call them e_1 and e_2. (It's cheating to name them both "0.") Then consider

$$e_1 + e_2 = e_1 \qquad (\text{since } e_1 \text{ is an integer and } e_2 \text{ is an identity element})$$

but also

$$e_1 + e_2 = e_2 \qquad (\text{since } e_2 \text{ is an integer and } e_1 \text{ is an identity element})$$

Therefore,

$$e_1 = e_2$$

But we had assumed that e_1 and e_2 were distinct (i.e., $e_1 \neq e_2$); hence we have our contradiction [namely, $e_1 \neq e_2$ (assumed) and $e_1 = e_2$ (derived)] and we can safely conclude that the additive identity element in the system of integers is unique.

(*Note:* This proof will be easily generalized later in Chapter 6 to the case where "integers" is replaced by an abstract system called a "group," and where the operation " + " is replaced by an abstract binary operation " ∗ " acting on the group.)

Example 6   Find a real solution to the following equation:

$$x^{12} - 4x^9 + 6x^6 - 4x^3 + 5 = 0$$

At first glance, this looks like a formidable task. In fact, this kind of problem often can lead to frustration. The source of difficulty in this particular case is the implicit assumption that a solution *really exists*. The way the problem is stated could lead many to believe that such is the case. Rather than trying numbers or some algorithmic procedure, suppose we start by assuming we already have a solution, and see where this takes us. Call the solution *r*. Then *r* must satisfy the equation, and so

$$r^{12} - 4r^9 + 6r^6 - 4r^3 + 5 = 0$$

Actually, it would appear that all we've accomplished is the replacement of an "*x*" with an "*r*." However, we might note the pattern of the exponents—all multiples of 3—and the coefficient arrangement 1, 4, 6, 4, 5, and ask ourselves if this looks like anything we've seen before. Certainly it appears to be related to a binomial expansion, $(a + b)^n$; the alternating signs hint at $(a - b)^n$; the coefficient arrangement cues $(a - b)^4$; and the lead term tells us to try $(r^3 - 1)^4$:

$$\left(r^3 - 1\right)^4 = r^{12} - 4r^9 + 6r^6 - 4r^3 + 1$$

Therefore, our original equation is really

$$\left(r^3 - 1\right)^4 + 4 = 0$$

from which $(r^3 - 1)^4 = -4$.

Since an even power of a real number can never be negative, we conclude that there is no such *r*; that is, the original equation has no real solution. If it had, we would be faced with the contradiction:

(a) For every real number *x*, every even power of *x* is nonnegative (known information).

(b) There exists a real number *x* and an even power of *x* that is negative (derived information).

Example 7   Prove that $\sqrt{2}$ is irrational.

What follows is the classical little proof of this assertion, and it exemplifies the power of indirect proof. The proof will be sketched here in a series of brief statements, and the reader should fill in the reasons.
Assume $\sqrt{2}$ is rational.
Then $\sqrt{2} = \dfrac{a}{b}$ where $b \neq 0$ and $\dfrac{a}{b}$ is in lowest terms.
So $b\sqrt{2} = a$.
Therefore, $2b^2 = a^2$.
Now *a* and *b* are positive integers each of which is different from 1; therefore, they have unique prime factorizations. It is easy to see that integers a^2 and b^2 have prime factori-

zations each containing an *even* number of primes (counting multiplicities of each repeated prime). But in the last equation above, $2b^2$ and a^2 are supposed to be equal; yet the integer $2b^2$ has an *odd* number of primes in its prime factorization (counting repeated primes), and the integer a^2 has an *even* number of primes in its prime factorization (counting repeated primes). This contradicts the Fundamental Theorem of Arithmetic, which asserts that every positive integer greater than 1 has a unique (up to order) prime factorization.

Assigned Problems

1. Construct, using indirect proof strategy, two-column logic proofs of the following symbolic arguments.

   (a) Prove: R

   Premises: $\begin{cases} \sim (P \land Q) \\ \sim R \to Q \\ \sim P \to R \end{cases}$

   (b) Prove: $\sim T$

   Premises: $\begin{cases} P \lor Q \\ T \to \sim P \\ \sim Q \lor R \\ \sim R \end{cases}$

   (c) Prove: $\sim (A \land D)$

   Premises: $\begin{cases} A \to (B \lor C) \\ B \to \sim A \\ D \to \sim C \end{cases}$

   (d) Prove: $\sim E \lor M$

   Premises: $\begin{cases} S \lor O \\ S \to \sim E \\ O \to M \end{cases}$

2. Prove via indirect proof: For any real numbers x, y, if $xy = 0$, then $x = 0$ or $y = 0$.
   (*Note:* This theorem is the basis for the solution of quadratic equations by the factoring method.)

3. Prove using indirect proof: If p is a prime number, then $\sqrt{p}$ is an irrational number.
   (*Hint:* Have you seen a statement like this before?)

4. Show that if A and B are sets in some universe U, then if $A \cap B \neq \varnothing$, then $A \neq \varnothing$.

5. Is it true that the sum of any positive real number and its reciprocal is always greater than or equal to 2? Show why or why not.

6. Prove in the system of real numbers that any nonzero real number has at most one multiplicative inverse (reciprocal).

7. Demonstrate in a proof by contradiction that for any natural number n,

$$\frac{n}{n+1} < \frac{n+1}{n+2}$$

8. Euclid, the great Greek geometer (330–275 B.C.), constructed a proof that given any nonempty finite set of primes, there is always a prime number that is not in the set. Essentially what Euclid did was to consider the number

$S = \{2, 3, 5, 7, \cdots\}$

Proof by contradiction

$(p_1 p_2 \ldots p_n) + 1$ where $p_1, p_2, \ldots p_n$ are the primes in the finite set. Construct an *indirect proof* (which is equivalent to Euclid's reasoning) of the statement "there are infinitely many prime numbers."

9. Prove by contradiction:

   (a) $\forall_{a,\, b \in \mathbf{R}^+}$ $(a + b \geq \sqrt{a^2 + b^2})$

   (b) $\forall_{n \in N}$ (if n^2 is even, then n is even)

10. Consider the parabola $y = x^2$ and the point $(1, 2)$. Prove that there is no tangent line to the curve containing this point.

11. We often exploit an analogy to finite cases when confronted with a problem involving infinitely many objects. Sometimes, however, the analogy breaks down, and only serves to confuse us. Consider, for example, the following situation: Visualize three of Santa's sacks labeled A, B, and C. Sacks B and C are empty, but sack A contains toys numbered $1, 2, 3, \ldots$ (assume infinitely many toys labeled by natural numbers). Assume Santa removes those toys numbered 1 through 10 from sack A and places them in sack B; then he removes toy number 1 from sack B and places it in sack C. After this process is completed, he takes toys numbered 11 through 20 and places them in sack B, then removes toy number 2 and places it in sack C. He continues the process in similar fashion, taking the next ten toys from sack A, placing them in sack B, then taking toy number 3 and placing it in sack C. Suppose he could continue like this until sack A is empty. How many toys would sacks B and C contain?

2.4   *Proof by Cases*

There are a number of instances in mathematics where it becomes necessary to examine several possibilities (cases) one-by-one to see what conclusion we might reach in each case, particularly to see if the conclusion in every case is the same. Some frequently occurring cases are:

$$x + \frac{1}{x} = \frac{x}{x}$$

1. x is positive or x is negative

2. $a < b$; $a = b$; $a > b$

3. n is even or n is odd

4. m is prime, $m = 1$, or m is composite

5. $A \cap B = \varnothing$ or $A \cap B \neq \varnothing$

6.

$x < 0$	$x < 0$	$x > 0$	$x > 0$	
and	and	and	and	$x = 0$; $y = 0$
$y < 0$	$y > 0$	$y < 0$	$y > 0$	

7. r is rational or r is irrational

8. G is finite or G is infinite

9. f is increasing; f is decreasing

10. $x \in A_1$; $x \in A_2$; $x \in A_3 \ldots x \in A_n$

11. $P \rightarrow Q$; $Q \rightarrow P$; $P \leftrightarrow Q$; neither $P \rightarrow Q$ nor $Q \rightarrow P$

12. m divides n; n divides m; neither; both

13. $\angle A$ is acute; $\angle A$ is right; $\angle A$ is obtuse (oblique)

14. $\displaystyle\sum_{i=0}^{\infty} a_i x^i$ converges; $\displaystyle\sum_{i=0}^{\infty} a_i x^i$ diverges

In the strategy of *proof by cases* we list all the possible alternatives that could occur in our domain of interest and try to show that the same conclusion is reached in every case. Let us examine the logical structure of *proof by cases*. First, we can simplify the set of alternatives by reducing it to two cases: either something is true or it is not ($R \vee \sim R$). Suppose we wish to deduce a conclusion C by itself from known information (original set of premises $P_1, P_2, P_3, \ldots P_n$). If we can deduce C from our known information $P_1, P_2, \ldots P_n$ *together with* an added premise R, and then if we can independently deduce C from the known information $P_1, P_2, \ldots P_n$ *together with* the added premise $\sim R$ (negation of the first added premise), then we can deduce C from the original premises alone. The foregoing statements are summarized in the *rule of proof by cases:*

> If a conclusion C can be derived independently from a statement R and from its negation $\sim R$ together with a set of premises, then C can be derived from the original premises alone.

As you can see, in the strategy of proof by cases we are adding premises. (*Note:* We are *not* adding a contradiction, since R is added, C is deduced; then $\sim R$ is added, C is deduced again.) Therefore, proof by cases is a form of conditional proof. Its logical basis resides in two tautologies: the *law of the excluded middle* ($R \vee \sim R$) and the *law of the dilemma*:

$$\{[(P \rightarrow R) \wedge (Q \rightarrow S)] \wedge (P \vee Q)\} \rightarrow (R \vee S)$$

in the instance where P is replaced by R, Q is replaced by $\sim R$, and both R and S are replaced by C,

$$\{[(R \rightarrow C) \wedge (\sim R \rightarrow C)] \wedge (R \vee \sim R)\} \rightarrow (C \vee C)$$

Since $C \vee C$ is logically equivalent to C itself (a tautology), we see that the outcome is the derivation of C alone resting on whatever known information we used $(P_1, P_2, \ldots P_n)$ and not dependent on either added premise $(R, \sim R)$.

It might be helpful to visualize *proof by cases* schematically as shown in Figure 2.1.

Note that proof by cases is really a method of circumventing a series of additional steps that would be required in a conditional proof: the deduction of $R \rightarrow C$ from the original premises, the deduction of $\sim R \rightarrow C$ from the original premises, adjunction of these two conditionals and a second adjunction with $(R \vee \sim R)$, the law of the dilemma, the conclusion $C \vee C$ by modus ponens, and finally C (deduced from the original premises).

Example 1   A logic proof.

Prove: $\sim P \vee Q$

Premise: $P \rightarrow Q$

We observe that there is an important tautology that asserts

$$(P \rightarrow Q) \leftrightarrow (\sim P \vee Q)$$

Here we are asked to verify the $(\leftarrow)$ part of that tautology. Obviously, we shall not invoke that particular tautology in the proof (it is a stronger statement than that we're trying to prove). Note that part of the proof is indented (where the *second* added premise leads to the desired conclusion) to separate it from the first part and to avoid any misinterpretation of a contradiction in steps (2) and (8).

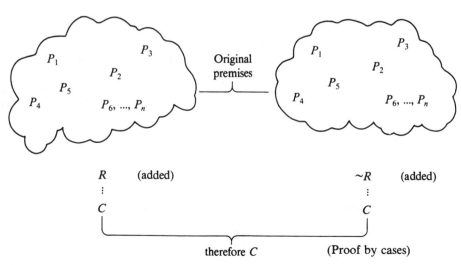

Figure 2.1

Proof	(1)	$P \to Q$	P
	(2)	$\sim Q$	P_a (first added premise)
	(3)	$(P \to Q) \leftrightarrow (\sim Q \to \sim P)$	Ins (contraposition)
	(4)	$\sim Q \to \sim P$	BR (1)(3)
	(5)	$\sim P$	MP (2)(4)
	(6)	$\sim P \to (\sim P \vee Q)$	Ins (simplification)
	(7)	$\sim P \vee Q$	MP (5)(6)
	(8)	Q	P_a (second added premise)
	(9)	$Q \to (\sim P \vee Q)$	Ins (simplification)
	(10)	$\sim P \vee Q$	MP (8)(9)
	(11)	$\sim P \vee Q$	Proof by cases (2)(7)(8)(10)

Example 2   For any integer n, $n^2 - n$ is always even.

The premises here include the information that n is an integer as well as all our known information about integers and mathematics relevant to this problem. We know that the integers fall into two classes: even or odd. Thus we can construct a proof by cases. Note that, in any mathematical proof by cases, there is an understood *or* between the cases.

Case 1: Suppose n is *even*. Then $n = 2k$ for some integer k and

$$n^2 - n = (2k)^2 - 2k$$
$$= 4k^2 - 2k$$
$$= 2(2k^2 - k)$$

an even integer since $2k^2 - k$ is an integer if k is an integer.

Case 2: Suppose n is *odd*. Then $n = 2k + 1$ for some integer k and

$$n^2 - n = (2k + 1)^2 - (2k + 1)$$
$$= 4k^2 + 4k + 1 - 2k - 1$$
$$= 4k^2 + 4k - 2k$$
$$= 2(2k^2 + 2k - k)$$

again, an even integer since $2k^2 + 2k - k$ is an integer if k is an integer.

In each case (and these were the only two possibilities, n even or n odd) we reached a conclusion that $n^2 - n$ is an even integer. Therefore, appealing to the method of proof by cases, we conclude $n^2 - n$ is always even when n is an integer.

Example 3   A different proof by cases for the statement in Example 2.

Prove: If n is an integer, then $n^2 - n$ is always even.

Note that $n^2 - n = n(n - 1)$, the latter being two consecutive integers. Now if n is even (case 1), then regardless of what $n - 1$ is (it's odd), the product $n(n - 1)$ must be even. On the other hand, if n is odd (case 2) then $n - 1$ must be even, and again the product $n(n - 1)$ is even. Thus, in either case, we have demonstrated that $n^2 - n$ is an even integer, if n is an integer.

Example 4   Rolle's theorem from calculus.

If a function f is continuous on a closed interval $[a, b]$, differentiable on an open interval (a, b), and if $f(a) = f(b)$, then $f'(c) = 0$ for at least one number c in (a, b).

This theorem has three main conditions on the function f: continuity over $[a, b]$, differentiability over (a, b), and equality at the endpoints $f(a) = f(b)$. It is probably best classified as an "existence" theorem, because we wish to show the *existence* of a number "c" somewhere in the interval (a, b) that makes the derivative zero.

Geometrically, what Rolle's theorem is saying is that under the conditions given above, the graph of the function f has a horizontal tangent somewhere between $(a, f(a))$ and $(b, f(b))$.

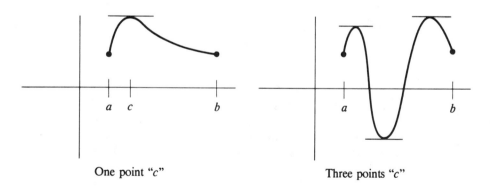

One point "c"   Three points "c"

A proof by cases is given next. The proof rests upon the observation that the function f must fall into at least one of the three categories defining the cases.

Case 1: $f(x) = f(a)$ for *all* $x \in (a, b)$. In this case, f is a constant function and therefore $f'(x) = 0$ for *every* $x \in (a, b)$.

Case 2: $f(x) > f(a)$ for *some* $x \in (a, b)$. In this case, the maximum value f takes on (a, b) must be greater than $f(a)$ or $f(b)$, and therefore it must occur at some interior point $c \in (a, b)$. Since f is differentiable over (a, b), we conclude $f'(c) = 0$. (A theorem in calculus tells us that extrema of functions can exist only at critical points—where the function has derivative zero or undefined—or at endpoints.)

Case 3: $f(x) < f(a)$ for *some* $x \in (a, b)$. This case is symmetric to case 2, except we make those observations about the minimum value of f. As in case 2, we conclude $f'(c) = 0$ for some number $c \in (a, b)$.

Thus, in every possible case we reach the same conclusion: $f'(c) = 0$ for some $c \in (a, b)$, providing the three conditions of Rolle's theorem are met.

Example 5   A theorem on absolute value: the triangle inequality. If a, b are real numbers, then

$$|a + b| \leq |a| + |b|$$

In words, the absolute value of a sum of any two real numbers is always less than or equal to the sum of their absolute values.

This theorem is called the *triangle inequality* because we can think of $|a|$, $|b|$, and $|a + b|$ representing magnitudes (lengths) of vectors (arrows indicating length and direction). We add two plane vectors by visualizing the terminal point of one connecting the initial point of the other, then drawing a vector (the sum vector) extending from the initial point of the first to the terminal point of the second.

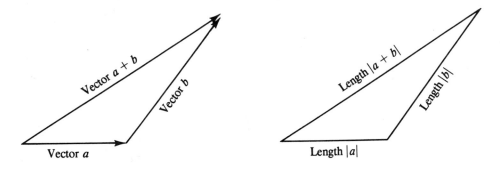

If the vectors are not collinear, then this arrangement forms a triangle, and since the shortest distance between two points is a straight line, we have

$$|a + b| < |a| + |b|$$

Equality occurs when the vectors are collinear and have the same direction (i.e., the "numbers" a and b are both positive or both negative).

Proof of Triangle Inequality (by Cases)

Case 1: $a \geq 0$ and $b \geq 0$

Here, $|a| = a$; $|b| = b$

and $|a + b| = a + b$

$$= |a| + |b|$$

$\therefore \quad |a + b| \leq |a| + |b|$

Case 2: $a < 0$ and $b < 0$

$$\text{Here, } |a| = -a; \ |b| = -b$$

$$\text{and } |a + b| = -(a + b)$$

$$= (-a) + (-b)$$

$$= |a| + |b|$$

$$\therefore \quad |a + b| \le |a| + |b|$$

Case 3: $a \ge 0$ and $b < 0$

$$\text{Here, } |a| = a; \ |b| = -b$$

Now we take two subcases inside case 3:

Subcase 3(a)	*Subcase 3(b)*
$(a + b) \ge 0$	$(a + b) < 0$
then $|a + b| = a + b$	then $|a + b| = -(a + b)$
$< a + (-b)$	$= (-a) + (-b)$
since $b < 0$	$\le a + (-b)$
$= |a| + |b|$	since $a \ge 0$
$\therefore \quad |a + b| < |a| + |b|$	$= |a| + |b|$
$\therefore \quad |a + b| \le |a| + |b|$	$\therefore \quad |a + b| \le |a| + |b|$

Thus (from our two-case argument inside case 3), in either case, we obtain $|a + b| \le |a| + |b|$.

Case 4: $a < 0$ and $b \ge 0$

This case is exactly symmetric to case 3, interchanging the roles of a and b. Thus, we again conclude

$$|a + b| \le |a| + |b|$$

The above four cases exhaust all possibilities for a and b (which affect the absolute values), hence we conclude

$$|a + b| \le |a| + |b| \text{ for all real numbers } a, b$$

Assigned Problems

1. Construct, using proof by cases, two-column logic proofs of the following symbolic arguments.

(a) Prove: Q

Premises: $\begin{cases} \sim R \to S \\ S \to (P \land Q) \\ R \to \sim T \\ T \lor Q \end{cases}$

(b) Prove: C

Premises: $\begin{cases} A \to B \\ \sim A \to C \\ B \to C \end{cases}$

2. Prove, by taking appropriate cases on x,

   (a) If $|x| < a$, then $-a < x < a$ (a, x are real numbers; can a be negative?)

   (b) $|x \cdot y| = |x| \cdot |y|$ for any real numbers x, y

3. Symbolize the argument below (M, L, H, P) and apply proof by cases to verify it.

   Prove: Logic is delightful.

   Premises: If mathematics is fun, then logic is delightful.
   If history is dull, then mathematics is fun.
   If physics is not absorbing, then logic is delightful.
   History is dull or physics is not absorbing.

4. The function f described by $f(x) = |x|$, $x \neq 0$ is differentiable, but the function g described by $g(x) = |x|$ is not differentiable. Why? Find a formula for f' using proof by cases.

5. Prove (by cases): If m is an integer, then $m^2 + m + 1$ is always odd.

6. There is a theorem in set theory that states, for any two sets A, B,

$$\overline{(A \cup B)} = \overline{A} \cap \overline{B}$$

Illustrate the truth of this statement by taking various cases on the relationship between A and B ($A \cap B = \emptyset$, $A \subseteq B$, etc.) and drawing Venn diagrams in each case. The symbol $\overline{S}$ refers to the universal complement of the set S, or the set of all elements in the universe that are *not* elements in S (see Chapter 3).

7. Verify the trigonometric identity

$$\sin(a + b) = \sin a \cos b + \sin b \cos a$$

You may use the trigonometric identity

$$\cos(a \pm b) = \cos a \cos b \mp \sin a \sin b$$

and any other identities you know. (*Hint:* This is not necessarily a proof by cases. You might think geometrically or use certain identities strategically.)

8. Heron's formula for the area of a triangle having side lengths a, b, c and semiperimeter s, where $s = (a + b + c)/2$, is

$$A^2 = s(s - a)(s - b)(s - c)$$

Check the validity of this formula in the two special cases where the triangle is (a) a right isosceles triangle; (b) an equilateral triangle. (This does not, of course, verify the formula in general, since we have not considered all possible cases.)

9. Every integer leaves a remainder of 0, 1, or 2 upon division by 3. In each case, respectively, the integer could be represented in the form $3k$, $3k + 1$, or $3k + 2$, where k is some integer. Using this information, construct a proof by cases that in any sequence of three consecutive integers, one is a multiple of 3.

10. Informal observation is all that this problem requires, but think in terms of cases in reasoning through the solution; that is, examine all the variation on chord AB. The two circles C_1 and C_2 are concentric. Let AB be a chord of C_1 that intersects C_2 in two points C and D. There is no instance of a chord AB being exactly twice the length of CD. If the radius of C_1 is 8, what are the possibilities for the radius of C_2?

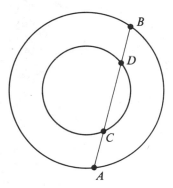

2.5   Existence Proofs

According to the James and James *Mathematics Dictionary*, an *existence theorem* is a theorem that asserts the existence of at least one object of some specified type, and an argument that establishes an existence theorem is called an *existence proof*. The key words to be aware of here are *at least one*; that is, in order to verify the existence of some object, we need only produce one example. In some cases, we need not even produce the object itself, but rather we might deduce its existence from known conditions.

Philosophically, there are different forms of existence. For example, when we assert that there exists a fish, we mean that somewhere in nature an object that ichthyologists call a fish can be physically observed. This kind of existence is *empirical* existence. In contrast, another kind of existence is *definitional* existence. That is, in mathematics, we are generally free to define into existence anything we wish, such as the idea of real number. (If, however, the associated system becomes self-contradictory, we reject the definition.) Note that a real number is empirically nonexistent (did you ever actually observe a real number as a physical entity?).

In this course we shall not be as concerned with empirical or definitional existence as we are with *deductive* existence. This latter form of existence deals with entities whose existence is not specifically indicated in the definitions or axioms that characterize the given system, but whose existence or nonexistence can be deduced from these definitions or axioms. For example, assuming we are familiar with the real numbers as an axiomatic system, is there a real number that equals its square? If we wish to prove that there (deductively) exists such a real number, we exhibit it and show it works. Note that to verify its existence, we need only produce one object or deductively force its existence, even though there may be more. (The latter issue leads to questions of uniqueness or nonuniqueness, a different question than existence.) If we wish to prove there does not exist such a real number, we have the burden of proving the property stated is not true for any real number. The latter is not an existence proof. The three statements below exemplify the foregoing three forms of existence:

1. There exists a highway from Chicago to New York (empirical).

2. There exists a line (definitional).

3. There exists a real root for the polynomial $x^2 - 3x + 2$ (deductive).

Statements of (deductive) existence may at first glance appear to be easy to prove, since all we need to produce is a single example. However, in mathematics, one encounters situations where finding even a single case is quite difficult, or even undetermined. Take, for example, the famous Fermat conjecture:

If $n > 2$, then the equation $x^n + y^n = z^n$ has no nontrivial solution in the integers.

The Fermat conjecture could be disproved (say in the case where $n = 5$) by showing a single triple of integers x, y, and z that satisfy the equation, but nobody has been able to disprove nor prove it. (Fermat himself constructed a proof of its truth for $n = 4$.)

You may recall, with varying degrees of frustration, the first time in a calculus course you were presented with the following:

Given any $\varepsilon > 0$, *produce a $\delta > 0$* so that $|x^2 - 9| < \varepsilon$ whenever $|x - 3| < \delta$.

This is the crux of proving, from the formal definition of limit, that

$$\lim_{x \to 3} (x^2 - 4) = 5$$

All such limit proofs are really existence proofs asking the question "given any ε, find a suitable δ that satisfies certain requirements" (a conditional statement involving certain absolute value inequalities).

The following examples provide a range of problems on existence.

Example 1   Show that the equation $6x^4 - 13x^3 - 6x^2 + 5x + 2 = 0$ has a real solution in the interval $[0, 1]$.

First we restate the problem. Let $f(x) = 6x^4 - 13x^3 - 6x^2 + 5x + 2$. We want to find a real root (zero) for f. Now we know that f is a polynomial function, and we recall that polynomial functions are continuous for all real numbers; that is, their graphs have no breaks. Using the analogy between algebraic representation and geometric representation, we might start thinking about the graph of this function over the interval $[0, 1]$. The function will have a zero if the graph crosses the x-axis. The graph will cross the x-axis if the function is positive-valued at one endpoint and negative-valued at the other. Thus we test $f(0) = 2$ and $f(1) = -6$, and we conclude the existence of a real root between 0 and 1. Note that we have deduced its existence without even producing the actual root (which, incidentally, is $2/3$). We have implicitly invoked the intermediate value theorem from calculus in our reasoning.

An alternate procedure would be to actually graph the function and visually see that the graph crosses the x-axis somewhere between 0 and 1. The graph of f follows.

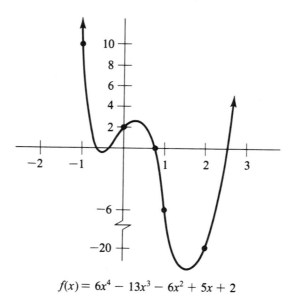

$$f(x) = 6x^4 - 13x^3 - 6x^2 + 5x + 2$$

Additionally, one could use numerical methods (e.g., bisection method) or a computer to obtain an approximate solution. Moreover, we might be lucky enough by trial and error to hit on the right solution. However, in many practical instances, it is helpful to know whether a solution exists before expending the time, energy, and perhaps money to determine it.

Example 2   Find a function $(R \rightarrow R)$ that is both odd and even.

We need to know the definition of odd and even functions before we can proceed here. A function is odd if $f(-x) = -f(x)$ for every $x \in \text{dom } f$; a function is even if $f(-x) = f(x)$

for every $x \in \text{dom } f$. The sine function is an example of an odd function, and the cosine function is an even function. Not all functions are either odd or even (e.g., $f(x) = x^2 + x^3$), but we want one that is both. Again, visualization may be of great assistance. There are graphical characteristics involving symmetry for even and odd functions. Even functions have graphs symmetric with respect to the y-axis; odd functions have graphs symmetric with respect to the origin. Suppose we try to sketch the graph of a function that is both even and odd. Start with point A in the first quadrant. The symmetry conditions force the existence of both A' and A'' on the graph of the function. But this makes it a nonfunction (vertical line test). We run into the same difficulty if we try to place A in *any* quadrant. This tells us something about the placement of A—it must be *on* the x-axis to avoid the nonfunction phenomenon, and there must be a symmetric point through the origin. Thus, one example of such a function is the zero-constant function $f(x) = 0$ for all $x \in R$, whose graph is the entire x-axis. Other examples would include functions whose graphs consist of point-sets on the x-axis that are symmetric with respect to the origin.

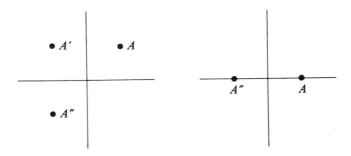

Example 3   Prove the denseness property of the system of rational numbers: Between any two distinct rational numbers a and b there is another rational number.

Since a and b are distinct, suppose (without loss of generality) that $a < b$. We conjecture that the arithmetic mean $(a + b)/2$ does the job, but we must show that it works.

Since $a < b$, then

$$a + b < b + b \quad \text{(add } b \text{ to both sides)}$$

$$\therefore \quad \frac{a + b}{2} < \frac{b + b}{2}$$

$$\therefore \quad \frac{a + b}{2} < b$$

Then, by a symmetric argument, since $a < b$, then

$$a + a < a + b \quad \text{(add } a \text{ to both sides)}$$

$$\therefore \quad \frac{a + a}{2} < \frac{a + b}{2}$$

$$\therefore \quad a < \frac{a + b}{2}$$

Therefore, we have

$$a < \frac{a + b}{2} < b$$

and the arithmetic mean of a and b is between a and b.

Thus, our existence proof is complete. We have produced a rational number that is between any two given rational numbers. (Note that this reasoning extends to prove that between any two rational numbers there are *infinitely many* rational numbers.)

Example 4   It can be proved (nice argument using backwards reasoning) that the geometric mean $\sqrt{ab}$ of any two nonnegative real numbers a and b is always less than or equal to their arithmetic mean. That is,

$$\text{if } a \geq 0 \text{ and } b \geq 0, \text{ then } \sqrt{ab} \leq \frac{a + b}{2}$$

Suppose we wish to find real numbers a, b for which equality in this statement holds. Clearly there is a trivial solution $a = b = 0$. That observation in itself is an existence proof and we are done. But suppose we explore a little further and ask if there are any *nontrivial* solutions, and if so, what conditions would have to be met?

We want

$$\sqrt{ab} = \frac{a + b}{2}$$

Thus,

$$2\sqrt{ab} = a + b$$

and

$$4ab = a^2 + 2ab + b^2$$

so

$$a^2 - 2ab + b^2 = 0$$
$$(a - b)^2 = 0$$

We conclude that $a = b$.

Thus, we have found that for nonnegative real numbers a and b, if $\sqrt{ab} = (a + b)/2$, then $a = b$.

Example 5   Assuming not both a and b are 0, does the complex number $a + bi$ have a multiplicative inverse?

If $b = 0$ and $a \neq 0$, then the desired inverse is $(1/a)$ in the system of real numbers. So, let us assume $b \neq 0$. We ask ourselves, "What is the role of a multiplicative inverse?" When $a + bi$ is multiplied by this inverse, we should expect as a product the identity element for multiplication $(1 + 0i)$ in the system of complex numbers. Thus, suppose the inverse exists; what would it have to be? Call it $c + di$. Then,

$$(a + bi)(c + di) = 1 + 0i$$

so

$$ac + (bc + ad)i - bd = 1 + 0i$$

Thus,

$$ac - bd = 1 \text{ and } bc + ad = 0 \qquad \text{(equating real and imaginary coefficients)}$$

Now, can we solve for c and d in terms of a and b? Since $bc = -ad$, if $b \neq 0$, then $c = (-ad/b)$. Substituting in the first equation and solving for d, we obtain (after some algebra)

$$d = \frac{-b}{a^2 + b^2}$$

Similarly, we note that if $b \neq 0$, then

$$bd = ac - 1$$

and

$$d = \frac{ac - 1}{b}$$

Substituting this into the second equation and solving for c, we obtain (after some algebra)

$$c = \frac{a}{a^2 + b^2}$$

Thus,

$$c + di = \left(\frac{a}{a^2 + b^2}\right) - \left(\frac{b}{a^2 + b^2}\right)i$$

Finally, we test this to make sure it does the job of a multiplicative inverse:

$$(a + bi) \cdot \left[\left(\frac{a}{a^2 + b^2}\right) - \left(\frac{b}{a^2 + b^2}\right)i\right]$$

$$= \left(\frac{a^2}{a^2 + b^2}\right) + \left[\left(\frac{ab}{a^2 + b^2}\right) - \left(\frac{ab}{a^2 + b^2}\right)\right]i + \left(\frac{b^2}{a^2 + b^2}\right)$$

$$= 1 + 0i$$

Assigned Problems

1. Show that the equation $x^4 - x^2 - 5 = 0$ has a real solution that lies in the interval between $x = 1$ and $x = 2$.

2. Show that the polynomial $y^{70} + 2y - 1$ has a zero in $[0, 1]$.

3. A common error made by first-year algebra students is to assume

$$\frac{1}{a + b} = \frac{1}{a} + \frac{1}{b}$$

It is easy to show that this equality is not true for all real numbers a and b. Are there *any* real numbers for which it is true? Explain.

4. Produce, in each case below, a real number for which the inequality is true.

   (a) $x^3 \leq x^2$

   (b) $\dfrac{1}{x} \geq x$

   (c) $\sqrt{x} \geq \tan^{-1}x \; (x \geq 0)$

   (d) $a^x \leq x$ for some $a > 0$, $a \neq 1$

   (e) $|\ln x| \geq x \; (x > 0)$

   (f) $\cosh x > e^x$

   Does visual thinking help on these exercises?

5. Prove: For any irrational number x, there exists another irrational number y such that $x \cdot y$ is rational.

6. Does the complex number $0 - 2i$ have a square root? If so, produce it; if not, show why not.

7. Prove:

$$\lim_{x \to 2} (5x + 4) = 14$$

   That is, given any $\varepsilon > 0$, show there exists a $\delta > 0$ such that $|(5x + 4) - 14| < \varepsilon$ whenever $|x - 2| < \delta$. (*Note:* In most cases, the choice of δ depends on ε; also, work backwards.)

8. Prove an integer exists that meets the following conditions:

   (a) Its digits have strictly increasing values (i.e., going from left to right).

   (b) The sum of its digits is 10.

   (c) The product of its digits is 24.

9. Produce a function whose domain consists of the set of all real numbers and whose range is the set of all positive integers.

10. Find a pair of nonsimple functions f and g such that

$$\lim_{x \to 2} f(x) \quad \text{and} \quad \lim_{x \to 2} g(x)$$

   do *not* exist, but

$$\lim_{x \to 2} [f(x) + g(x)]$$

   does exist. (*Nonsimple* means the functions do not take a given value more than once.)

2.6   *Mathematical Induction*

Mathematical induction is a specific, direct form of deductive proof that is used in verifying generalizations about *integers*. It is different from other forms of direct proof in several ways, and it is different from the heuristic process we call "inductive reasoning." We shall first explore these differences, and then discuss the proof technique of *mathematical induction*.

Inductive reasoning (or reasoning by induction) is a heuristic strategy that involves making generalizations based upon observation and experimentation with a series of particular instances. For example, suppose we start drawing lines at random in a plane (with no two lines parallel and no three intersecting in a point), and each time we add a line, we ask, "Into how many separate regions is the plane divided?" We observe that if there were no lines ($n = 0$), then there would be one region—the plane itself; if there were one line ($n = 1$), we have two regions; two lines ($n = 2$) produce four regions; three lines ($n = 3$) produce seven regions. To organize our experimental observations, we make a table where n means the number of lines, and P_n means the number of plane regions created by n lines:

n	P_n
0	1
1	2
2	4
3	7
4	11
⋮	⋮

Next we look at the table to see if there is any regularity or pattern in the numbers for P_n. Can we predict how many regions there would be if there were any number (n, in general) of lines drawn? Suppose, by some technique or sheer luck, we came up with the explicit formula

$$P_n = \frac{n^2 + n + 2}{2}$$

Have we demonstrated that this formula actually produces the number of plane regions created by n lines *for any nonnegative integer n*? Certainly not, since our formula is based on observation of perhaps only five or six special cases. It is a *guess*, a *conjecture*; and it must be verified in general if we are to assert that it holds for all $n \geq 0$.

The foregoing example illustrates reasoning by induction, or *pattern-search*. The problem solver explores special cases to see if there is a pattern; if he or she discovers a pattern, he or she makes a *conjecture* (guess). *It must be stated emphatically here that the result of inductive reasoning is a conjecture.* No one is forced logically to agree with a conjecture, although the more cases that support it tend to increase our confidence that it may be correct. However, examination of any number of special cases *never* proves a conjecture in general. At the same time, a conjecture may be disproved with a single counterexample. For example, the expression

$$n^2 + n + 41$$

yields values of 43, 47, 53, 61, and 151 for trial cases where n is 1, 2, 3, 4, and 10 respectively. These observations may lead us to the conjecture, for all positive integers n:

$$n^2 + n + 41 \text{ is a prime number}$$

Is this conjecture true?

Thus, *inductive reasoning* (pattern-search) is different from *deductive reasoning* (proof-making). Deductive reasoning involves proceeding from a hypothesis (premise) to a conclusion through a finite sequence of steps, each step being formally justified by an axiom, definition, theorem, statement accepted as true, or rule of logical inference. The result of deductive reasoning is a proof. Anyone who accepts the axioms, definitions, theorems, and agrees with the logic must by necessity accept the result of a deductive proof as true. In contrast, inductive reasoning, being based on a finite number of observations, generates conjectures that may be true or false, and the reasoning process does not verify the result. After a conjecture is made, we may wish to attempt to verify that the conjecture is true for all cases (integers) and it is here that we need a deductive method to construct a proof. *Mathematical induction* is such a deductive method. The relationship between inductive reasoning, conjecture, and deductive reasoning is summarized in Figure 2.2.

Although mathematical induction is a proof technique, a form of deductive reasoning, it differs in certain ways from the "general" proof techniques—direct, conditional, indirect, cases. First, mathematical induction is restricted to one class of mathematical statements—generalizations about integers (e.g., "prove for all positive integers n...," "show for all $n \leq -1$...," "prove that this statement...holds for all integers"). Moreover, it is an algorithmic technique; that is, it can be used without a great deal of creativity. Certain steps to follow

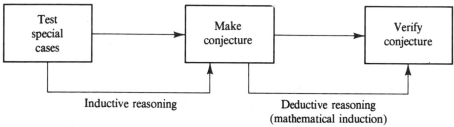

Figure 2.2

will be described to outline the technique, and if these steps are followed carefully, then there is usually not much difficulty in constructing the proof. Unlike direct or conditional proof, where the proof-maker has the problem of *inventing* the proof, mathematical induction provides a rough-structured framework in which to operate. To further simplify our discussion of this important technique, we shall restrict our generalizations to conjectures involving only positive (natural numbers) or nonnegative integers.

Mathematical induction, as a proof technique, is based on the following major theorem of number theory, called *The Principle of Mathematical Induction* (*PMI*):

> *If* N is the set of natural numbers   ($N = \{1, 2, 3, \ldots\}$)
>   and $M \subseteq N$ having the following properties:
>   (a) $1 \in M$
>   (b) for each $n \in N$, $n + 1 \in M$ whenever $n \in M$
>   (i.e., if $n \in M$, then $n + 1 \in M$)
> *then* $M = N$.

This theorem says that if a subset of the natural numbers contains the number "1" and for any natural number in the subset, it also contains the successor (the next one), then it must contain *all* the natural numbers. Usually, property (a) in the theorem statement is called the *starting point*, the "if" part of property (b) is called the *inductive hypothesis*, and all of property (b) is called the *inductive step*. For the sake of simplicity, the discussion here will be restricted to the positive integers starting at 0 or 1 and upward. It should be noted, however, that the technique of mathematical induction extends to situations involving infinite subsets of the set of integers, starting at any point and proceeding in either the positive or negative direction.

Associated with the above theorem (Principle of Mathematical Induction), the proof technique we call *mathematical induction* is outlined next.

- We have a statement P_n that we wish to show is true for all natural numbers n. (It probably arose by investigating special cases and making the conjecture "P_n".)

- We define a set $M = \{n \mid n \in N \text{ and } P_n \text{ is true}\}$. (If we can show $M = N$, this would be equivalent to showing that M contains *all* natural numbers; i.e., the statement P_n is true for *all* natural numbers.)

- To show $M = N$, we do three things; namely,

   (i) show $1 \in M$ (starting point; show P_1 is true)
   (ii) assume $n \in M$ (inductive hypothesis; assume P_n is true)
   (iii) show $n + 1 \in M$ (inductive step complete; using assumed fact that P_n is true, deduce that P_{n+1} must be true)

- This done, we can invoke the PMI to conclude $M = N$, and consequently assert the statement "P_n is true for all natural numbers n."

We illustrate the technique of mathematical induction with several examples.

Example 1   Prove: $\forall_{n \in N} \left(\sum_{i=1}^{n} i = \dfrac{n(n+1)}{2} \right)$

(The statement "$\sum_{i=1}^{n} i = \dfrac{n(n+1)}{2}$" is the statement "$P_n$")

Proof   Define $M = \{ n \mid n \in N \text{ and } \sum_{i=1}^{n} i = \dfrac{n(n+1)}{2} \}$

Claim: $M = N$

To show this:

(i) Is $1 \in M$? Yes, since $\sum_{i=1}^{1} i = 1 = \dfrac{1+1}{2}$ (P_1 is true)

(ii) Assume $n \in M$; that is, $\sum_{i=1}^{n} i = \dfrac{n(n+1)}{2}$ for any specific n (inductive hypothesis; P_n assumed to be true)

(iii) We show $n + 1 \in M$; that is, $\sum_{i=1}^{n+1} i = \dfrac{(n+1)[(n+1)+1]}{2}$ (prove P_{n+1} true)

$$\sum_{i=1}^{n+1} i = \left(\sum_{i=1}^{n} i \right) + (n+1) \qquad \text{(group together the first } n \text{ terms of the sum so that the inductive hypothesis can be used)}$$

$$= \left(\frac{n(n+1)}{2} \right) + (n+1) \qquad \text{(inductive hypothesis/substitution)}$$

$$= \frac{n(n+1) + 2(n+1)}{2} \qquad \text{(algebra)}$$

$$= \frac{(n+1) \cdot (n+2)}{2} \qquad \text{(algebra)}$$

$$= \frac{(n+1)[(n+1)+1]}{2} \qquad \text{(algebra)}$$

- Since $1 \in M$, and whenever $n \in M$, then $n + 1 \in M$; by the PMI, $M = N$.
- The statement P_n, or $\sum_{i=1}^{n} i = \dfrac{n(n+1)}{2}$, holds for *all* natural numbers $n = 1, 2, 3, \dots$.

We can paraphrase the technique of mathematical induction in other ways than using a proof involving sets. One such way is outlined below:

- We wish to show that a statement P_n is true for all integers $n \geq a$ (here a is the starting point).

- We *show* the statement is true when $n = a$.

- We *assume* the statement is true for any specific integer $n = k$ (where $k \geq a$) and we *show* the statement is true for the next integer $n = k + 1$.

- We conclude that the statement is true for all $n \geq a$.

Example 2   Prove: for all positive integers n, $(1 + x)^n \geq (1 + nx)$ (where x is assumed to be a real number and $x \geq -1$).

We note that since $x \geq -1$, the expression $1 + x \geq 0$. In this example, we shall discard some of the formality of a mathematical induction proof and not cast the proof in the form of statements about sets. Rather we shall concentrate on the statements themselves in the structure of mathematical induction. We now begin the proof by mathematical induction.

Proof   The statement (P_n) of concern here is

$$(1 + x)^n \geq (1 + nx)$$

We must show two things: (1) that this statement is true when $n = 1$ (P_1 is true), and (2) assuming it is true when $n = k$ (k some positive integer; P_k assumed true), use that to show it is true when $n = k + 1$ (the next positive integer; show P_{k+1} is true).
First we note that

$$(1 + x)^1 \geq 1 + 1x$$

is obviously true, since equality holds.
Next we assume

$$(1 + x)^k \geq 1 + kx \text{ for some positive integer } k \quad \text{(inductive hypothesis)}$$

and try to deduce

$$(1 + x)^{k+1} \geq 1 + (k + 1)x$$

Now

$$
\begin{aligned}
(1 + x)^{k+1} &= (1 + x)^k (1 + x) \\
&\geq (1 + kx)(1 + x) \quad \text{(using the inductive hypothesis)} \\
&= 1 + kx + x + kx^2 \\
&= 1 + (k + 1)x + kx^2
\end{aligned}
$$

We note that since $kx^2 \geq 0$ for all $k \geq 1$ and for all $x \geq -1$, then

$$\left(1 + (k + 1)x + kx^2\right) \geq \left(1 + (k + 1)x\right)$$

Hence, we have deduced

$$\left[(1 + x)^{k+1}\right] \geq \left[1 + (k + 1)x\right]$$

which was to be proved.

We conclude, by the PMI, that the statement

$$(1 + x)^n \geq 1 + nx \qquad x \geq -1; \; x \in R$$

is true for all positive integers n.

Example 3   Prove: For any nonnegative integer n, $n^3 - n$ is divisible by 3.

Proof   We start with $n = 0$. $0^3 - 0 = 0$, and 0 is divisible by 3 since $0 = 3 \cdot 0$ (b is divisible by a means there exists an integer c such that $b = a \cdot c$).

Next, suppose for $n = k$,

$k^3 - k$ is divisible by 3

and we wish to use that assumption to show that the divisibility statement is true when $n = k + 1$; that is, "Is $(k + 1)^3 - (k + 1)$ divisible by 3?"

$$(k + 1)^3 - (k + 1) = k^3 + 3k^2 + 3k + 1 - k - 1$$
$$= (k^3 - k) + 3(k^2 + k)$$

Now we observe that we have assumed, in the inductive hypothesis, that $k^3 - k$ is divisible by 3, and the term $3(k^2 + k)$ clearly is a multiple of 3 (hence divisible by 3); therefore, the sum of these two expressions

$$(k^3 - k) + 3(k^2 + k) \text{ is divisible by 3}$$

The observation that if two numbers are each divisible by 3, then their sum (difference) is divisible by 3 is an easy consequence of distributive properties.

We thus conclude that for every nonnegative integer n,

$n^3 - n$ is divisible by 3.

Example 4   Show that for any nonnegative integer $m \geq 0$ and for every finite set A having m elements, A has 2^m subsets.

Proof   We start at $m = 0$. If a set A has 0 elements, it must be the empty set, and the empty set has exactly one subset—itself. Thus,

If $n(A) = 0$, then A has $2^0 = 1$ subset.

Now assume that whenever a finite set A has k elements, it will have 2^k subsets. That is, assume

If $n(A) = k$, then A has 2^k subsets.

What is the effect of adding one more element to the set? We would like to be able to give a convincing argument that when A has $k + 1$ elements, it will have

$$2^{k+1} \text{ subsets}$$

One way to accomplish this is to visualize a set with k elements having all 2^k of its subsets lined up. The first subset might be the empty set, then there would be a singleton set for each element (k singletons), then the sets having two elements, and so on until finally the set itself would be listed. As we have assumed, there would be 2^k subsets total. Now take an additional ($k + 1$st) element that was not in the original set, and imagine it being added to the set. The set now would have $k + 1$ elements. How many subsets would it have? Well, all the 2^k subsets we had listed could be left unchanged—they are still subsets of the $(k + 1)$ element set. But we could, *in addition*, form precisely 2^k *new* subsets by tossing that $k + 1$st new element into every one of the existing subsets. (When we put the $k + 1$st element into the empty set, we'd have a singleton set we didn't have before, etc.) Thus, the totality of subsets for the $(k + 1)$ element set would be

$$2^k + 2^k = 2 \cdot 2^k$$
$$= 2^{k+1} \text{ subsets}$$

Appealing to the PMI, we have shown

$$\forall_{m \geq 0} \quad [\text{if } n(A) = m, \text{ then } A \text{ has } 2^m \text{ subsets}]$$

The reader might note from the foregoing four examples that the first dealt with a general sum, the second was an inequality, the third a divisibility proof, and the fourth was a counting argument. All dealt with generalizations involving integers, but they point out some of the latitude and flexibility of proof by mathematical induction. That is, there can be general statements about integers that occur in the context of many different mathematical backgrounds.

Assigned Problems

1. Construct, as special cases, partial sums of n odd positive integers where $n = 1, 2, 3, \ldots$. That is, consider

$$1 = 1$$
$$1 + 3 = 4$$
$$1 + 3 + 5 = 9$$

(a) Make a conjecture about the general sum of the first n odd numbers

$$\sum_{i=1}^{n} (2i - 1) \quad \text{or} \quad 1 + 3 + 5 + \cdots + (2n - 1)$$

(b) Verify your conjecture in a proof by mathematical induction.

2. How many diagonals are there in a convex polygon having n sides ($n \geq 3$)? Verify your conjecture by mathematical induction.

3. There are n persons at a party ($n \geq 2$). Each person shakes hands with every other person. How many handshakes are there? Have you seen a problem like this before? Verify your conjecture using mathematical induction.

4. Prove, for any natural number n,

$$\frac{1}{1^2} + \frac{1}{2^2} + \frac{1}{3^2} + \cdots + \frac{1}{n^2} \leq \frac{2n}{n+1}; \text{ that is, } \forall_{n \in N} \sum_{i=1}^{n} \frac{1}{i^2} \leq \frac{2n}{n+1}$$

5. Verify by mathematical induction

$$\sum_{i=1}^{n} (2i - 1)^3 = n^2(2n^2 - 1) \text{ for all } n \geq 1$$

6. Show that for every $k \in N$,

$$2^k < 2^{k+1}$$

7. For any collection of n sets in a universe U,

$$\overline{\left(\bigcup_{i=1}^{n} A_i \right)} = \bigcap_{i=1}^{n} \overline{A_i}$$

That is, the complement of the generalized union of n sets is a set that is the same as the generalized intersection of the individual set complements. (See Chapter 3 for a discussion of complements of sets.) This is a generalization of De Morgan's law, stated below for two sets:

$$\overline{(A_1 \cup A_2)} = \overline{A_1} \cap \overline{A_2}$$

Verify, by mathematical induction, the generalized De Morgan's law for n sets, $n \geq 2$.

8. Prove: $\displaystyle\forall_{n \in N} \sum_{i=1}^{n} i^3 = \left[\frac{n(n+1)}{2} \right]^2$

9. Show for every $n \geq 1$,

$$\sum_{i=1}^{n} i! i = (n + 1)! - 1$$

(Recall that $n! = n(n - 1)(n - 2) \ldots 1$ and $(n + 1)! = (n + 1)n!$.)

10. Prove (via mathematical induction):

$$\forall_{n \in N} D_x^n(x^n) = n!$$

That is, the nth derivative of the nth power function is $n!$.

2.7   Mathematical Induction, Part II

In the preceding section, we stated the Principle of Mathematical Induction (PMI) and applied its associated proof strategy to a variety of problems involving generalizations about integers. It is noteworthy that the PMI is really a theorem about natural numbers, and therefore the basis step in proofs that appeal directly to the PMI involves demonstrating that the desired statement is true in the case where $n = 1$. Therefore, we make an inductive hypothesis that the statement is true for $n = k$, where k is some arbitrary, unspecified natural number greater than or equal to 1, and try to show the statement is true for the next natural number, $k + 1$.

It is easy to see how the Principle of Mathematical Induction can be extended to cases where the basis step is $n = a$ (where a is *any* natural number), and the strategy proceeds stepwise to verify the desired statement for all natural numbers $n \geq a$. All we need do is define the set M to include the numbers $1, 2, \ldots a - 1$, then show that $a \in M$, and proceed with standard mathematical induction after that. In fact, we can create variations in the strategy of mathematical induction that allow us to start at any integer and proceed stepwise in either direction. In this section, we shall examine some of these "stepwise" variations as well as another (equivalent) form of the PMI, called "strong" induction. We do this without proof, but encourage the reader to develop an intuitive feeling for the close relationship between these variations and the original PMI, and particularly a recognition of situations to which these strategies may apply.

Stepwise Variations on Mathematical Induction

Variation #1

Wish to prove:
    P_n is true for every $n \geq a$,
        where a is *any* integer.
(1) Show P_n is true for $n = a$.
(2) Assume P_n is true for $n = k$,
        where $k \geq a$.
(3) Show P_n is true for $n = k + 1$.

Variation #2

Wish to prove:
    P_n is true for every $n \leq b$,
        where b is *any* integer.
(1) Show P_n is true for $n = b$.
(2) Assume P_n is true for
        $n = k$ (where $k \leq b$).
(3) Show P_n is true when $n = k - 1$.

Variation #3

Wish to prove:
    P_n is true for every nonnegative
        multiple of j, $j \geq 1$.
(1) Show P_n is true for $n = 0$.
(2) Assume P_n is true for $n = k$,
        where $k \geq 0$ and k is a
        multiple of j.
(3) Show P_n is true for $n = k + j$.

Variation #4

Wish to prove:
    P_n is true for *all* integers n.
(1) Show P_n is true for $n = a$
        (where a is any specific integer).
(2) Assume P_n is true for $n = k$,
        where k is *any* integer.
(3a) Show P_n is true for $n = k - 1$.
(3b) Show P_n is true for $n = k + 1$.

Following are several examples that involve application of these variations on mathematical induction.

Example 1   Prove $\forall_{n \geq 5} \; n^2 < 2^n$

Proof   (by Variation #1)
Our starting point is $n = 5$. $5^2 = 25$ and $2^5 = 32$. Thus the statement is true for $n = 5$. Next, assume $k^2 < 2^k$ for some $k \geq 5$ (inductive hypothesis). We need to show that the statement is true for $n = k + 1$. That is,

$$(k + 1)^2 < 2^{k+1}$$

Since $(k + 1)^2 = k^2 + 2k + 1$ and $2^{k+1} = 2 \cdot 2^k$, where $k^2 < 2^k$ (inductive hypothesis), then $2k^2 < 2^{k+1}$ and it suffices to show that $k^2 + 2k + 1 < 2k^2$, or equivalently, $2k + 1 < k^2$. This latter statement is true for $k \geq 5$. (Actually for $k \geq 3$; check the solutions to the quadratic inequality $x^2 - 2x - 1 > 0$). Therefore, we have proved that $n^2 < 2^n$ for all $n \geq 5$.

Example 2   Prove $\forall_{n \leq 0} \; e^n \leq 1 - n$

Proof   (by Variation #2)
The statement is true for $n = 0$, since $e^0 = 1$ and $1 - 0 = 1$. Assume the statement is true for $n = k$, where $k \leq 0$. That is, assume $e^k \leq 1 - k$. Now consider the situation for $n = k - 1$. We need to show that

$$e^{k-1} \leq 1 - (k - 1) = 2 - k$$

Since $e^{k-1} = e^k \cdot e^{-1}$ and $e^k < 1 - k$ by inductive hypothesis and $e^{-1} < 1$, then

$$e^{k-1} < 1 \cdot (1 - k) = 1 - k$$

and since $1 - k < 2 - k$, we have $e^{k-1} < 2 - k$. Therefore

$$\forall_{n \leq 0} \; e^n \leq 1 - n$$

The next example illustrates a proof using standard mathematical induction, but with an algebraic twist that assists us in making the inductive step.

Example 3   Prove that if x and y are any real numbers and if n is any positive integer, then

$$x - y \text{ divides } x^n - y^n$$

Proof   For $n = 1$, $x - y$ divides $x^1 - y^1$, so we have the basis step. Now, assume that $x - y$ divides $x^k - y^k$ for some $k \geq 1$. We wish to show that $x - y$ also divides $x^{k+1} - y^{k+1}$. We express

$$x^{k+1} - y^{k+1} = x \cdot x^k - y \cdot y^k \qquad \text{(algebra)}$$

$$= xx^k - yx^k + yx^k - yy^k \quad \text{(addition and subtraction of the term } yx^k)$$

$$= (x - y)x^k + y(x^k - y^k)$$

Since $x - y$ divides $x^k - y^k$, by inductive hypothesis, and $x - y$ divides itself, then $x - y$ divides the sum. Therefore, $x - y$ divides $x^{k+1} - y^{k+1}$, and our theorem is proved.

Example 4   Prove that $2^n - 1$ is divisible by 3 for every *even* integer $n \geq 2$.

Proof   (by Variation #3), with $j = 2$.
For $n = 2$, $2^2 - 1 = 3$, which is surely divisible by 3. Assume that $2^k - 1$ is divisible by 3, where k is an *even* integer ($k \geq 2$). Next, we show that the statement is true for the *next even* integer, $k + 2$. That is, we wish to prove

$2^{k+2} - 1$ is divisible by 3

To do this we rewrite $2^{k+2} - 1$ in an equivalent form which allows us to access the inductive hypothesis (a standard technique in divisibility proofs).

$$2^{k+2} - 1 = 2^2 \cdot 2^k - 1$$
$$= 4(2^k - 1) + 3 \quad \text{(an equivalent form)}$$

Now 3 divides $2^k - 1$, by inductive hypothesis, and 3 divides itself. Therefore, since 3 divides both summands, it divides the sum. Thus, we have verified our desired result for all even integers.

Sometimes "stepwise" mathematical induction is quite difficult to apply, or the problem situation does not fit well with the strategy of bridging from one integer to another. Sometimes, there is no apparent connection between the nth case and the $(n + 1)^{st}$ case for the desired property. An alternate form of induction is needed. There is still another form of mathematical induction which is worthwhile knowing. It is mathematically equivalent to the PMI, and we shall refer to it here as the PSI, or Principle of Strong Induction.

Principle of Strong Induction

If N is the set of natural numbers ($N = \{1, 2, 3, \dots\}$)
and $M \subseteq N$ having the following properties:
(a) $1 \in M$
(b) for each $n \in N$, n is also in M whenever $1, 2, 3, \dots n - 1 \in M$,
then $M = N$

Basically, the only difference between the *strong* (PSI) and *standard* (PMI) forms of mathematical induction is in the inductive hypothesis. In the strong form, we assume the statement we wish to prove is true for all natural numbers up to and including $n - 1$, and the inductive step involves verification for n. In standard induction, we assume the statement is true for some unspecified natural number $n \geq 1$, and deduce its truth for the next one, $n + 1$. Cast in the form of a proof strategy, strong induction says:

> We wish to prove:
>
> P_n is true for every $n \geq 1$

(1) Show P_n is true for $n = 1$.

(2) Assume P_k is true for all k such that $1 \leq k \leq n - 1$.

(3) Show P_n is true.

The next example demonstrates an application of strong induction.

Example 5   Every natural number $n \geq 2$ has a prime factorization. That is, $n = p_1 \cdot p_2 \cdot p_3 \cdots p_k$, where p_i is a prime ($i = 1, 2, \ldots k$).

Proof   (by strong induction on n)
The basis step is for $n = 2$. Since 2 is prime, it has a prime factorization (namely, itself). Assume that the theorem is true for all natural numbers k, where $2 \leq k \leq n - 1$. We will show it is also true for n. Let t be a natural number such that $2 \leq t \leq n$. If $t < n$, then t has a prime factorization by the inductive hypothesis. If $t = n$, then either t is prime (in which case n is prime and we're done) or t is composite. If t is composite, then $t = a \cdot b$ where $a, b \neq 1$ and $a, b \neq t$. Thus $2 \leq a < t$ and $2 \leq b < t$, and the inductive hypothesis applies to a and b. Therefore, a and b have prime factorizations, and the product of these factorizations is a prime factorization of t. In every case, however, we obtain t as a product of primes. Therefore, we have proved that n has a prime factorization for all $n \geq 2$.

It is interesting to think about how we might have taken the inductive step had we tried to apply a standard mathematical induction proof for the theorem in Example 5. Knowing that one integer has a prime factorization doesn't really provide much leverage to deduce that the next integer also has a prime factorization.

The variations stated earlier for standard induction also hold for strong induction. This means we can initiate our proof at any integer b (positive or negative), make an assumption that the statement holds for all integers between (and including) b and k—

$$\{ b, b + 1, b + 2, \ldots k \} \quad \text{if} \quad k \geq b, \text{ or}$$

$$\{ k, k + 1, k + 2, \ldots b \} \quad \text{if } k \leq b$$

and then try to demonstrate the case for $k + 1$ or $k - 1$ respectively, depending on the direction in which we wish to proceed.

We conclude this discussion with another example involving strong induction.

Example 6   **Prove:**   For every odd natural number n,

$$1 + 3 + 5 + \cdots + n = \frac{(n + 1)^2}{4}$$

If we try to set up a standard proof by mathematical induction, we run into some difficulties. The key here is to use the strong form of mathematical induction, and select the "truth set" M with some forethought. We call the statement to prove, P_n.

Proof   Define $M = \{n \in N \mid n$ is even, or n is odd and P_n holds$\}$. We wish to show $M = N$. First, $1 \in M$ since 1 is odd and $1 = \dfrac{(1 + 1)^2}{4}$. Next, suppose $n \in N$ and

$$1, 2, 3, \ldots, n \in M$$

We wish to show that $(n + 1) \in M$. Either $n + 1$ is even or $n + 1$ is odd. If $n + 1$ is even, then $(n + 1) \in M$ by the definition of M, and we're done. If $n + 1$ is odd, then so is $n - 1$. But since $(n - 1) \in M$, we know that

$$1 + 3 + 5 + \cdots + (n - 1) = \frac{n^2}{4}$$

Adding $n + 1$ to both sides, we obtain

$$1 + 3 + 5 + \cdots + (n - 1) + (n + 1) = \frac{n^2}{4} + (n + 1)$$

$$= \frac{n^2 + 4n + 4}{4}$$

$$= \frac{(n + 2)^2}{4}$$

$$= \frac{[(n + 1) + 1]^2}{4}$$

Therefore, $(n + 1) \in M$, and $M = N$ by PSI.

Assigned Problems

1. (a) For which natural numbers n is the statement $n^3 < 2^n$ true?

    (b) Verify, by mathematical induction, the truth of your conjecture. (*Hint*: you will need to use your lower bound on n in the proof.)

2. Suppose P_n is the statement "$n^2 + 5n + 1$ is even," where n is a natural number.

    (a) Prove $\forall_{n \in N} [P_n \rightarrow P_{n+1}]$.

    (b) Does your proof in (a) verify that the statement P_n is true for all natural numbers? Why or why not? Comment on the structure of mathematical induction with respect to this situation.

3. Prove: for every odd integer $n \geq 3$,

$$\prod_{i=2}^{n}\left[1 + \frac{(-1)^{i}}{i}\right] = 1$$

(The symbol $\prod$ means generalized *product*, just as Σ means generalized sum.)

4. Prove: for every integer $n \geq 4$, $2^n < n!$

5. Prove that every positive integer greater than 1 has a prime factor.

6. Prove: for all positive integers $n \geq 2$,

$$\sum_{i=1}^{n} \frac{1}{\sqrt{i}} > \sqrt{n}$$

7. (a) Show by mathematical induction that 6 divides $n^3 - n$ for all negative integers n.

(b) Show by mathematical induction that 6 divides $n^3 - n$ for all nonnegative integers n.

(c) What conclusion do you reach from (a) and (b)?

No (d) If you had verified (b) first, would (a) have had to be proved by mathematical induction? Explain your answer.

No (e) Is it necessary to prove (a) and (b) by mathematical induction? Explain.

8. A recursive, or inductive, definition for a sequence includes stating the first term (or first several terms) of the sequence, and then defining the other members in terms of previous ones. The famous Fibonacci sequence $1, 1, 2, 3, 5, 8, 13, 21, 34, \ldots$ is defined recursively as follows:

$$\begin{cases} f_1 = 1 \\ f_2 = 1 \\ f_{n+2} = f_{n+1} + f_n \quad \text{for} \quad n \geq 1 \end{cases}$$

(a) Compute f_{10}, f_{13}, and $f_{n+3} - f_{n+1}$.

(b) Prove: $\forall_{n \geq 1} \sum_{i=1}^{n} f_i = f_{n+2} - 1$.

9. (a) Use a proof by strong induction to show that the nth Fibonacci number, f_n, is an integer.

(b) Prove, using standard induction, that the sum of the first n odd-indexed Fibonacci numbers is f_{2n}. That is, $f_1 + f_3 + f_5 + \cdots + f_{2n-1} = f_{2n}$.

(c) Make a conjecture about the sum of the first n even-indexed terms of the Fibonacci sequence, and verify your conjecture by standard mathematical induction.

10. Prove, by strong induction, that the nth Fibonacci number f_n is given by

$$f_n = \frac{\left(\dfrac{1 + \sqrt{5}}{2}\right)^n - \left(\dfrac{1 - \sqrt{5}}{2}\right)^n}{\sqrt{5}}$$

This gives us an explicit (although quite tedious) formula for calculating Fibonacci numbers. *Hint*: in your proof, the following observations may be helpful:

$$\left(\frac{1 + \sqrt{5}}{2}\right) + \left(\frac{1 - \sqrt{5}}{2}\right) = 1; \qquad \left(\frac{1 + \sqrt{5}}{2}\right) \cdot \left(\frac{1 - \sqrt{5}}{2}\right) = -1$$

$$\left(\frac{1 + \sqrt{5}}{2}\right)^2 = \frac{3 + \sqrt{5}}{2}; \qquad \left(\frac{1 - \sqrt{5}}{2}\right)^2 = \frac{3 - \sqrt{5}}{2}$$

2.8   *Overgeneralization and Counterexample*

To disprove a general (universally quantified) statement such as

$$\forall_{x \in R}(x < x^2)$$

we are required to prove that its negation is true. That is,

$$\exists_{x \in R}(x \geq x^2)$$

To prove the last statement, it suffices to produce one example of such a real number (e.g., $1/3$). Such an example is called a *counterexample* for the original (general) statement. A counterexample is simply a single object that demonstrates that a generalization is *not* true. If an individual claims that all college students are egotistical, and you can find one college student who is altruistic, then you have disproved that individual's generalization about college students. Thus, in order to construct "disproofs," you must be able to do two things:

1. Formulate accurately the negation of the generalization you are attempting to disprove. (This formulation will be some existentially quantified statement.)

2. Produce a counterexample. (That is, satisfy the existence statement with an example.)

You may wonder why, after all the work it takes to construct proofs, mathematicians would be interested in disproofs. There are numerous instances in mathematics where an individual investigates statements without knowing at the outset whether they are generally true or false. For example, if the statement under investigation is some generalization, the investigator may try many special cases. If the generalization holds in each of the individual special cases, this certainly does *not* validate the generalization, but it does serve to build one's faith that the generalization *might* be true. It becomes more plausible with each example for which it holds. Thus, the investigator might try to *construct a proof* that the generalization is true. Of course, if the generalization is actually false, no valid proof exists, and all attempts at proof should necessarily lead to dead ends. (Occasionally, one individual may come up with a "proof," and another with a "counterexample." Since a valid proof and counterexample cannot peacefully coexist, either the proof has some logical or mathematical flaw, or the counterexample does not faithfully represent the conditions involved, or perhaps both. This is another reason why it is so important to have good command of the underlying logic.) After various attempts at proof, the investigator may return to testing special cases in search of a counterexample. Usually it is helpful to test *extreme* cases as likely candidates for counterexamples. Also, it is helpful to stay away from situations having special properties when searching for counterexamples. That is, if the generalization is about integers, and you suspect it is false, it might be unwise to try 0 or 1 because these numbers play a special role in the system of integers—additive and multiplicative identity elements. On the other hand, there are numerous situations where 0 makes a perfectly fine counterexample because of divisibility properties. As a further example, if some generalization about polygons is under suspicion, testing only regular polygons could lead to "pseudo" support for the generalization when, in fact, it may be false for some bizarre polygon.

Besides the time and effort saved in searching for nonexistent proofs, there is another distinct benefit to disproving generalizations. Often the original generalization may be too "strong." That is, it may cover too much ground (e.g., all real numbers) when in fact it is true for a more restricted space (e.g., all integers). Sometimes a counterexample helps a problem solver discover that "all is not lost." Rather than simply rejecting the proposed generalization when a counterexample is produced, the problem solver might ask if the statement could be "weakened" so that it remains a generalization, but over some reduced space, or with some additional condition(s) imposed. For example, suppose an assertion "all continuous functions are differentiable" is made. After looking around, one might discover the absolute value function as a counterexample (it's continuous everywhere but not differentiable at the origin). This could lead to a restatement of the generalization such as "all continuous functions whose graphs have no corners are differentiable." Subsequent tests may produce a counterexample to that—the cube root function is continuous everywhere, its graph has no corners, but it does have a twist at the origin where a vertical tangent exists, and is not differentiable there (see Figure 2.3). Eventual considerations may lead to considerably weakened statements (e.g., All polynomial functions are differentiable), which nevertheless may be true in general and still worthy of consideration. The reader may recall from calculus that

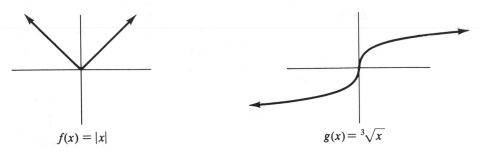

$$f(x) = |x|$$ $$g(x) = \sqrt[3]{x}$$

Figure 2.3

continuity $\nrightarrow$ differentiability

but

differentiability $\rightarrow$ continuity

This says that differentiability is a stronger condition on functions than continuity. (If a function is differentiable, it is continuous; but the converse is not true.)

As another example, consider the generalization:

For all real numbers x, if $x \leq 1$, then $x^2 \leq x$.

A little reflection (perhaps even a reflection of the positive branch of the $y = x^2$ graph in the y-axis) will show that $x = -3$ is a counterexample. Note that the logical form of the statement is

$$\forall_x (\text{if } P(x), \text{ then } Q(x))$$

and the negation of this statement (which $x = -3$ makes true) is

$$\exists_x (P(x) \wedge \sim Q(x))$$

That is, there exists an x (namely -3) such that $x \leq 1$ ($-3 \leq 1$) and whose square is neither less than nor equal to itself. This observation, however, should not lead to a total rejection of the generalization, since in a slightly weakened form it is true; namely:

For all nonnegative real numbers x,

if $x \leq 1$, then $x^2 \leq x$

or, equivalently, for all real numbers x,

if $0 \leq x \leq 1$, then $x^2 \leq x$

The remainder of this section will be devoted to a series of examples illustrating *overgeneralization*. Overgeneralization occurs when an overzealous pattern searcher discovers a pattern among finitely many cases and states or implies that the pattern holds for all (when in fact, it doesn't, and it can be exploded via a counterexample). This does *not* mean that the reader should be timid about making conjectures; mathematics has advanced because of many "fearless" conjectures. However, the reader should be cautioned that there are numerous instances in mathematics where patterns seem to exist and really don't. Thus, while you should not be fearful of making conjectures (remember that's exactly what they are—conjectures, guesses), and in order to convince everyone (including yourself) completely in the generalization, you must verify it deductively. Another important lesson we learn from overgeneralization and counterexample is that we should not be afraid to believe (especially when that belief is enhanced by many supporting cases), but we must be prepared to change our belief in the face of contradictory evidence.

Example 1   Consider the expression $2^n + 3$.

Suppose we evaluate this expression for various natural number values for n:

$$\text{If } n = 1, 2^1 + 3 = 5$$
$$\text{If } n = 2, 2^2 + 3 = 7$$
$$\text{If } n = 3, 2^3 + 3 = 11$$
$$\text{If } n = 4, 2^4 + 3 = 19$$

After these four examples, it may appear that the expression $2^n + 3$ always generates prime numbers when n is a natural number. That statement would be an overgeneralization, however, since the next case ($n = 5$) disproves it:

$$\text{If } n = 5, 2^5 + 3 = 35 \quad \text{(not prime)}$$

Example 2   Is the expression $n^2 + n + 11$ a prime generator for n any integer?

Suppose we test some cases.

$$\text{If } n = 0, 0^2 + 0 + 11 = 11 \quad \text{(prime)}$$
$$\text{If } n = 1, 1^2 + 1 + 11 = 13 \quad \text{(prime)}$$
$$\text{If } n = 2, 2^2 + 2 + 11 = 17 \quad \text{(prime)}$$
$$\text{If } n = 3, 3^2 + 3 + 11 = 23 \quad \text{(prime)}$$
$$\text{If } n = 4, 4^2 + 4 + 11 = 31 \quad \text{(prime)}$$

The first five cases all generate primes; suppose we try some negative integers:

$$\text{If } n = -1, (-1)^2 + (-1) + 11 = 11 \quad \text{(prime)}$$
$$\text{If } n = -2, (-2)^2 + (-2) + 11 = 13 \quad \text{(prime)}$$

$$\text{If } n = -3, (-3)^2 + (-3) + 11 = 17 \quad \text{(prime)}$$

It appears we are producing symmetric results to those already obtained by choosing negative numbers. Keep testing:

$$\text{If } n = 5, 5^2 + 5 + 11 = 41 \qquad \text{(prime)}$$

$$\text{If } n = 6, 6^2 + 6 + 11 = 53 \qquad \text{(prime)}$$

$$\text{If } n = 7, 7^2 + 7 + 11 = 67 \qquad \text{(prime)}$$

$$\text{If } n = 8, 8^2 + 8 + 11 = 83 \qquad \text{(prime)}$$

$$\text{If } n = 9, 9^2 + 9 + 11 = 101 \qquad \text{(prime)}$$

$$\text{If } n = 10, 10^2 + 10 + 11 = 121 \quad \text{(not prime)}$$

Thus, $n = 10$ is a counterexample! Actually, an insightful individual may have spotted an easy counterexample right off by wondering about the constant "11" in the expression $n^2 + n + 11$. If n is any multiple of 11, for example 11 itself, then each term of the expression contains a factor of 11 and the expression splits into two integers, neither of which is 1.

Example 3   Take any circle and consider n points on the circle. Draw chords joining all n points in such manner that no three chords intersect in the same point (if that happens, adjust the location of the points on the circle). Now count the number of regions interior to the circle for various values of $n = 1, 2, 3, \ldots$.

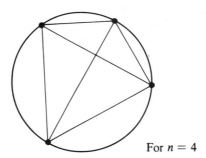

For $n = 4$

We arrange our observations in a table. The variable n represents the number of points on the circle, and R_n represents the number of regions separated by the chords interior to the circle.

n	R_n
1	1
2	2
3	4
4	8
5	16
⋮	⋮

Do you have a conjecture? Does the apparent generalization hold for all n?

Example 4   (a) Compute some values of the function

$$f(n) = \tfrac{1}{6}(18 - 14n + 9n^2 - n^3) \qquad n = 1, 2, 3,\ldots$$

$$f(1) = 2$$

$$f(2) = 3$$

$$f(3) = 5$$

$$f(4) = 7$$

$$f(5) = ?$$

The first four cases appear to produce primes, but the fifth produces 8. Thus, $f(5)$ is a counterexample to the assertion that the range of this function contains only prime numbers.

(b) Compute some values of the function

$$g(n) = (n - 1)(n - 2)(n - 3)(n - 4) + 2^{n-1}$$
$$g(1) = 1$$
$$g(2) = 2$$
$$g(3) = 4$$
$$g(4) = 8$$
$$\text{but } g(5) = 40 \qquad \text{(and not the expected 16)}$$

Example 5   Do your recall the modification on Euclid's proof that there exist infinitely many prime numbers, which you were asked to do in Section 2.3, Problem 8? Briefly, it was an indirect proof in which we assume there are only finitely many primes $p_1, p_2, p_3, \ldots p_n$, and then construct the number

$$(p_1 p_2 p_3 \cdots p_n) + 1$$

showing this number is both prime and not prime (a contradiction). Suppose we were to carefully examine numbers having the form

$$(p_1 p_2 p_3 \cdots p_n) + 1$$

where the p_i's are primes.

$$Q_1 \quad 2 + 1 = 3$$
$$Q_2 \quad 2 \cdot 3 + 1 = 7$$
$$Q_3 \quad 2 \cdot 3 \cdot 5 + 1 = 31$$
$$Q_4 \quad 2 \cdot 3 \cdot 5 \cdot 7 + 1 = 211$$
$$Q_5 \quad 2 \cdot 3 \cdot 5 \cdot 7 \cdot 11 + 1 = 2311$$

All of these are primes! Is it true that *all* numbers of the form $(p_1 p_2 \cdots p_n) + 1$ are prime if $p_1, p_2, \ldots p_n$ are primes? We test Q_6:

$$Q_6 \quad 2 \cdot 3 \cdot 5 \cdot 7 \cdot 11 \cdot 13 + 1 = 30031$$

It turns out 30031 is not prime since

$$30031 = 59 \cdot 509$$

and we have our counterexample.

Example 6   An *algorithm* is a finite set of unambiguous steps or instructions, which when executed for a given set of initial conditions (input), produces a corresponding result (output), and terminates in a finite amount of time. For example, applying the quadratic formula

$$x = \frac{-b \pm \sqrt{b^2 - 4ac}}{2a}$$

to solve a quadratic equation

$$ax^2 + bx + c = 0 \quad (a \neq 0)$$

illustrates the use of a rather popular algorithm: memorize the formula, punch in the appropriate coefficients, carry out the indicated operations, and out will come the solution(s) to the quadratic equation.

Another standard algorithm is the one for extracting square roots. You may recall there exists a long, involved method something like: mark off the original number in groups of two digits from each side of the decimal point, estimate the square root of the number represented by the first grouping, place that number above the grouping and off to the side, square it, subtract, etc.... Fortunately, in the age of calculators we no longer need such devices. However, let us propose at least an examination of the following *square root algorithm* in case our calculator fails.

Step 1: Take the original number and add 100.
Step 2: Drop the units digit from the number obtained in step 1.
Step 3: Divide the result of step 2 by 2.

We illustrate the process with a table:

Root	+100	Drop Units	÷2		Root	+100	Drop Units	÷2
$\sqrt{49}$	149	14	7		$\sqrt{144}$	244	24	12
$\sqrt{64}$	164	16	8		$\sqrt{169}$	269	26	13
$\sqrt{81}$	181	18	9		You guessed it! The next one fails...			
$\sqrt{100}$	200	20	10		$\sqrt{196}$	296	29	$14\frac{1}{2}$
$\sqrt{121}$	221	22	11					

(You may also have wondered why the table started at $\sqrt{49}$.) However, being the dauntless mathematicians we are, we can "patch up" the problem and revise our algorithm:

Step 1: Add 200.
Step 2: Drop units digit.
Step 3: Divide by 3.
Step 4: Add 1.

So we obtain

$$\sqrt{196} \qquad 396 \qquad 39 \qquad 13 \qquad ⑭$$

Does our revision always work? Or can you find a counterexample?

Example 7 This one is a beauty! It's another square root algorithm designed especially for nice large numbers and consists of two steps:

Step 1: Take first two digits.
Step 2: Add 25.

We start with $\sqrt{1681}$. Since the pattern continues a while, we shall tabulate our results. Watch the pattern in the first two digits of each number and especially the pattern of squares in the last two digits of each number. These numbers are successive perfect squares!

Root	First Two Digits	+ 25	=	Result
$\sqrt{1681}$	16	+ 25	=	41
$\sqrt{1764}$	17	+ 25	=	42
$\sqrt{1849}$	18	+ 25	=	43
$\sqrt{1936}$	19	+ 25	=	44
$\sqrt{2025}$	20	+ 25	=	45
$\sqrt{2116}$	21	+ 25	=	46
$\sqrt{2209}$	22	+ 25	=	47
$\sqrt{2304}$	23	+ 25	=	48
$\sqrt{2401}$	24	+ 25	=	49
$\sqrt{2500}$	25	+ 25	=	50
$\sqrt{2601}$	26	+ 25	=	51
$\sqrt{2704}$	27	+ 25	=	52
$\sqrt{2809}$	28	+ 25	=	53
$\sqrt{2916}$	29	+ 25	=	54

$\sqrt{3025}$	30	$+25$	$=$	55
$\sqrt{3136}$	31	$+25$	$=$	56
$\sqrt{3249}$	32	$+25$	$=$	57
$\sqrt{3364}$	33	$+25$	$=$	58
$\sqrt{3481}$	34	$+25$	$=$	59
$\sqrt{3600}$	36	$+25$	$=$	61!!

Our pattern of extracting square roots explodes *after 19 successive cases*! The moral of the story: mathematics has some beautiful patterns and generalizations, but it also has some beautiful counterexamples!

Assigned Problems

1. A calculus student claims that there are no negative solutions to the equation

$$x^3 - 3x^2 + x + 4 = 0$$

Is this assertion true? Why or why not?

2. An eleventh-grade student evaluated an expression of the form

$$\frac{\log A}{\log B}$$

by "reducing," as follows:

$$\frac{\log A}{\log B} = \frac{\cancel{\log} A}{\cancel{\log} B}$$

$$= \frac{A}{B}$$

which further reduced to $\frac{2}{3}$, the correct result.

(a) Does this always work? If so, show why; if not, give a counterexample.

(b) What must A and B have been to produce the correct answer obtained by the student?

3. Beginning algebra students often make the following kinds of errors:

(a) $\dfrac{1}{a + b} = \dfrac{1}{a} + \dfrac{1}{b}$ (for all real a, b (nonzero))

(b) $\sqrt{a + b} = \sqrt{a} + \sqrt{b}$ (for all positive reals a, b)

(c) $\dfrac{a + b}{a} = \dfrac{1 + b}{1}$ (for all $a \neq 0$, b reals)

(d) $(a + b)^2 = a^2 + b^2$ (for all real a, b)

In each situation, find a counterexample to illustrate that the statement is false, and then see if you can modify the statement or the universe to make it true under different circumstances.

4. Find counterexamples for each of the following:

(a) $\displaystyle\sum_{i=1}^{n} x_i^2 = \left(\sum_{i=1}^{n} x_i \right)^2$ (where x_i is a real number, n a positive integer)

(b) $\displaystyle\sum_{i=1}^{n} x_i y_i = \left(\sum_{i=1}^{n} x_i \right)\left(\sum_{i=1}^{n} y_i \right)$ (where x_i, y_i are real numbers, n a positive integer)

5. Observe the cases given below, especially the pattern of numerators in each proper fraction of the results (circled). Is this an overgeneralization (to assume that the counting numbers appear in order in those numerators)?

Given $\sqrt[3]{1}$, consider $1 \div 2 = 0\dfrac{①}{2}$.

Given $\sqrt[3]{8}$, consider $8 \div 3 = 2\dfrac{②}{3}$.

Given $\sqrt[3]{27}$, consider $27 \div 4 = 6\dfrac{③}{4}$.

Given $\sqrt[3]{64}$, consider $64 \div 5 = 12\dfrac{④}{5}$.

Given $\sqrt[3]{125}$, consider $125 \div 6 = 20\dfrac{⑤}{6}$.

$\vdots$ $\vdots$

6. Recall that if f and g are functions, the *composite* function $f \circ g$ is defined by

$$(f \circ g)(x) = f(g(x)), \forall_{x \in \text{dom } g} \text{ and where range } g \subseteq \text{domain } f$$

Two questions:

(a) Is the composite function meaningful for *any* two functions f and g?

(b) For meaningful compositions, is function composition a commutative operation?

7. For each of the following situations, if the equation is true, prove it; if false, produce a counterexample.

(a) $\displaystyle\sum_{i=1}^{n} \frac{i^2}{2i+1} = \frac{n(n+1)(n+2)}{8(n^2+1)}$

(b) $\displaystyle\sum_{i=1}^{n} \frac{1}{i(i+1)} = \frac{n}{n+1}$

8. For each of the following statements, assume all variables are real variables, and find a counterexample to disprove the statement. Visual thinking (using graphs of known functions) may help in some cases.

(a) $xy \geq x + y$

(b) $\dfrac{1}{x} \leq x$

(c) $|x + y| = |x| + |y|$

(d) $a^x \geq 1$ for $a > 0$ and $x \geq 0$

(e) $|\ln x| \leq x$ for all $x > 0$

(f) If $x \leq y$, then $\dfrac{1}{x} \geq \dfrac{1}{y}$

(g) $f^{-1}(x) = \dfrac{1}{f(x)}$

(h) If $\sin x = 1$, then $x = \pi/2$

9. Discuss the truth of the following statements:

(a) If a function f has a finite domain (finitely many elements in its domain), then the inverse function f^{-1} always exists.

(b) Let a, b, c be three integers. If a divides c and b divides c, then a divides b or b divides a.

10. Definition: A function f has a limit L at a if and only if for every $\varepsilon > 0$ there is a $\delta > 0$ such that for every $x \in \mathrm{dom}\, f$

$$\text{If } 0 < |x - a| < \delta, \text{ then } |f(x) - L| < \varepsilon.$$

Claim: The function $f(x) = \sin \dfrac{1}{x}$ at $a = 0$ has a limit of 0.

Prove or disprove this claim.

3

Set Theory

3.1 Sets and Operations

Of all the basic ideas of modern mathematics, two are most fundamental —the concept of *set* and the concept of *function*. In this section, we shall discuss the basic ideas about sets and the operations on sets that form new sets. It is assumed that you are already familiar with many of the basic notions about sets, hence these will be treated briefly. Our emphasis will be on operating with sets.

Our discussion here will be informal; that is, we shall not delve into axiomatic set theory. However, a little later we shall be proving theorems about sets.

The term *set* in colloquial usage as well as in mathematical language is a concept or idea that refers to any collection of objects. Other terms that generally refer to the same idea are *class*, *collection*, *family*, or *aggregate*. (We shall not use the term *group*, since this has a definite meaning in mathematics different from that of *set*.)

The objects or members of a set are called *elements*, and if we adopt the notational convention that capital letters $(A, B, C, \ldots)$ refer to sets and small letters $(a, b, c, \ldots)$ refer to elements, then we use the notation

$a \in A$ means "a is an element of A"

Whenever we talk about sets there will be some large class (in the back of our mind) from which we take elements and form sets. This large class will be called the *domain of discourse* or simply the *universe*, and denoted U.

There are two fundamental axioms about sets that help clarify the meaning and formation of sets. They are called the *Axiom of Extent* and the *Axiom of Specification*.

Axiom of Extent

Let A and B be sets.

$$A = B \leftrightarrow \forall_x \, (x \in A \leftrightarrow x \in B)$$

This axiom says that *a set is completely determined by its members*. It is sometimes taken as a definition of set equality. In other words, two sets A and B are equal if and only if every element of A is also an element of B, and (conversely) every element of B is also an element of A. We shall use this axiom later as the basis of a *method for proving two sets are equal*. This method is called the *pick-a-point process*. That is, to show two sets are equal, we will arbitrarily pick any element (or point) x in the first set and show that it is necessarily an element in the second set; then we will arbitrarily pick any element (or point) x in the second set (this x having no relationship with the first one selected), and show that it must be an element of the first set. These two activities completed, we will conclude that the two sets are equal.

Axiom of Specification

Let U be any universe of discourse, x a variable on U, B a set of elements in U (B could be U itself), and $P(x)$ some predicate (statement or condition involving x). There exists a set A such that

$$\forall_x \ (x \in A \leftrightarrow x \in B \text{ and } P(x))$$

We denote the set A by

$$A = \{x \mid x \in B \text{ and } P(x)\}$$

The axiom of specification says that we can pick a set and a property and build a new set. This is why the notation for A given above is sometimes referred to as *set-builder notation*.

The two axioms above are stated only to clarify the nature of sets. We will not use them to deduce theorems since our development here is informal.

Example 1 Let U be the set of all integers, B the set of even integers, and $P(x)$ the statement "x is prime." Then the axiom of specification says there exists a set—we'll call it A—defined by

$$A = \{x \mid x \in B \text{ and } x \text{ is prime}\}$$

That is, any element in A must be an integer (member of U) that is even (member of B) and prime (have property P). As you know, the set A could have been specified by listing its elements, in which case

$$A = \{2\}$$

In view of the axioms of extent and specification, we have several different ways of describing the elements of any given set:

(a) by listing them explicitly,

(b) by some verbal descriptive statement, or

(c) by using set-builder notation.

A *well-defined set* is a set A together with a description or specification such that given any element x from our universe of discourse U, we can tell whether or not the element is in the set A. That is, we can judge the statement $x \in A$ to be true or false.

Example 2   Let U be the set of all natural numbers, O the set of odd natural numbers. Consider the three descriptions of the set A below:

1. Listing: $A = \{1, 3, 5, 7, 9\}$

2. Verbal: $A = \{$all odd natural numbers from 1 to 9, inclusive$\}$

3. Set-builder: $A = \{x \mid x \in O \text{ and } x < 10\}$

Note that there are, in common usage, a variety of ways to describe sets. For example, if the set is not finite, we might list the elements and infer a pattern (if one exists):

$$A = \{2, 4, 6, 8, 10, \dots\}$$

This is not mathematically precise, since it is an easy matter to define functions that generate *any* next element in the sequence. However, when it is intuitively clear which set we are describing, it is common practice to list elements inferring a pattern in this manner. Also, we might specify some mathematical condition, like

$$B = \{x \mid x \in R \text{ and } |x - 3| < 2\}$$

or

$$C = \{x \mid x^2 - 5x + 6 = 0\}$$

These latter are implicit specifications because they do not produce a direct, explicit view of the set's elements. Nevertheless, they describe well-defined sets since given any object from the universe in mind we can tell whether it has set membership in each case. An example of a *not* well-defined set would be

$$T = \{x \mid x \text{ will be President of the United States in the year 2000}\}$$

Several cautionary remarks about sets follow.

1. When listing the elements of a set, *we do not include duplications*. That is, an element may qualify for membership in a set by way of several different criteria (e.g., when x is even or x is less than 10, the number 8 qualifies twice), but we do not list it more than once in the same set description. For example, the sets

$$\{1, 1, 2, 3\} = \{1, 2, 3\}$$

We are only interested in whether a given element is in a given set, not how many times it may qualify for membership.

2. When listing the elements of a set, *order is immaterial*. That is,

$$\{1, 2, 3\} = \{3, 1, 2\}$$

There are occasions in mathematics where order of elements within a set is important. For such situations, the concept of *ordered set* is introduced. However, in general set theory, unless otherwise specified, elements can be reordered or rearranged in a listing that describes a set without changing the set itself.

3. A *singleton* set is a set with exactly one element in it; the *empty* set is the set with no elements in it. We denote the empty set by $\varnothing$. We must be extremely careful to distinguish between such things as a singleton set and the element in a singleton set as well as the empty set and a singleton set containing as its only "element" the empty set. *Note:*

$$\{a\} \text{ is } not \text{ the same as } a$$

The symbol on the left refers to a singleton set whose only element is a; the symbol on the right is a symbol for an element a. On the one hand we have a *set*; on the other we have an *element*. In this case, they are different entities, although, in some cases, sets themselves are elements of other sets. Also note:

$$\{\varnothing\} \text{ is } not \text{ the same as } \varnothing$$

The symbol on the left refers to a singleton set whose only element happens to be the empty set. (It is perfectly all right for sets to be elements of other sets.) The symbol on the right is the symbol for the empty set—the set with no elements. On the one hand we have a set with something in it; on the other hand we have a set with nothing in it. They are different entities.

We next introduce a particular *relation* between sets—the *subset relation*.

Subset Relation

Let A and B be sets. A is a *subset* of B, denoted $A \subseteq B$, if and only if

$$\forall_x (\text{if } x \in A, \text{ then } x \in B)$$

In other words, every element of A is also an element of B. We can represent this situation pictorially by using a Venn diagram (named after the English logician John Venn, 1834–1923), as shown in Figure 3.1.

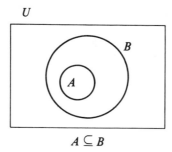

$A \subseteq B$

Figure 3.1

It is worthwhile to see what it means for a set to *not* be a subset of another set. For this, we simply negate the definitional statement:

$A \not\subseteq B$ means "$\exists_x(x \in A$ and $x \notin B)$"

This says there is at least one element (of the universe U) that is in A but *not* in B (see Figure 3.2).

Note that $A \subseteq B$ is *not* a set; it is a *relationship* between sets A and B. It is a statement, and we can judge it true or false.

Following are some remarks about subsets.

1. Every set is a subset of itself ($\forall_A(A \subseteq A)$).

2. The empty set is a subset of every set ($\forall_A(\varnothing \subseteq A)$).

3. Any given set is always a subset of the universe of discourse under consideration ($\forall_A(A \subseteq U)$).

4. Given a set A, the subsets $\varnothing$ and A itself are called the *trivial* subsets of A. All other subsets of A (if there are any) are called *nontrivial* subsets of A. Also, every subset of A except the set A itself is called a *proper* subset

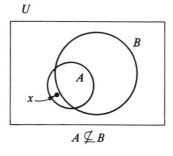

$A \not\subseteq B$

Figure 3.2

of A (denoted by the relational symbol $\subset$). That is, $(B \subseteq A$ and $B \neq A)$ if and only if $B \subset A$.

Another way of viewing proper subsets is as follows. A is a proper subset of B if and only if A is a subset of B and there is at least one element of B that is not in A.

5. In view of the definition of subset, the *Axiom of Extent* can be restated as follows:

> Let A and B be sets.
>
> $A = B \leftrightarrow (A \subseteq B$ and $B \subseteq A)$

In other words, to prove that two sets are equal, show that the first is a subset of the second, and then show that the second is a subset of the first.

6. If A is a subset of B, then the relation between A and B is one of inclusion or containment (A is included or contained within B). On the other hand, the relation between B and A is sometimes referred to as the *superset* relation. That is, B is a superset of A if A is a subset of B. In this case, we say that B includes or contains A.

7. If $7 \in A$, note that it is incorrect to say "$7 \subseteq A$." This is again the element/set distinction. It *is* correct to say "$\{7\} \subseteq A$."

Since the major number systems will be used quite frequently as examples in this course, we list the standard symbols for them below:

N (natural numbers)	$\{1, 2, 3, 4, \ldots\}$
W (whole numbers)	$\{0, 1, 2, 3, \ldots\}$
I (integers)	$\{\ldots -3, -2, -1, 0, 1, 2, 3, \ldots\}$
Q (rationals)	$\left\{\dfrac{a}{b} \middle| a, b \in I \text{ and } b \neq 0\right\}$
R (reals)	$\{$all numbers named by finite and infinite decimals$\}$
C (complex)	$\{a + bi | a, b \in R \text{ and } i^2 = -1\}$

We turn now to ways of creating new sets from old ones.

Operations on Sets

There are four basic operations on sets—union, intersection, universal complement, and relative complement (or set difference).

Union

Let A and B be two sets. We define a set called the *union of A and B*, symbolized by $A \cup B$, by

$$A \cup B = \{x | x \in A \text{ or } x \in B\}$$

Thus, the union of two sets is the set of all elements in the universe belonging to one or the other, or both. We form the union in a mechanical sense by "putting the two sets together" and eliminating any redundant elements. Thus, if

$$A = \{2, 3, 4\}$$
$$B = \{4, 8, 9, 10\}$$

then

$$A \cup B = \{2, 3, 4, 8, 9, 10\}$$

The different possibilities for the union of two sets are pictured in Figure 3.3 (shaded sets).

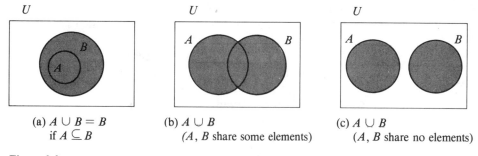

(a) $A \cup B = B$ (b) $A \cup B$ (c) $A \cup B$
    if $A \subseteq B$ *(A, B share some elements)*   *(A, B share no elements)*

Figure 3.3

Intersection

Let A and B be two sets. We define a set called the *intersection of A and B*, symbolized by $A \cap B$, by

$$A \cap B = \{x | x \in A \text{ and } x \in B\}$$

Here, the intersection of two sets is the set of all elements in the universe belonging to *both* sets. We form the intersection by looking for all elements common to or shared by both sets. For example, if A and B are the same sets as those listed above for set union, then

$$A \cap B = \{4\}$$

since the element 4 is the only element shared by both A and B. Venn diagrams for set intersections (shaded) are given in Figure 3.4.

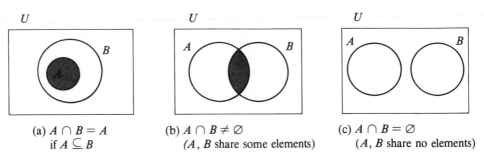

(a) $A \cap B = A$
if $A \subseteq B$

(b) $A \cap B \neq \emptyset$
(A, B share some elements)

(c) $A \cap B = \emptyset$
(A, B share no elements)

Figure 3.4

In the third Venn diagram, where $A \cap B = \emptyset$, we say that the sets A and B are *disjoint*.

It should be carefully noted by the reader that there is a striking analogy between the roles of the logical connectives $\vee$ (or) and $\wedge$ (and), with respect to operating on sentences to form new sentences called *disjunctions* and *conjunctions*, and the roles of set operation symbols $\cup$ (union) and $\cap$ (intersection), with respect to operating on sets to form new sets called *unions* and *intersections*. This is no accident.

One of the most powerful reasoning tools in mathematics is called *reasoning by analogy*, where the problem solver sees a similar situation, translates a problem into that situation, solves it in the different context, and translates the solution back to the original context. This translation process happens frequently with sets and sentences where there is an almost perfect analogy.

Universal Complement

Given a set A and a universe of discourse U, the *universal complement* of A is a set denoted by $\overline{A}$, defined by

$$\overline{A} = \{ x \mid x \notin A \}$$

In other words, the universal complement of set A (sometimes called *absolute complement*, or simply *complement*) is the set of all elements in the universe *except* those in A. For example, if the universe were the set of integers, A the set of even integers, then $\overline{A}$ would be the set of all odd integers. A Venn diagram for set complement is given in Figure 3.5.

Consistent with the analogy suggested above, the counterpart of *universal complementation* (of a set) is *negation* (of a sentence). Note also that universal complementation is a *unary* set operation, that is, it deals with only one set. You take one set and do something with it (complementation) to form another set, just as you take one sentence and negate it to form another sentence (its negation).

Complementation (or forming a set complement) is different than forming unions or intersections, the latter being operations that are examples of *binary*

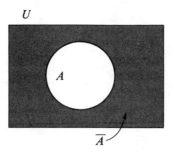

Figure 3.5

operations on sets, requiring two sets to form a third (the union or the intersection). The next set operation is another example of a binary operation on sets.

Relative Complement

Given two sets A and B, the *relative complement* of B with respect to A is a set, denoted $A - B$ and defined by

$$A - B = \{x \mid x \in A \text{ and } x \notin B\}$$

Thus, the relative complement of one set with respect to another (sometimes called the *set difference*) is the set of all elements that are in the first set *and not* in the second set. Appealing to the analogy between sets and sentences, the words "and not" translate into "$\cap \overline{B}$," hence we might expect

$$A - B = A \cap \overline{B}$$

This will indeed turn out to be the case, and we shall prove it later. Venn diagrams illustrating various possibilities for relative complements are shown in Figure 3.6. Note that neither set has to be a subset of the other to form the set $A - B$ in a meaningful way.

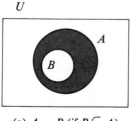

(a) $A - B$ (if $B \subseteq A$)

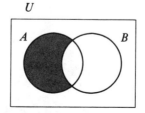

(b) $A - B$
(A, B share some elements)

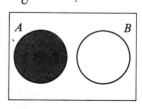

(c) $A - B = A$
(A, B share no elements)

Figure 3.6

Example 3   Suppose $U = \{1, 2, 3, 4, \ldots 10\}$.

$$A = \{\text{all multiples of 3 in } U\}$$

$$B = \{\text{all primes in } U\}$$

$$C = \{x \mid x \in U \text{ and } x \geq 5\}$$

Then the following statements are all true:

- $A \cap B = \{3\}$
- $A \cup B = \{2, 3, 5, 6, 7, 9\}$
- $\overline{A} = \{1, 2, 4, 5, 7, 8, 10\}$
- $B - C = \{2, 3\}$
- $\overline{(A - B)} = \{1, 2, 3, 4, 5, 7, 8, 10\}$
- $\overline{(B \cup C)} = \{1, 4\}$
- $\overline{\varnothing} = U$
- $\overline{U} = \varnothing$
- $\overline{\overline{B}} = \{2, 3, 5, 7\} = B$
- $B \cap \overline{C} = \{2, 3\} = B - C$

To summarize this discussion about sets and operations, we tabulate the various analogous correspondents in three systems—sets, sentences (logic), and arithmetic (integers). The analogy is not perfect; for example, sentence $\vee$ distributes over $\wedge$, and vice-versa, but arithmetic $\cdot$ distributes over $+$, but *not* vice-versa. However, it is so close that we use this analogy frequently to translate back and forth between similar systems, being careful to consider only meaningful statements.

Sets	*Sentences*	*Arithmetic*
$\cup$	($\vee$) or	$+$
$\cap$	($\wedge$) and	$\cdot$
$\subseteq$	($\rightarrow$) if … then	$\leq$
$\overline{A}$	($\sim$) not	$-$
$=$	($\leftrightarrow$) if and only if	$=$
$\varnothing$	(F) false	0
U	(T) true	1

Assigned
Problems

1. Give a rational argument, based on logic, why the empty set should be considered as a subset of every set.

2. Is the following an acceptable alternate definition of subset?

$(A \subseteq B)$ if and only if (A does not contain any element that is not also an element of B).

Why or why not?

3. Using the sets U, A, B, C given in Example 3, list the elements of each of the following sets.

(a) $\bar{A} \cap B$

(b) $\overline{(A \cap B)}$

(c) $(A \cup B) - (A \cap B)$

(d) $B - A$

(e) $\bar{C} - \bar{B}$

(f) $\overline{(A \cup C)} \cap B$

(g) $A \cap (B \cup C)$

(h) $(A \cap B) \cup (A \cap C)$

4. Take the five-element set $M = \{a, b, c, d, e\}$, and in orderly fashion, list all the subsets of M. That is, start by listing all 0-element subsets, then list all singleton subsets of M, then all subsets of M having two elements, and so on.

(a) How many subsets does M have totally?

(b) Do you recognize any pattern in the numbers of subsets having 0, 1, 2, 3, 4, and 5 elements? Explain.

5. In each situation below, a pair of sets is given. Decide first if the sets are equal. If not, decide if one of the sets is a subset of the other and explain briefly.

(a) $C = \{$all constant functions$\}$
$A = \{$all functions with derivative 0 everywhere$\}$

(b) $T = \{e^n | n \in I\}$
$L = \{x | x \in R \text{ and } \ln x \in I\}$

(c) $C = \{$all continuous functions$\}$
$D = \{$all differentiable functions$\}$

(d) $C = \{$all continuous functions$\}$
$P = \{$all polynomial functions$\}$

6. A general three-set Venn diagram is shown below for sets A, B, and C. Using the same kind of diagram in each case below, shade in the following sets.

(a) $A \cap (\bar{B} \cup C)$

(b) $\bar{A} \cap B \cap C$

(c) $A - (B \cup C)$

(d) $\overline{(B \cup C)}$

(e) $B \cup (A \cap \bar{C})$

(f) $(B \cup A) \cap (B \cup \bar{C})$

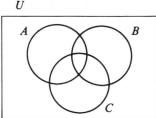

7. In set algebra, the *symmetric difference* of two sets A and B, symbolized by $A * B$, is defined by

$$A * B = (A - B) \cup (B - A)$$

(a) Draw a Venn diagram for $A * B$.

(b) Draw a Venn diagram for $(A * B) * C$ where A, B, and C are sets each sharing some elements.

(c) Draw a Venn diagram for $A * (B * C)$ where A, B, and C are sets, each sharing some elements.

(d) What is $A * A$? Explain briefly.

(e) What is $A * \varnothing$? Explain briefly.

(f) What is $A * U$? Explain briefly.

8. For each of the following statements, provide the analogous statement in the system noted.

(a) In arithmetic: if $a \leq b$, then $a + c \leq b + c$. (analogue, in sets)

(b) In sets: if $A = \varnothing$ or $B = \varnothing$, then $A \cap B = \varnothing$. (analogue, in arithmetic)

(c) In logic: $P \wedge (Q \vee R) \leftrightarrow (P \wedge Q) \vee (P \wedge R)$. (analogue, in sets)

(d) In arithmetic: $a \cdot 0 = 0$. (analogue, in sets)

(e) In logic: If $P \vee Q$ is false, then P is false and Q is false. (analogue, in arithmetic)

9. A four-set Venn-type diagram is shown below for the sets A, B, C, and D. Recall the *lexicographical* technique for finding all combinations of T's and F's in a truth table. Develop an analogous technique for labeling all 16 distinct regions in the diagram by forming a table with headings A, B, C, D

and using complement symbols when a given set is not included. For example, $\overline{A} \cap \overline{B} \cap C \cap D$ would mean do not include A, do not include B, include C, and include D—or, outside A, outside B, inside C, and inside D. Using your technique, label each region in the diagram.

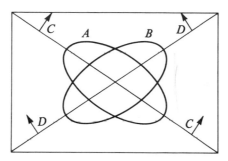

10. Suppose a student says, for any two sets A and B,

   If $A \cap B = A \cup B$, then $A \subseteq B$.

   Is he/she correct or incorrect? Discuss briefly.

3.2 Counting

The primitive idea of *counting* rests upon a mathematical concept we call *one-to-one correspondence*. Suppose a shepherd in ancient times wished to keep track of his sheep, making certain that all that were let out from the pen in the morning actually returned in the evening (so he would know how many were missing). To accomplish this, he might have a pouch full of stones. As the sheep left one-by-one in the morning, he could place one stone in the pouch for each sheep; in the evening, he could reverse the process, taking one stone out of the pouch for each sheep entering the pen. Any discrepancy between sheep and stones would indicate something was wrong. The shepherd had developed a method for *counting* his sheep, and his method *did not require numbers*. In mathematics, we call this kind of matching process a *one-to-one correspondence*. It is a special kind of function associating the elements of two sets A and B (e.g., sheep and stones) in such manner that for every element of A, there is exactly one correspondent in B, and for every element in B, there is exactly one correspondent in A. This is the basis of counting, and when we see a small child counting the ants on a sidewalk "on his fingers," that child has developed a sense of one-to-one correspondence just like the shepherd, only he is using fingers instead of stones.

We say two sets are *equivalent* if they can be placed in one-to-one correspondence with one another. Note that equality $\rightarrow$ equivalence, but not conversely. Also, there are usually many different one-to-one correspondences between two sets if there is one (depending on how many elements are in the sets). Furthermore, if two sets can be placed in one-to-one correspondence with a given set, then there exists a one-to-one correspondence between them.

Example 1   Find a one-to-one correspondence between the two sets

$$A = \{1, 2, 3\} \quad \text{and} \quad B = \{a, b, c\}$$

To do this, we construct an arrow diagram to connect corresponding elements.

$$
\begin{array}{cc}
A & B \\
1 & \longleftrightarrow a \\
2 & \longleftrightarrow b \\
3 & \longleftrightarrow c
\end{array}
$$

The above diagram illustrates one one-to-one correspondence between A and B. Another might be

Actually, it is quite easy to see that there are six different one-to-one correspondences between A and B. The element 1 can be matched with any of three choices (a, b, or c). Once this is done, 2 has exactly two possible correspondents. Once 1 and 2 are assigned correspondents, then the correspondent for 3 is uniquely determined.

We use the notion of one-to-one correspondence to distinguish between *finite* and *infinite* sets, and to make comparisons between the numbers of elements in finite sets.

Finite Set

A *finite set A* is a set that is either the empty set or that can be placed in one-to-one correspondence with the set $\{1, 2, 3, \ldots k\}$ for some natural number k. If $A = \varnothing$, then we say A has 0 elements, and denote this by

$$n(A) = 0$$

If A is in one-to-one correspondence with $\{1, 2, 3, \ldots k\}$, then we say A has k elements, denoted by

$$n(A) = k$$

Infinite Set

An *infinite set* is a set which is not finite (i.e., a nonempty set for which there exists no one-to-one correspondence with a set $\{1, 2, 3, \ldots k\}$ for any natural number k).

There are different kinds of infinite sets, and the notion of one-to-one correspondence is used once again to distinguish them. An infinite set that can be placed in one-to-one correspondence with the set of natural numbers (or positive integers) is said to be *countably infinite* (or denumerably infinite). For example, the set of even positive integers can be placed in one-to-one correspondence with the set of positive integers

$$E = \{2, 4, 6, 8, 10, \ldots 2k \ldots\}$$
$$\updownarrow \ \updownarrow \ \updownarrow \ \updownarrow \ \updownarrow \qquad \nearrow$$
$$N = \{1, 2, 3, 4, \ 5, \ \ldots k \ldots\}$$

via the correspondence $2k \leftrightarrow k$. This happens to point out a paradoxical characteristic of infinite sets that distinguishes them from finite sets; namely, every infinite set can be placed in one-to-one correspondence with some *proper subset* of itself.

If an infinite set *cannot* be placed in one-to-one correspondence with the set of positive integers $\{1, 2, 3, \ldots\}$, then it is said to be *uncountably infinite* or *nondenumerably infinite*. It turns out that the set of rational numbers is countably infinite.

Two theorems concerning these observations, made by the famous set theorist Georg Cantor (1845–1918), follow.

Theorem   A

The set of positive rational numbers is countably infinite (i.e., it can be placed in one-to-one correspondence with the set of positive integers).

Proof: (due to Cantor, who used a diagonalization method)
Arrange the positive rational numbers in a configuration having infinitely many rows, each having infinitely many entries, by placing all rational numbers with numerator "1" in the first row (where the denominators vary over 1, 2, 3, . . .), and next, all rational numbers with numerator "2" in the second row (varying denominators), and continue producing rows in this fashion. There will be various redundancies (e.g., $\frac{1}{1}, \frac{2}{2}, \ldots$), but if *this* set can be placed into 1–1 correspondence with the positive integers, then so can an infinite subset of it. (We use the facts, without proof, that any infinite subset of a countable set is countable, and that the union of any two countable sets is countable). Note the array.

$$
\begin{array}{cccccc}
\frac{1}{1} & \frac{1}{2} & \frac{1}{3} & \frac{1}{4} & \frac{1}{5} & \frac{1}{6} & \cdots \\[4pt]
\frac{2}{1} & \frac{2}{2} & \frac{2}{3} & \frac{2}{4} & \frac{2}{5} & \frac{2}{6} & \cdots \\[4pt]
\frac{3}{1} & \frac{3}{2} & \frac{3}{3} & \frac{3}{4} & \frac{3}{5} & \frac{3}{6} & \cdots \\[4pt]
\vdots & \vdots & \vdots & \vdots & \vdots & \vdots &
\end{array}
$$

Now, we set up the correspondence as follows. Think of the entries $\frac{a}{b}$ as ordered pairs (a, b). Label the first entry $(\frac{1}{1})$ "1." Thereafter, count diagonally from upper right to lower left, as shown in the next table.

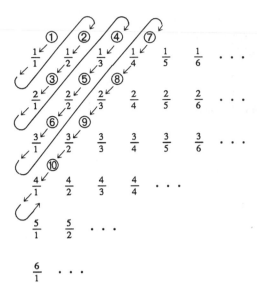

In this manner, we can assign to each positive rational number a unique positive integer. Therefore, the preceding set is countably infinite, and by implication, the set of positive rationals is countably infinite. In fact, the specific count function is given by: If $\frac{a}{b}$ is any positive rational number,

$$f\left(\frac{a}{b}\right) = \frac{1}{2}(a + b - 2)(a + b - 1) + a$$

Note that $f\left(\frac{2}{3}\right) = \frac{1}{2}(2 + 3 - 2)(2 + 3 - 1) + 2 = 8$, and "8" is the positive integer assigned to $\frac{2}{3}$ by this diagonalization process. You might check to see, for example, that the rational number $\frac{3}{6}$ is assigned "31." Finally, if you agree that the union of two countable sets is countable, and that we could adapt the preceding diagonalization process to the nonpositive rationals (to show that they are countable), then it follows that the entire set of rational numbers is countably infinite.

Theorem   B

The set of real numbers in the open interval $(0, 1)$ is *uncountably* infinite.

Proof: We wish to prove that the set of real numbers in $(0, 1)$ cannot be placed in one-to-one correspondence with the set of positive integers. Our proof will be *indirect*. Suppose these real numbers were *countable*. Then we could index them, using positive integers, as r_1, r_2, r_3, and so on. Now, the decimal expansion of every real number in $(0, 1)$ can be represented in binary notation as an infinite string of "0" and "1" digits. Suppose we create a tabular array of all real numbers in $(0, 1)$ (hypothetically):

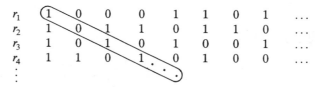

Now, supposedly *all* real numbers in $(0, 1)$ are in this table. If we take the main diagonal elements (left to right) and form a new real number r by reversing the binary digits (if a "1" appears, change it to a "0"; if a "0" appears, change it to a "1"). This real number $(0100\ldots)$ disagrees in at least one digit with every real number in the table. It differs from r_1 in the first decimal place, from r_2 in the second decimal place, and, in general, from r_k in the kth decimal place. Therefore, it does not appear in the table. This contradicts our assumption that *every* real number in $(0, 1)$ appears in the table (and is indexed by some positive integer). Therefore, the set of real numbers in $(0, 1)$ is uncountable. It follows that the set of all real numbers is uncountably infinite, if you agree that a countable union of uncountably infinite sets is an uncountably infinite set. (This will be given as an exercise.)

As mentioned earlier, we use the notion of one-to-one correspondence to make comparisons between finite sets. How can we tell if a finite set A has fewer or more elements than a finite set B? We can try to match the two sets in one-to-one correspondence and if one set (say A) matches in one-to-one correspondence with a *proper subset* of the other (B), then we say

$$n(A) < n(B)$$

This comparison goes hand-in-hand with order comparison among positive integers. We usually define the order relation $(<)$ by

$$\forall_{a,\, b \in I^+} \left[(a < b) \leftrightarrow \left(\exists_{c \in I^+}(a + c = b) \right) \right]$$

In other words, if $n(A) = a$ and $n(B) = b$, and A is in one-to-one correspondence with a *proper* subset of B, then let C be the subset of B "left over" unmatched with A, and let $n(C) = c$.

We summarize some of these ideas in the following two theorems.

Theorem   1

If $n(A) = k$, $n(B) = m$, and $A \subset B$ (A is a proper subset of B), then $m > k$.

Proof: Following is a direct proof of the statement. Since $n(A) = k$, then A is in one-to-one correspondence with the set $\{1, 2, 3, \ldots k\}$. Thus, we can denote the elements of A by $a_1, a_2, a_3, \ldots a_k$. (This is an example of *indexing*, to be discussed later in this section.) Since A is a proper subset of B, there exists at least one element of B (possibly more) that is not in A. Denote the element(s) of B that are not in A by $b_1, b_2, \ldots b_j$ (note that j is at least 1). Now arrange the elements of B in order $\{a_1, a_2, \ldots a_k, b_1, b_2, \ldots b_j\}$, and set up a correspondence with the set $\{1, 2, 3, \ldots k, k + 1, k + 2, \ldots k + j\}$. Thus $n(B) = k + j$. But $m = k + j$ and $j \geq 1$ so $m > k$.

Theorem 2

The set I^+ of positive integers is infinite.

Proof: We prove this statement by the *indirect* method. Assume I^+ is finite. We know $1 \in I^+$, so $I^+ \neq \varnothing$. Thus I^+ can be placed in one-to-one correspondence with the set $K = \{1, 2, 3, \ldots k\}$, for some k. So I^+ has k elements. But we also know that K is a proper subset of I^+, since I^+ contains the integer $k + 1$ whenever it contains the integer k (a property of I^+).

What we have, then, is a proper subset K of I^+ and I^+ itself each having k elements. This is a contradiction to Theorem 1 above, which says the number of elements in a finite superset must be strictly greater than the number of elements in a proper subset. Therefore, our assumption that I^+ is finite is false; hence I^+ must be infinite.

Following is a list of some of the basic counting properties of *finite* sets.

F1 A is a nonempty finite set if and only if $n(A)$ is a positive integer. This means

$$\forall_{\text{finite set } A,\ A \neq \varnothing} \exists_{k \in I^+} \text{ such that } n(A) = k$$

and, conversely,

$$\forall_{k \in I^+} \exists_{\text{finite set } A,\ A \neq \varnothing} \text{ such that } n(A) = k$$

F2   If A and B are finite sets, then $A \cup B$ is finite (the union of any two finite sets is a finite set).

F3   If $A \subseteq B$ and B is a finite set, then A is a finite set (any subset of a finite set is a finite set).

F4 $A = \varnothing$ if and only if $n(A) = 0$.

F5 A is a singleton set if and only if $n(A) = 1$.

F6   If A and B are finite sets, then

$$n(A \cup B) = n(A) + n(B) - n(A \cap B)$$

F7   If A and B are finite disjoint sets, then

$$n(A \cup B) = n(A) + n(B)$$

F8   If A and B are finite sets and $B \subseteq A$, then

$$n(A - B) = n(A) - n(B)$$

To verify property F6, suppose $n(A \cup B) = m$, $n(A) = r$, $n(B) = s$, and $n(A \cap B) = t$. We wish to prove $m = r + s - t$. Suppose we index the elements in $A \cap B$ by $\{c_1, c_2, \ldots c_t\}$. Then A could be indexed by $\{a_1, a_2, \ldots a_{r-t}, c_1, c_2, \ldots c_t\}$ and B could be indexed by

$\{b_1, b_2, \ldots b_{s-t}, c_1, c_2, \ldots c_t\}$. Now $A \cup B$ contains the elements

$$\{a_1, a_2, \ldots a_{r-t}, c_1, c_2, \ldots c_t, b_1, b_2, \ldots b_{s-t}\}$$

Hence $A \cup B$ has $(r - t) + t + (s - t)$ elements. That is, $n(A \cup B) = r + s - t$, which was to be proved.

Some of the other properties (e.g., F2, F3, F7, and F8) can be verified using similar reasoning.

We turn now to the process of indexing.

Indexing

Quite frequently in mathematics, we need to deal with more objects or sets than we have letters in the English (or Greek) alphabet. Suppose we wish to consider a system of 50 sets (or for that matter, a general system of k sets). If we try to label them via capital letters of the English alphabet, A, B, C, $\ldots$, we'll run out of available letters at 26, and moreover, that kind of labeling is somewhat confusing. One way out of this impasse is to *index* the objects or sets with which we are dealing. We do this by taking some set Δ where elements can be ordered in a systematic fashion and using its elements in one-to-one correspondence as subscripts for the correspondent objects or sets we wish to index. The set Δ is called an *indexing set*, and the process of assigning its elements to the objects or sets we have in mind is called *indexing*. One of the more standard indexing sets is the set of positive integers $I^+ = \{1, 2, 3, 4, \ldots n \ldots\}$. If the system we wish to index is finite or countably infinite, the positive integers serve as a very natural indexing set. For example, a collection of 50 sets could be denoted easily by

$$A_1, A_2, A_3, \ldots A_{50}$$

Similarly, a general family of k sets could be denoted by

$$A_1, A_2, A_3, \ldots A_k$$

Additionally, a *countably* infinite collection of sets might be denoted $\{A_i | i \in I^+\}$. If, however, the collection of objects or sets we wish to index is uncountably infinite, we may wish to use some general indexing set Δ that has sufficiently many elements to use as indices. In this case, the collection would be the set $\{A_\delta | \delta \in \Delta\}$.

We can use the idea of indexing to generalize the set operations of union and intersection.

Generalized Union of Sets

Suppose $A_1, A_2, A_3, \ldots A_n$ is an indexed finite collection of sets. The *generalized union* of all these sets, denoted by

$$\bigcup_{i=1}^{n} A_i$$

is the set

$$\bigcup_{i=1}^{n} A_i = A_1 \cup A_2 \cup A_3 \cup \cdots \cup A_n$$

Suppose $\{A_i | i \in I^+\}$ is a countably infinite collection of sets. Then the generalized union of these sets is symbolized by

$$\bigcup_{i=1}^{\infty} A_i = A_1 \cup A_2 \cup \cdots \cup A_i \cup \cdots$$

Generalized Intersection of Sets

Suppose $A_1, A_2, A_3, \ldots A_n$ is an indexed collection of sets. The *generalized intersection* of all these sets, denoted by

$$\bigcap_{i=1}^{n} A_i$$

is the set

$$\bigcap_{i=1}^{n} A_i = A_1 \cap A_2 \cap A_3 \cap \cdots \cap A_n$$

For a countably infinite collection of sets $\{A_i | i \in I^+\}$, the generalized intersection would be

$$\bigcap_{i=1}^{\infty} A_i = A_1 \cap A_2 \cap \cdots \cap A_i \cap \cdots$$

In the case of *uncountably* infinite collections of sets, if Δ is an indexing set and the collection is

$$\{A_\delta | \delta \in \Delta\}$$

then the generalized union and intersection are denoted by

$$\bigcup_{\delta \in \Delta} A_\delta \text{ and } \bigcap_{\delta \in \Delta} A_\delta$$

respectively. Therefore,

$$\left(x \in \bigcup_{\delta \in \Delta} A_\delta \right) \leftrightarrow (\exists_{\delta \in \Delta})(x \in A_\delta)$$

and

$$\left(x \in \bigcap_{\delta \in \Delta} A_\delta \right) \leftrightarrow (\forall_{\delta \in \Delta})(x \in A_\delta).$$

Example 2   Find $\bigcup_{i=1}^{\infty} A_i$ and $\bigcap_{i=1}^{\infty} A_i$, where for each positive integer i, $A_i = \{1, 2, 3, \ldots i\}$. To get a feeling for this problem, suppose we look at some special cases for the A_i's.

$$A_1 = \{1\}$$
$$A_2 = \{1, 2\}$$
$$A_3 = \{1, 2, 3\}$$
$$\vdots$$
$$A_n = \{1, 2, 3, \ldots n\}$$

Thus,

$$\bigcup_{i=1}^{\infty} A_i = A_1 \cup A_2 \cup \cdots$$
$$= \{1\} \cup \{1, 2\} \cup \{1, 2, 3\} \cup \cdots$$
$$= \{1, 2, 3, \ldots\}$$
$$= I^+$$

Also,

$$\bigcap_{i=1}^{\infty} A_i = A_1 \cap A_2 \cap \cdots$$
$$= \{1\} \cap \{1, 2\} \cap \{1, 2, 3\} \cap \cdots$$
$$= \{1\}$$

We have seen that two sets A and B are called *disjoint* if and only if $A \cap B = \varnothing$. Now we introduce a related idea for indexed collections of sets—*pairwise disjoint sets*. Suppose Δ is some indexing set and $\{A_\delta | \delta \in \Delta\}$ is an indexed collection of sets. Then the sets in this collection are said to be *pairwise disjoint* if and only if $A_\delta \cap A_\gamma = \varnothing$ whenever $\delta \neq \gamma$.

Example 3   The integers 8, 15, and 77 are called relatively prime to each other (even though they are not necessarily prime themselves) since they share no common prime factors. Consider their sets of prime factors:

$$A_1 = \{2\} \qquad A_2 = \{3, 5\} \qquad A_3 = \{7, 11\}$$

These three sets are pairwise disjoint since

$$A_1 \cap A_2 = \varnothing$$
$$A_1 \cap A_3 = \varnothing$$
$$A_2 \cap A_3 = \varnothing$$

It is worthy to note that if $\mathscr{C}$ is any collection of pairwise disjoint nonempty sets, the existence of an indexing set for $\mathscr{C}$ is a consequence of a celebrated mathematical assumption called the *Axiom of Choice* (or Zermelo's Axiom, which is an equivalent statement of the Axiom of Choice).

Axiom of Choice

If $\mathscr{C}$ is any collection of pairwise disjoint nonempty sets, then $\exists_{a\ set\ B}$ such that

$$\forall_{set\ A\ \in\ \mathscr{C}}\ B \cap A \text{ is a singleton set}$$

This means that B is a set that has a single member in common with each set in the entire collection $\mathscr{C}$. Another way of saying this is that we can form the set B by choosing, in turn, exactly one member from each of the sets in $\mathscr{C}$. Thus, B could be used to index the collection of sets $\mathscr{C}$ regardless of the relative "size" of $\mathscr{C}$.

A final concept we shall discuss in this section is the notion of *power set* of a finite set.

Power Set

If A is any set, then the set that has as its *elements* all the *subsets of A* is called the *power set* of A, and denoted

$$\mathscr{P}(A)$$

In other words,

$$\mathscr{P}(A) = \{ B | B \subseteq A \}$$

Consider, for example, the power sets of several finite sets.

$$\mathscr{P}(\varnothing) = \{\varnothing\}$$

$$\mathscr{P}(\{x\}) = \{\varnothing, \{x\}\}$$

$$\mathscr{P}(\{x, y\}) = \{\varnothing, \{x\}, \{y\}, \{x, y\}\}$$

$$\mathscr{P}(\{x, y, z\}) = \{\varnothing, \{x\}, \{y\}, \{z\}, \{x, y\}, \{x, z\}, \{y, z\}, \{x, y, z\}\}$$

Note carefully that the power set of a finite set has *sets* as its elements. When working with power sets, one must be careful with the notation, making sure to distinguish between set and element (e.g., $\{a\}$ is not the same as a; $\{\varnothing\}$ is not the same as $\varnothing$), and to use set braces whenever sets are being listed.

From the foregoing samples of power sets for zero-, one-, two-, and three-element sets, we see that the power sets have one, two, four, and eight elements, respectively. These observations lead to the next theorem.

Theorem   3

If $n(A) = k$, then $n(\mathscr{P}(A)) = 2^k$. In other words, if a finite set has k elements, then it has 2^k subsets (its power set has 2^k elements). A formal proof of this theorem would be by mathematical induction (on k) (see Section 2.6). This was given in earlier discussion on that proof technique. The proof given below is a rather informal sketch of one mode of reasoning. It is sometimes helpful to reason this way in counting arguments.

Proof: Suppose A has k elements; then we can index the elements of the set A by $a_1, a_2, a_3, \ldots a_k$. Now consider how the possible subsets might be formed. Considering each element a_i in turn, there are exactly two choices—either the element is in a given subset or it is not. We use a tree diagram to lay out the possible subsets. The count of each row is given by the numbers in the right-hand column. (This is what would have to be verified formally by mathematical induction.)

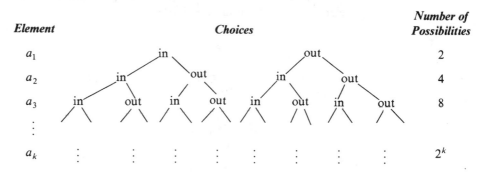

Element	Choices	Number of Possibilities
a_1	in / out	2
a_2	in out / in out	4
a_3	in out in out in out in out	8
$\vdots$		
a_k		2^k

Because of two choices at each stage, each row doubles the number of possibilities of the prior row. Thus, the kth row would exhibit 2^k different branches through the tree—each branch being a different subset of the original set A.

1. Given the following pairs of sets, in how many ways can a one-to-one correspondence be set up between them? Draw arrow diagrams to illustrate each correspondence. Assume, in the sets containing letters of the alphabet, that these are only distinct letters (symbols) and not variables on some set.

   (a) $\{1\}, \{a\}$

   (b) $\{1, 2\}, \{a, b\}$

   (c) $\{1, 2, 3\}, \{a, b, c\}$

   (d) $\{1, 2, 3, 4,\}, \{a, b, c, d\}$

2. (a) As a generalization of Problem 1, suppose A is a finite set having k elements. Then A can be placed in one-to-one correspondence with the set $\{1, 2, 3, \ldots k\}$ (or with itself, for that matter). In how many different ways can such one-to-one correspondences be arranged? Prove your conjecture.

(b) When we ask "In how many ways can a finite set be placed in one-to-one correspondence with itself," what mathematical idea are we dealing with?

3. Verify property F2: The union of two finite sets is a finite set.

4. Verify property F3: Any subset of a finite set is finite.

5. Verify property F8: If A and B are finite sets and $B \subseteq A$, then $n(A - B) = n(A) - n(B)$.

6. (a) Extend property F6 to the situation with three finite sets A, B, and C. In other words, develop a counting formula for the number of elements in the union of three finite sets.

$$n(A \cup B \cup C) =$$

(b) Generalize the "counting of a union" formulas to any generalized union of k sets. That is, develop a counting formula for the total number of elements in the union of k sets:

$$n\left(\bigcup_{i=1}^{k} A_i \right) =$$

7. Suppose Δ is an indexing set and $\Delta = \{r | r$ is a real number and $r > 0\}$. Suppose also that $B_r = \{x | x$ is a real number and $0 \le x < r\}$. Determine the sets

$$\bigcup_{r \in \Delta} B_r \quad \text{and} \quad \bigcap_{r \in \Delta} B_r$$

8. For each positive integer n, let A_n be the set of points in the plane defined by

$$A_n = \{(x, y) | y - x^{2n} \ge 0\}$$

Describe the region of the plane given by $\bigcup_{n=1}^{\infty} A_n$.

9. If $A = \{a, b\}$,

(a) Write out the elements of $\mathscr{P}(A)$.

(b) Write out the elements of $\mathscr{P}(\mathscr{P}(A))$.

(c) Generalize: If A is a finite set and $n(A) = k$, then what is $n(\mathscr{P}(\mathscr{P}(A)))$?

10. A *Cartesian product* of two sets A and B is the set of all ordered pairs (a, b) in which the first component is selected from A and the second component from B. The Cartesian product is denoted $A \times B$ (see Section 4.1).

$$A \times B = \{(a, b) | a \in A \text{ and } b \in B\}$$

Also, two ordered pairs (a, b) and (c, d) are equal if and only if they are equal componentwise. That is, $(a, b) = (c, d) \leftrightarrow (a = c \text{ and } b = d)$.

(a) If $n(A) = 5$ and $n(B) = 4$, what is $n(A \times B)$?

(b) If $A = \{1, 2\}$ and $B = \{x\}$, list the elements in $\mathscr{P}(A \times B)$.

(c) If $n(A) = a$ and $n(B) = b$, what is $n(\mathscr{P}(A \times B))$?

(d) Suppose $A = \{1, 2\}$, $B = \{x\}$, and $C = \{\Delta\}$.
   (i) List the elements of $(A \times B) \times C$.
   (ii) List the elements of $A \times (B \times C)$.
   (iii) Is Cartesian product an associative operation on sets? Why or why not?

3.3   Cardinality

We have seen that infinite sets can be distinguished by "relative size" into two different types—those which are countably infinite versus those which are uncountably infinite. In Section 3.2, Cantor's arguments led to conclusions that the set of rational numbers is countably infinite and that the set of real numbers is uncountably infinite. The question arises, "In what ways can we distinguish between various degrees of infinity?" In this section we shall attempt informally to shed some light on that question. The theory which resolves questions about "infinite numbers" is quite complex, incomplete even to the present day, and considerably beyond the level of this course. It is largely due to Cantor, and his attempts to answer these kinds of questions inspired much of the foundations of modern mathematics. To begin, we need the concept of *cardinality*, or the *cardinal number of a set*.

Cardinality

Let A and B be sets. We say that A and B *have the same cardinality* (*or the same cardinal number*) if and only if A and B are equivalent. That is, there is a one-to-one correspondence between A and B.

Thus a nonempty set A is finite if it has the same cardinal number as $\{1, 2, 3, \ldots n\}$, for some n. The cardinal number of the empty set is defined to be zero. A set is countably infinite if it has the same cardinal number as N (the set of natural numbers).

To proceed, we need several definitions concerning properties of functions. A more detailed look at functions will be taken in Chapter 5. For now, it shall suffice to provide several intuitive ideas and properties of functions as a working vocabulary in which to discuss the idea of cardinality. Let f be a function from one set A to another set B. We denote this by $f: A \to B$, and we say, "f maps A into B." The set A is called the *domain* of the function and B is called the *image space* of the function. The elements in A are called "preimages" of f. The set of all function values of elements in the domain is called the *range* of the function. The elements in the range are called "images" of f. The function f is said to map A *onto* B if for every $b \in B$ there exists an $a \in A$ such that $f(a) = b$. That is, f is onto if and only if range f = image space of f. The function is *one-to-one* (1–1) if for every $a_1, a_2 \in A$, if $f(a_1) = f(a_2)$, then $a_1 = a_2$. That is, whenever two images are equal, their preimages are equal. Finally, a function that maps A one-to-one and onto B is called a *one-to-one (1-1) correspondence* from A to B.

In Chapter 5 we shall study specific properties of 1–1 and onto functions, the existence of their inverses, and compositions of such functions in more detail so that we could justify the following three characteristics of cardinality. We state them here, and defer their proofs until later.

(1) Any set A has the same cardinality as itself. (There exists a 1–1 function mapping A onto itself.)

(2) For any two sets A and B, if A has the same cardinality as B, then B has the same cardinality as A. (If there is a 1–1 function mapping A onto B, then there is a 1–1 function mapping B onto A.)

(3) For any three sets A, B, and C, if A has the same cardinality as B and B has the same cardinality as C, then A has the same cardinality as C. (If there is a 1–1 function mapping A onto B and a 1–1 function mapping B onto C, then there is a 1–1 function mapping A onto C.)

From earlier results, we see that the cardinal number of any nonempty finite set is its count, whereas the infinite sets of natural numbers N, whole numbers W, integers I, and rational numbers Q all have the same cardinality. Still other infinite sets—the set of real numbers R and $(0, 1)$ have the same cardinality which is different from that of N, W, I, and Q. The cardinal number of the natural numbers we denote by $\aleph_0$ (aleph-null), and the cardinal number of the real numbers we denote by c. The question as to whether there exist any other cardinal numbers for infinite sets of real numbers has been proved to be unprovable! Cantor attempted, but failed, to prove that every infinite set of real numbers has cardinality $\aleph_0$ or cardinality c—this is the famous Continuum Hypothesis. In 1963, the logician Paul Cohen proved that the continuum hypothesis is undecidable on the basis of the standard axioms of set theory. Thus, if we accept the continuum hypothesis, then we accept the idea that within the set of real numbers and all its subsets, there are only two levels of infinity—that of the natural numbers and that of the real numbers. If we reject the continuum hypothesis, then we must accept the notion that there are infinite sets of real numbers with cardinality different from $\aleph_0$ and c, even though we have no such examples. In any event, we have here an unprovable statement which has shaken the foundations of logic and set theory, and which has divided mathematicians into different philosophical camps. Following are several theorems about countable and uncountable sets. The first is a major result (called the Schröder-Bernstein Theorem), which we state without proof.

Theorem 1   *[Schröder-Bernstein]*

Let A and B be sets such that there is a 1–1 function from A into B and a 1–1 function from B into A. Then A has the same cardinality as B.

Note that the Schröder-Bernstein Theorem does not require 1-1 correspondences (1-1 *and* onto functions), but only 1-1 functions. That is, we can show two sets have the same cardinality if we can produce either a 1-1 onto function from one set to the other (definition of equivalence) or 1-1 functions both ways between the two sets (Schröder-Bernstein). The following theorem illustrates an application of the latter procedure.

Theorem   2

If a and b are distinct real numbers, the cardinality of the open interval (a, b) is the same as the cardinality of the closed interval $[a, b]$.

Proof: It is not too difficult to show that if $a < c < d < b$ then $[a, b]$ has the same cardinality as $[c, d]$ (see problem 1). Now, we apply the Schröder-Bernstein Theorem, showing there exist 1-1 functions both ways between $[a, b]$ and (a, b). First the identity function is 1-1 from (a, b) into $[a, b]$. Second, if $a < c < d < b$, then since $[a, b]$ has the same cardinality as $[c, d]$, there must be some 1-1 correspondence from $[a, b]$ onto $[c, d]$. But this function maps $[a, b]$ 1-1 *into* (a, b). Therefore, by the Schröder-Bernstein Theorem, $[a, b]$ and (a, b) have the same cardinality.

If we think graphically and know a little about (continuous) functions, it is not difficult to establish some interesting and unexpected results. As an example, we show that the unit open interval $(0, 1)$ is equivalent to all of R.

Example 1   Show that $(0, 1)$ has the same cardinality as the set of all real numbers, R.

All we need is a 1-1 function f which maps the open interval $(0, 1)$ (continuously) onto R. Consider the graph shown. The function

$$f(x) = \frac{x - \frac{1}{2}}{x(x - 1)}$$

does the job. (There are many other such 1-1 onto functions.)

The next theorem provides us with an interesting fact about the cardinalities of sets and their power sets. Recall that if S is any set, then the power set of S, denoted $\mathscr{P}(S)$, is the set of all subsets of S. For example, if S is a finite set having 5 elements, then $\mathscr{P}(S)$ is also finite, and it has 32 elements (all subsets of S). We know, in general, that any finite set having n elements has 2^n subsets. It is easy to see that if S is any finite set, then the cardinality of S is different than the cardinality of $\mathscr{P}(S)$. However, in the infinite cases, that fact is not so readily apparent.

Theorem   3

If S is any set, then S and $\mathscr{P}(S)$ are not equivalent. That is, a set never has the same cardinality as its own power set.

Proof: (by cases on S)

Case 1. $S = \varnothing$.

Then $\mathscr{P}(S) = \{\varnothing\}$, and so $n(S) = 0$ and $n(\mathscr{P}(S)) = 1$. Therefore S and $\mathscr{P}(S)$ do not have the same cardinal number.

Case 2. $S \neq \varnothing$.   (Indirect proof).

Suppose S and $\mathscr{P}(S)$ were equivalent. Then there would be a one-to-one function f mapping S onto $\mathscr{P}(S)$—associating elements of S with subsets of S in such a manner that for each subset of S there exists an element of S which f maps to that subset. Define a subset A of S by

$$A = \{x \mid x \in S \quad \text{and} \quad x \notin f(x)\}$$

That is, A is the set of all elements in S which are *not* members of their image subset under f. Whether $A = \varnothing$ or not, A is a subset of S, by the Axiom of Specification, and since f maps S onto $\mathscr{P}(S)$, there must exist an element $x \in S$ such that $f(x) = A$. Now, what about x? If $x \in A$, then by definition of A, $x \notin f(x) = A$ (a contradiction). On the other hand, if $x \notin A$, then we have $x \in f(x) = A$, and so $x \in A$ by the definition of A (again, a contradiction). We conclude that there can exist no such function f; hence S and $\mathscr{P}(S)$ do not have the same cardinal number.

The proof of Theorem 3 contains reasoning reminiscent of the Russell Paradox (see page 10). This is precisely why in mathematics we do not consider the "set of all sets" to be a well-defined set. If we did, we would run into the same kind of contradiction when we consider A to be the set of all sets that are not elements of themselves. Either possibility—$A \in A$ or $A \notin A$—produces a contradiction.

One consequence of Theorem 3 is that the set of natural numbers N has different cardinality than its power set $\mathscr{P}(N)$. This means (if we accept the continuum hypothesis) that $\mathscr{P}(N)$ must have cardinality c or "larger." The next theorem settles the question, and is a classical application of the Schröder-Bernstein Theorem.

Theorem   4

In N is the set of natural numbers and $\mathscr{P}(N)$ is its power set, then $\mathscr{P}(N)$ has the same cardinality as the set of real numbers, R.

Proof: Since we know from Example 1 that the open interval $(0,1)$ has the same cardinality as R, it suffices to show here that $(0,1)$ has the same cardinality as $\mathscr{P}(N)$. To do this, we will produce 1–1 functions from $(0,1)$ into $\mathscr{P}(N)$ and from $\mathscr{P}(N)$ into $(0,1)$ and appeal to the Schröder-Bernstein Theorem. As in the Cantor proof of the uncountability of $(0,1)$, let every element in $(0,1)$ be written in its binary expansion form, $.a_1a_2a_3\ldots$, where each a_i is either a "0" or a "1." (Disallow the infinite string of all "1s.") Now, given any real number $x \in (0,1)$, suppose $x = .a_1a_2a_3\ldots$ and define a set $A = \{n \mid n \in N$ and $a_n = 1\}$. For example, if $x = .101010\ldots$ (alternating 1s and 0s), then $A_x = \{1,3,5,\ldots\}$, or the set of odd natural numbers. For each x in $(0,1)$, A_x is a subset of N and thus $A_x \in \mathscr{P}(N)$. Now define a function $f\colon (0,1) \to \mathscr{P}(N)$ by $f(x) = A_x$. By the uniqueness of binary representation of real numbers, f is a 1–1 mapping.

Next, let $B \in \mathscr{P}(N)$, so $B \subseteq N$. We now wish to associate B with some real number in $(0,1)$. To do this, use the natural numbers in B to form a decimal representation of a real number in $(0,1)$ as follows: Let $.b_1b_2b_3\ldots$, be a decimal string of "3s" and "5s" such that $b_j = 3$ if $j \in B$ and $b_j = 5$ if $j \notin B$. Thus, for every set $B \in \mathscr{P}(N)$ we obtain a real number in $(0,1)$ with decimal representation a string of 3s and 5s, depending on whether the natural-number position of each digit is an element of B or not. Since decimal representation is unique, two different subsets of N cannot map to the same real number in $(0,1)$, and we have a 1–1 function from $\mathscr{P}(N)$ into $(0,1)$. Therefore, by the Schröder-Bernstein Theorem, $\mathscr{P}(N)$ has the same cardinality as $(0,1)$, and by implication, as the set of real numbers, R.

Another consequence of Theorem 3 is that the cardinality, c, of the set of real numbers, R, is different from the cardinality of $\mathscr{P}(R)$, the power set of R. This gives rise to other cardinal numbers besides $\aleph_0$ and c. Note that this does not contradict the continuum hypothesis, which asserts that there are no sets of *real numbers* with cardinality other than $\aleph_0$ or c. We shall not deal with extended cardinal numbers here. The formal study of such entities belongs to a branch of mathematics known as "*transfinite arithmetic.*"

We conclude this section with a series of observations that can be deduced from other results (including the Axiom of Choice and the Well-Ordering Principle, see pp. 128, 207). These are presented without proof to summarize some important points in our brief treatment of cardinality.

(1) There can exist a one-to-one correspondence between two infinite sets, even if one is a proper subset of the other.

(2) Any infinite subset of a countably infinite set is countably infinite.

(3) Any union of a countably infinite collection of countably infinite sets is a countably infinite set.

(4) Any Cartesian product of a nonempty finite collection of countably infinite sets is a countably infinite set.

(5) Any infinite set (countable or uncountable) has a countably infinite subset.

(6) All intervals of real numbers—$[a, b]$, $[a, b)$, $(a, b]$, (a, b), (a, ∞), $[a, \infty)$, $(-\infty, b)$, $(-\infty, b]$—(where $a < b$) have the same cardinal number, c, which is the cardinal number of the set of real numbers R.

(7) There are many different levels of infinity from $\aleph_0$ to c, to the cardinality of $\mathcal{P}(R)$, to the cardinality of $\mathcal{P}(\mathcal{P}(R))$, and so on.

Assigned Problems

1. Prove that if a, b, c, and d are real numbers such that $a < b$ and $c < d$, then the closed interval $[a, b]$ has the same cardinal number as the closed interval $[c, d]$. (*Hint:* Think graphically and define a continuous 1–1 function from $[a, b]$ onto $[c, d]$.)

2. Prove that $\left(-\dfrac{\pi}{2}, \dfrac{\pi}{2}\right)$ has the same cardinal number as the entire set of real numbers, R.

3. In the proof of Theorem 3 we took two cases—$S = \emptyset$ or $S \neq \emptyset$. For the second case ($S \neq \emptyset$), we implicitly assumed, without explicitly calling attention to it, that $S \neq \emptyset$. Where exactly in the proof of Case 2 did we use the assumption that $S \neq \emptyset$?

4. Let S be the unit square with vertices $(-1, 1)$, $(1, 1)$, $(1, -1)$, and $(-1, -1)$. Let C be the unit circle, $C = \{(x, y) \mid x^2 + y^2 = 1\}$. Using a diagram to illustrate your reasoning, show that S and C have the same cardinality.

5. Complete the following proof that the interior and boundary of the unit square joining the four points $(0, 0)$, $(1, 0)$, $(1, 1)$, and $(0, 1)$ can be mapped by a 1–1 function into the interval $[0, 1]$, thereby showing that $[0, 1] \times [0, 1]$ has the same cardinality as $[0, 1]$.

   Take any point (x, y), where $0 \leq x \leq 1$ and $0 \leq y \leq 1$. Then x and y have infinite decimal expansions,

   $$x = .a_1 a_2 a_3 \ldots$$
   $$y = .b_1 b_2 b_3 \ldots$$

   Now form the real number z by alternating the digits of x and y, respectively

   $$z = .a_1 b_1 a_2 b_2 a_3 b_3 \ldots$$

   (You continue the argument from this point.)

6. Assuming that a finite or countable union of countable sets is countable, and that an infinite subset of a countable set is countable, prove that if A is an uncountable set and B is a countable set, then $A - B$ must be an uncountable set.

3.4   *Classification Problems*

The ideas of elementary set theory, especially the forms of representation such as Venn and tree diagrams, have some interesting application in *classification* problems. Whenever we try to sort, categorize, or classify objects in disjoint categories using two, three, four, or more attributes, we have a classification problem. Quite frequently we may be given a collection of information (e.g., numbers of objects having various classifications) and play a detective game using our logical principles to deduce the number of objects in other categories.

A simple classification used in art and photography is the classification of colors into color families. There are three *primary* colors—red, blue, and yellow. The intersections of these primary colors taken two at a time are called the *secondary* colors—violet, orange, and green. The intersection of all three primaries is *tertiary* color—brown. This classification of colors is illustrated in the three-way Venn diagram in Figure 3.7.

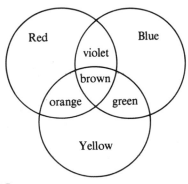

Figure 3.7

An application in medicine serves next as an illustration of a counting problem solved through classification. There are three proteins (antigens) in human blood, symbolized by A, B, and Rh. The presence or absence of A, B antigens determines *blood type*; the presence or absence of Rh determines *Rh factor*. Symbols for blood-type/Rh factor combinations are given in the following tables.

Antigen	*Blood Type*	*Antigen*	*Rh Factor*
A and B	AB	Rh present	+
A only	A	Rh absent	−
B only	B		
neither A nor B	O		

For example, AB$^+$ blood indicates the presence of all three antigens, while B$^-$ indicates the absence of A, presence of B, and absence of Rh. In set theoretic symbols, AB$^+$ can be regarded as $A \cap B \cap Rh$, while B$^-$ can be regarded as $\overline{A} \cap B \cap \overline{Rh}$. Since there are three attributes (the three antigens), we have eight disjoint classifications. We can represent these classifications in a three-way Venn diagram (see Figure 3.8) with "truth table" classification, where T means presence of the antigen and F means absence.

A	B	Rh	Blood Category
T	T	T	AB$^+$
T	T	F	AB$^-$
T	F	T	A$^+$
T	F	F	A$^-$
F	T	T	B$^+$
F	T	F	B$^-$
F	F	T	O$^+$
F	F	F	O$^-$

Now suppose an experiment is run on 100 people. Various numbers of samples are taken and the presence of these antigens in each sample are noted. The raw data from certain classifications appear in the following table.

No. of Samples	Antigens Present
50	A
52	B
40	Rh
20	A and B
13	A and Rh
15	B and Rh
5	A, B, and Rh

The experimenters are particularly interested in three categories:

1. How many samples were A$^-$?

2. How many samples were O$^+$?

3. How many samples were O$^-$?

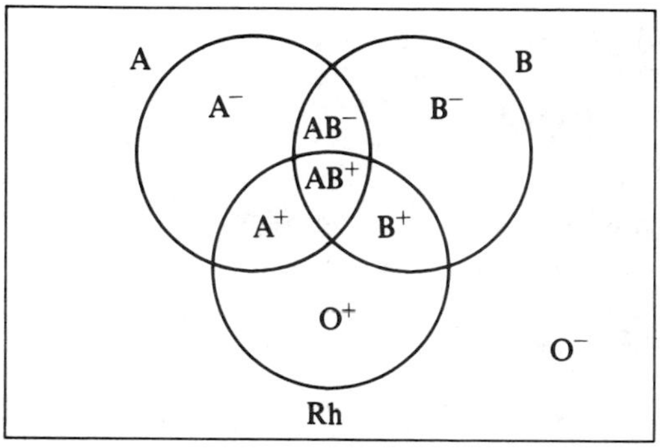

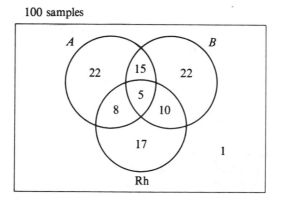

Figure 3.8

To answer these questions, our strategy will be to draw a Venn diagram (see Figure 3.9) and use all the information in the data table to help us deduce the number of samples in each disjoint category. As a start, the information that 50 samples had antigen A present is not much help, because the A circle really contains four disjoint classifications—$A^-(A \cap \overline{B} \cap \overline{Rh})$, $AB^-(A \cap B \cap \overline{Rh})$, $AB^+(A \cap B \cap Rh)$, and $A^+(A \cap \overline{B} \cap Rh)$. Therefore, it is probably more efficient to start with the smallest category (maximum number of intersections) and work outward. We know five samples contained A, B, and Rh, so we can place 5 in the triple intersection. Once that is done, the two-way intersections are easy because each involves the three-way intersection and one more disjoint category. Proceeding this way, we determine numbers of samples for each of the eight disjoint categories.

From the completed classification diagram, we can quickly answer the three questions, and our problem is solved.

Figure 3.9

Number of A⁻ samples: 22
Number of O⁺ samples: 17
Number of O⁻ samples: 1

As another example, suppose a manufacturer has 50 mattresses in stock, and he compiles the following list while taking inventory:

14   king-size

21   foam interior

6   plastic covers and foam interiors but not king-size

28   do not have plastic covers

16   king-size, or foam interiors, without plastic covers

6   king-size and plastic covers

7   king-size and foam interiors

3   king-size but neither foam interiors nor plastic covers

A week later, the manufacturer would like to fill some orders, and he wants to know how many mattresses have the following characteristics:

- king-size, foam interior, and plastic-covered

- foam interior only

- foam interior or plastic-covered, but not king-size

In attempting to solve this problem, we first note from the inventory that there are three categories of mattresses—king-size, foam interior, and plastic covered, but the three-way intersection is not apparent. We look to see if any information can be used to fill in a disjoint category. Surveying the data carefully, we recognize several pieces of information that are quite helpful:

3 king-size but neither foam interior nor plastic cover (a disjoint category)

14 king-size

From these two pieces of information we deduce that 11 mattresses must be in the king-size and foam interior or king-size and plastic cover category. That is, if K, F, and P represent the three categories, we know that

$$n[(K \cap F) \cup (K \cap P)] = 11$$

But other given information tells us

$$n(K \cap F) = 7$$
$$n(K \cap P) = 6$$

Putting this information together and recalling the counting formula for the union of two sets, we deduce that

$$n(K \cap F \cap P) = 2$$

This is a critical piece of information! It's the "pivot" we needed to lead us to other conclusions. We can now solve quite easily for the remaining categories. Our diagram is shown in Figure 3.10. Therefore, there are

2 king-size, foam interior, and plastic-covered

8 foam interior only

24 foam interior or plastic-covered but not king-size

Occasionally it is convenient to represent classifications in different ways to help visualize relationships. The Venn diagram is perhaps most convenient because there are regions available to fill with numbers, and numbers of objects in related classifications can be added or subtracted with ease. In addition to the Venn diagram, there are also truth table or membership table schematics, as well as tree diagrams. To illustrate these three methods of visualization, we will use a four-way classification scheme. Suppose all the women on campus are concerned with four categories of men—intelligent (I), socially concerned (S), honest (H), and good looking (G). For any particular woman (suppose we call her "Sue"), we can sort these four categories as shown at the top of page 142.

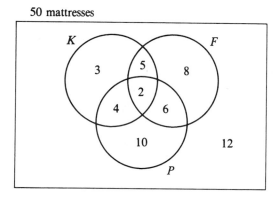

Figure 3.10

Truth Table

I	S	H	G	Number of men Sue places in the category
T	T	T	T	
T	T	T	F	
T	T	F	T	
T	T	F	F	
T	F	T	T	
T	F	T	F	
T	F	F	T	
T	F	F	F	
F	T	T	T	
F	T	T	F	
F	T	F	T	
F	T	F	F	
F	F	T	T	
F	F	T	F	
F	F	F	T	
F	F	F	F	

Lexicographical arrangement produces 16 distinct categories in which T means the characteristic is present and F means the characteristic is absent (according to Sue's perception of campus men).

Tree Diagram (Figure 3.11)

The four levels of the tree represent the four characteristics (or attributes) of interest. A symbol (e.g., H) means that characteristic is present (e.g., honest); a symbol with a bar (e.g., $\overline{H}$) means that characteristic is absent (e.g., not honest). Reading from top to bottom, there are 16 distinct branches, each branch representing one of the 16 disjoint categories. The squares at the bottom could be filled in with numbers of men having each particular combination of attributes (according to Sue's point of view).

Venn Diagram (Figure 3.12)

Here is a Venn diagram showing the 16 distinct regions that are possible when four sets intersect (see Problem 9, Section 3.1). Each region would have a number placed in it that would correspond to Sue's perception of the number of men in the category.

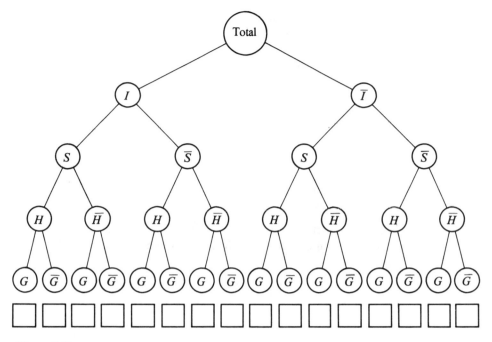

Figure 3.11

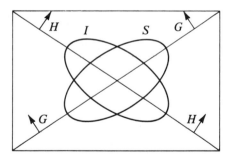

Figure 3.12

Assigned Problems

1. (a) In a set of 20 people, $n(A) = 15$, $n(B) = 14$, and $n(A \cap B) = 13$. What is $n(A \cup B)$?

(b) In a set of 30, $n(P) = 18$, $n(Q) = 12$, $n(P \cup Q) = 21$. Find $n(P \cap Q)$.

(c) Twenty-five items are sorted into two classifications A and B such that $n(A) = 23$ and $n(B) = 24$. List the possible range of values for $n(A \cup B)$, and also for $n(A \cap B)$. (An item need not belong to either classification, or it could belong to both.)

2. In a health survey of 100 children, the numbers having defective eyes, hearing, or teeth were as follows.

defective eyes	20
defective hearing	17
defective teeth	26
defective eyes and teeth	9
defective hearing and teeth	8
defective eyes or teeth and defective hearing	12
defective eyes and teeth or defective eyes and hearing or defective teeth and hearing	18

(a) How many had defective eyes but not defective teeth or defective hearing? $E \cap (\overline{T \cup H})$

(b) How many had defective hearing and defective teeth but not defective eyes? $(H \cap T \cap \overline{E}$

(c) How many had no defects in eyes, hearing, or teeth? $\overline{E} \cap \overline{H} \cap \overline{T}$

3. In a certain high school every student takes at least one course in science (biology, chemistry, physics) as a graduation requirement. Of a graduating class of 80, the number of those who took all three sciences was 8. Other enrollment figures indicated:

biology and chemistry	20
chemistry and physics	25
chemistry	40
biology only	16
physics only	15

How many took (a) biology, (b) physics?

4. Consider three sets P, Q, and R and the following information:

$$n(P \cap Q \cap R) = 3$$
$$n(Q \cap R) = 7$$
$$n(P \cap R) = 9$$

$$n((P \cup R) \cap Q) = 9$$
$$n(Q) = 13$$
$$n(P \cup Q) = 24$$

Determine $n((P \cap \overline{Q}) \cap \overline{R})$.

5. A doctor has 41 patients. Every patient who smokes (S) has a respiratory disease (R). Twenty of the smokers are also tall (T). Eight of those patients who have respiratory disease are not tall. However, a total of 30 patients have respiratory disease. One patient is neither tall nor has respiratory disease, but five nonsmokers have respiratory disease.

   (a) How many tall patients do not have respiratory disease?

   (b) How many patients have respiratory disease, but are neither smokers nor tall?

   (c) How many tall nonsmokers have respiratory disease?

6. The positive integer 252 has a prime factorization of

   $$2^2 \cdot 3^2 \cdot 7$$

   Suppose we wish to determine all the divisors of 252 in an orderly and systematic fashion. This essentially would involve all combinations of divisors of 2^2, 3^2, and 7. For example, the divisors of 2^2 are 1, 2, and 2^2. Construct a tree diagram that lays out all such combinations of divisors and such that when we multiply the numbers down any branch, we produce a divisor of 252. How many distinct positive integer divisors of 252 are there (including 1)? List all of them.

7. Generalize Problem 6 to the following situation. If n is any positive integer, suppose its prime factorization looks like

   $$n = p_1^{a_1} \cdot p_2^{a_2} \cdot p_3^{a_3} \cdot \cdots \cdot p_t^{a_t}$$

   (a) How many distinct positive integer divisors will n have?

   (b) Verify your answer to question (a) by constructing a proof by mathematical induction on t.

8. Consider all the positive integers from 1 to 100. Suppose we wish to separate them along the four cases of divisibility by 2, 3, 5, or 7. Construct a four-way Venn diagram, and in each of the 16 disjoint regions, place the *number* of positive integers (between 1 and 100, inclusive) that fit that characteristic. Let A, B, C, and D be the sets of multiples of 2, 3, 5, and 7, respectively. Use the Venn diagram for four sets given in this section, replacing I, S, H, and G with A, B, C, and D, respectively.

9. A survey was made of 9000 households to find what types of detergents each had used within the last three years. The following tabulation was made:

   3224 had used brand A

   3571 had used brand B

   5656 had used brand C

   1820 had used both A and B

   2376 had used both A and C

   2476 had used both B and C

   1545 had used A, B, and C

   It was then reported that 8305 households used at least one of these three products.

   (a) Is this figure correct?

   (b) How many households used exactly two of these detergents?

   (c) How many households used exactly one?

10. A high school class is working the following problem: "Analyze the data below and find out how many students are in the school. Assume that every student in the school takes at least one of the courses."

    28 take English

    23 take French

    23 take German

    12 take English and French

    11 take English and German

    8 take French and German

    5 take all three languages

    One student analyzes the data as follows:

    $28 + 23 + 23 = 74$ students   (if students took only one course)

    but

    $12 + 8 + 11 = 31$ taking two courses

    $31 - 5 = 26$ taking more than one course

    So $74 - 26 = 48$ students in the school. Is this student correct in the method used and the result? Explain why or why not.

3.5   *Algebra of Sets*

We have seen that sets are usually represented by capital letters $A, B, C, \ldots$, the universe of discourse by U, the empty set by $\varnothing$, and operations on sets by such symbols as $\cup, \cap, \overline{A}, A - B$. Just as there are basic tautologies in logic that describe the *algebra of sentences*, there are basic properties and equalities among sets that describe the *algebra of sets*. We list the following as selected theorems from the algebra of sets. The reader should study these 33 theorems carefully to understand not only their structure and meaning, but also how to use them to manipulate expressions involving sets (to produce equivalent expressions). In the next section, we shall see how to formally prove these and other theorems involving sets. For now, the reader might construct Venn diagrams to "get a feeling" for what these theorems are saying. It should be clear, however, that the drawing of Venn diagrams does not constitute a proof of a given theorem.

Set Theorems

S1	$A \cup B = B \cup A$	(commutativity for union)
S2	$A \cap B = B \cap A$	(commutativity for intersection)
S3	$(A \cup B) \cup C = A \cup (B \cup C)$	(associativity for union)
S4	$(A \cap B) \cap C = A \cap (B \cap C)$	(associativity for intersection)
S5	$A \cap (B \cup C) = (A \cap B) \cup (A \cap C)$	(distributivity for intersection over union)
S6	$A \cup (B \cap C) = (A \cup B) \cap (A \cup C)$	(distributivity for union over intersection)
S7	$A \cup \overline{A} = U$	(complementation-union)
S8	$A \cap \overline{A} = \varnothing$	(complementation-intersection)
S9	$A \cup A = A$	(idempotent law-union)
S10	$A \cap A = A$	(idempotent law-intersection)
S11	$(\overline{A \cup B}) = \overline{A} \cap \overline{B}$	(De Morgan's law)
S12	$(\overline{A \cap B}) = \overline{A} \cup \overline{B}$	(De Morgan's law)
S13	$(A \cup B) \cap B = B$	(absorption law)
S14	$(A \cap B) \cup B = B$	(absorption law)
S15	$A - B = A \cap \overline{B}$	(relative complementation)

Set Theorems   (continued)

S16	$A \cap U = A$	(properties of universe)
S17	$A \cup U = U$	
S18	$A \cap \varnothing = \varnothing$	(properties of empty set)
S19	$A \cup \varnothing = A$	
S20	$\overline{U} = \varnothing$	(complement of universe)
S21	$\overline{\varnothing} = U$	(complement of empty set)
S22	$\overline{\overline{A}} = A$	(double complementation)
S23	$(A = B) \leftrightarrow (A \subseteq B \text{ and } B \subseteq A)$	(double inclusion)
S24	$(A \cap B) \subseteq A$	(subset intersection)
S25	$A \subseteq (A \cup B)$	(union subset)
S26	$(A \subseteq B) \leftrightarrow (\overline{B} \subseteq \overline{A})$	(subset complement)
S27	$[(A \subseteq C) \text{ and } (B \subseteq C)] \leftrightarrow [(A \cup B) \subseteq C]$	(union of subsets)
S28	$[(A \subseteq B) \text{ and } (A \subseteq C)] \leftrightarrow [A \subseteq (B \cap C)]$	(subset of intersection)
S29	$(A \subseteq B) \leftrightarrow [(\overline{A} \cup B) = U]$	
S30	$(A \subseteq B) \leftrightarrow [(A \cap B) = A]$	(properties of subsets)
S31	$(A \subseteq B) \leftrightarrow [(A \cup B) = B]$	
S32	$[A - (B \cup C)] = [(A - B) \cap (A - C)]$	(De Morgan's law for relative complements)
S33	$[A - (B \cap C)] = [(A - B) \cup (A - C)]$	(De Morgan's law for relative complements)

Example 1   Prove theorem S30:

$$(A \subseteq B) \leftrightarrow [(A \cap B) = A] \text{ for all sets } A \text{ and } B$$

Proof   Since we are required to verify a biconditional ($\leftrightarrow$), we break the proof down to two parts —one for each conditional.

Part 1 ($\rightarrow$)

Assume $A \subseteq B$; we show $A \cap B = A$. To do this, we will invoke the Axiom of Extent, showing first that if x is any element of $A \cap B$, then it must be an element of A, and secondly, if x is any element of A, then it must be an element of $A \cap B$.

Let $x \in A \cap B$

then $x \in A$ and $x \in B$ (definition $\cap$)

$\therefore$ $x \in A$ (logic)

Next, let $x \in A$

then since $A \subseteq B$ (assumption)

we have $x \in B$ (definition $\subseteq$)

$\therefore$ $x \in A \cap B$ (definition $\cap$)

$\therefore$ $A \cap B = A$ (Axiom of Extent)

$\therefore$ If $A \subseteq B$, then $A \cap B = A$ (conditional proof)

Part 2 ($\leftarrow$)

Assume $A \cap B = A$; we show $A \subseteq B$. To do this, we will let x represent any arbitrary element in A and try to show it must also be in B. Then we will have proved $A \subseteq B$ by invoking the definition of subset.

Let $x \in A$

then, since $A = A \cap B$ (assumption)

we have $x \in A \cap B$ (definition $=$)

thus, $x \in A$ and $x \in B$ (definition $\cap$)

$\therefore$ $x \in B$ (logic)

$\therefore$ $A \subseteq B$ (definition $\subseteq$)

$\therefore$ If $A \cap B = A$, then $A \subseteq B$ (conditional proof)

As a final conclusion to both parts 1 and 2, we see that

$$(A \subseteq B) \leftrightarrow [(A \cap B) = A]$$ holds for all sets A and B.

Example 2   Prove theorem S31:

$$(A \subseteq B) \leftrightarrow [(A \cup B) = B]$$

Proof   Again, we verify a biconditional by showing that the two conditionals hold simultaneously.

Part 1 ($\rightarrow$)

Assume $A \subseteq B$; we show $A \cup B = B$

Let $x \in A \cup B$

then $x \in A$ or $x \in B$ (definition $\cup$)

Now we do a little *proof by cases*:

Case 1: Suppose $x \in A$

then since $A \subseteq B$ (assumption)

we have $x \in B$ (definition of subset)

Case 2: Suppose $x \in B$

then clearly $x \in B$ (logic)

$\therefore$ In either case we have $x \in B$

$\therefore$ $x \in B$ (proof by cases)

Next assume $x \in B$

then $x \in A$ or $x \in B$ (logic)

so $x \in A \cup B$ (definition $\cup$)

$\therefore$ we have $A \cup B = B$ (Axiom of Extent)

Thus, we have shown

If $A \subseteq B$, then $A \cup B = B$ (conditional proof)

Part 2 ($\leftarrow$)

Assume $A \cup B = B$; we show $A \subseteq B$

Let $x \in A$

then $x \in A$ or $x \in B$ (logic)

so $x \in A \cup B$ (definition $\cup$)

$\therefore$ $x \in B$, since $A \cup B = B$ (assumption)

$\therefore$ $A \subseteq B$ (definition of $\subseteq$)

$\therefore$ If $A \cup B = B$, then $A \subseteq B$ (conditional proof)

Thus, the result of both Parts 1 and 2 is the theorem

$$(A \subseteq B) \leftrightarrow [(A \cup B) = B]$$

Examples 1 and 2 (Theorems S30 and S31) have the effect of saying that the three statements

1. $A \subseteq B$

2. $A \cap B = A$

3. $A \cup B = B$

are logically equivalent. This is sometimes called the *consistency principle* in set theory. Any time one of these statements is holding, we know the other two statements also hold. We know these statements are logically equivalent because we have shown (1) ↔ (2) and (1) ↔ (3). It follows from a logic tautology that (2) ↔ (3).

An alternate method of proving the equivalence of three statements such as the above is to prove (1) → (2) → (3) → (1). This latter method is known as the *round-robin* technique for proving equivalence (↔) of statements. To apply it, we verify a series of conditionals that start with any statement, include each statement only once, and end with the same statement as the starting one.

It is also noteworthy that the relationship between two sets, their intersection, their union, the universe, and the empty set can be visualized by the diagram in Figure 3.13, where line segments connect subsets (lower) with supersets (upper).

Figure 3.13 summarizes the content of theorems S24 and S25 as well as the fact that the empty set is a subset of every set which is, in turn, a subset of the universe.

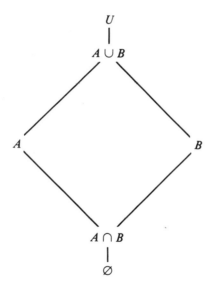

Figure 3.13

Assigned Problems

1. In a Venn diagram, sets are displayed in "general position" when every combination of intersections of the sets involved and their complements that might be nonempty is assigned a region in the diagram. Sketch Venn diagrams with sets in general position for the set theorems S4–S6, S11–S15, and S26–S33.

2. Using appropriate symbols for generalized unions or generalized intersections (of finite collections of sets), restate in correct generalized form the following set theorems: S5, S6, S11, S12, S32, and S33.

3. In each situation below, use a distributive law to find an equal set.

(a) $R \cap (S \cup T)$ (e) $(B \cap A) \cup (B \cap C)$

(b) $S \cup (R \cap T)$ (f) $(T \cup R) \cap (T \cup S)$

(c) $(Z \cup Y) \cap X$ (g) $(W \cap P) \cup (H \cap P)$

(d) $(M \cap N) \cup P$ (h) $(B \cup D) \cap (C \cup D)$

4. Simplify each of the following by applying set theorems and known information.

(a) $C \cup B \cup \overline{C}$ (f) $\overline{[(\overline{C \cup \varnothing}) \cup C]}$

(b) $(C \cap U) \cup \overline{C}$ (g) $(X \cap \overline{W}) \cup \overline{W}$

(c) $(C \cup \overline{B}) \cap (\overline{C} \cup \overline{B})$ (h) $(A \cap B) \cup (A \cup \overline{B})$

(d) $(C \cup U) \cap \overline{C}$ (i) $[(H \cap K) \cup (H \cap \overline{K}) \cup (\overline{H} \cap K)] \cap K$

(e) $\overline{(A \cap U)} \cup \overline{A}$

5. Simplify each of the following expressions involving sets.

(a) $(A \cup T) \cup \overline{A}$

(b) $A \cup (T \cup A)$

(c) $(\varnothing \cap Z) \cup W$

(d) $(A \cap R) \cap \overline{A}$

(e) $(U \cup R) \cap S$

(f) $(A \cap \varnothing) \cap U$

(g) $(A \cap \varnothing) \cup U$

(h) $[A \cap (B \cup \overline{A})] \cup B$

(i) $\overline{(A \cap \varnothing)} \cup U$

(j) $\overline{(A \cup B)} \cup \overline{B}$

(k) $[(D \cup T) \cap (D \cup \overline{T})] \cap (D \cup T)$

(l) $(X \cap U) \cup \overline{X}$

(m) $\overline{[(\overline{C \cap \varnothing}) \cap C]}$

(n) $(A \cap B) \cup (\overline{A} \cup B)$

(o) $(C \cup D) \cap (\overline{C} \cap D)$

(p) $[(A \cap B) \cup (\overline{A} \cup B)] \cup [(B \cap A) \cup (\overline{B} \cup A)]$

(q) $[(\overline{K} \cup T) \cap (\overline{K} \cup \overline{T})] \cup (R \cap \overline{K})$

6. Prove the following statement:
>   If $A \subseteq B$, then $A \cap \overline{B} = \emptyset$ for any sets A and B.

7. Prove the following set theorem:
>   S27 $[(A \subseteq C) \text{ and } (B \subseteq C)] \leftrightarrow [(A \cup B) \subseteq C]$
>   for any sets A, B, and C.

3.6   *Proofs of Set Theorems*

There are two standard deductive procedures for proving that two sets are equal: (a) the *pick-a-point* process, and (b) deductive derivation using known set theorems.

The Pick-a-Point Process

This method of proving two sets are equal rests on the Axiom of Extent. Suppose we wish to demonstrate that $A = B$. The Axiom of Extent says
Let A, B be sets

$$A = B \quad \text{if and only if} \quad (\forall_x (x \in A \leftrightarrow x \in B))$$

We start our proof by selecting any arbitrary element of A, call it x. This is called "picking the point x." Now A and B in this discussion are simply symbols for sets. Here they represent some combination of sets built up from unions, intersections, complementation, or other set operations. Therefore, by letting x represent any element of A, this means that we are assuming that the defining conditions for A hold for x (e.g., if A is an intersection of two sets, by assuming $x \in A$ we are assuming x is an element of both sets). Next we show that under these conditions and those that define B, x must also be an element of B. That being accomplished, we reverse the argument. Now assume x (no relationship with the first x, just a symbol for a general representative) is an element of B, and assuming that x meets the conditions for membership in B, we show x also meets the conditions for membership in A. We have therefore shown: for any x, if $x \in A$, then $x \in B$ *and* if $x \in B$, then $x \in A$. This is equivalent to the biconditional $\forall_x (x \in A \leftrightarrow x \in B)$. Hence, by the Axiom of Extent, the two sets A and B are equal.

Note that the form of argument used in the *pick-a-point* process is also equivalent (by the definition of the subset relation) to proving $A \subseteq B$ and $B \subseteq A$. Thus, by the double inclusion set theorem (S23), the sets A and B are equal. For this reason, proof of equality of sets by the pick-a-point process is sometimes known as *proof by double inclusion*. In this course we will write proofs by double inclusion more frequently than referring to the Axiom of Extent because the former will reinforce the ideas of subset and set equality.

Examples of proofs via the pick-a-point process appeared in Section 3.4 (Algebra of Sets), where set theorems S30 and S31 were verified formally. Those two proofs were further complicated by the fact that the theorems were biconditionals. Thus, in addition to having to prove two sets equal, there was the necessity to argue both conditionals ($\rightarrow$) and ($\leftarrow$). The following example further illustrates the pick-a-point method.

Example 1   Prove Theorem S12 (De Morgan's law):

$$\overline{(A \cup B)} = \overline{A} \cap \overline{B}$$

The statement is a simple assertion that two sets are equal. We use the pick-a-point method of double inclusion to show first

$$\overline{(A \cup B)} \subseteq \overline{A} \cap \overline{B}$$

and then

$$\overline{A} \cap \overline{B} \subseteq \overline{(A \cup B)}$$

Proof   ($\subseteq$)   Assume $x \in \overline{(A \cup B)}$

$\therefore$	$x \notin (A \cup B)$	(definition of complement)
$\therefore$	$\sim [x \in (A \cup B)]$	(meaning of $\notin$; notation)
$\therefore$	$\sim [x \in A$ or $x \in B]$	(definition $\cup$)
$\therefore$	$x \notin A$ and $x \notin B$	(logic/De Morgan's law)
$\therefore$	$x \in \overline{A}$ and $x \in \overline{B}$	(definition of complement)
$\therefore$	$x \in \overline{A} \cap \overline{B}$	(definition $\cap$)
$\therefore$	If $x \in \overline{(A \cup B)}$, then $x \in \overline{A} \cap \overline{B}$	(conditional proof)
$\therefore$	$\overline{(A \cup B)} \subseteq \overline{A} \cap \overline{B}$	(definition subset)

($\supseteq$)   Next, assume $x \in \overline{A} \cap \overline{B}$

$\therefore$	$x \in \overline{A}$ and $x \in \overline{B}$	(definition $\cap$)
$\therefore$	$x \notin A$ and $x \notin B$	(definition complement)
$\therefore$	$\sim [x \in A$ or $x \in B]$	(logic/De Morgan's law)
$\therefore$	$\sim [x \in A \cup B]$	(definition $\cup$)
$\therefore$	$x \notin (A \cup B)$	(meaning of $\notin$; notation)

∴ $x \in \overline{(A \cup B)}$ (definition complement)

∴   If $x \in \overline{A} \cap \overline{B}$, then $x \in \overline{(A \cup B)}$ (conditional proof)

∴ $\overline{A} \cap \overline{B} \subseteq \overline{(A \cup B)}$ (definition subset)

Therefore, we conclude by double inclusion (S23) that

$$\overline{(A \cup B)} = \overline{A} \cap \overline{B} \text{ for all sets } A, B$$

Several comments on the above proof are in order here. First, note that the steps in the ($\supseteq$) part were exactly in the reversed order of the steps in the ($\subseteq$) part. This happens in some, but certainly not all, proofs. This proof happened to be nicely *symmetric* with respect to the order of the steps. Secondly, note the heavy reliance on *definitions* used to justify the steps. This is characteristic of certain proofs in mathematics, especially those that involve going all the way back to primitive notions such as axioms (e.g., the Axiom of Extent) or fundamental theorems (e.g., the double inclusion theorem). Knowing definitions is very important in mathematical reasoning, because the prover follows a pattern of thinking which involves *asking himself lots of questions*, each of which cues further steps in the proof. For example, a typical thinking pattern in the first part of the proof above might go as follows:

- I want to show $\overline{(A \cup B)} \subseteq (\overline{A} \cap \overline{B})$.

- How can I do that?

- What does it mean to be a subset? (recall definition)

- OK, let x represent any arbitrary element of $\overline{(A \cup B)}$.

- Now, what does that mean? (recall definition)

- Aha! $x \notin A \cup B$.

- But I'd like to show $x \in \overline{A} \cap \overline{B}$. (start working backwards from goal)

- This involves an intersection . . . . (recall definition)

- OK, I want $x \in \overline{A}$ and $x \in \overline{B}$.

- Which means $x \notin A$ and $x \notin B$. (recall definition)

- Is the statement $x \notin (A \cup B)$ logically equivalent to the statement $x \notin A$ and $x \notin B$?

- Let's see . . . $x \notin (A \cup B)$ is the same as $\sim [x \in A \cup B]$.

- Now I can use the definition of union to obtain $\sim [x \in A \text{ or } x \in B]$.

- And there's a tautology in logic about negating "or" sentences.

- It's the De Morgan tautology (strange coincidence?!) etc.

Finally, it is significant to note how frequently the analogy between sets and sentences is used. Every tautology of logic gives rise to a corresponding theorem about sets. If you exploit this analogy, you can use what you know about logic to assist your thinking about sets, and vice-versa.

It turns out that the pick-a-point process for proving two sets are equal is an effective (but inefficient) method; it exploits definitions and the logic/set theory analogy, but it is rather long and at times arduous. A more elegant way of proving set theorems is to search one's memory for known set theorems that may apply to the situation at hand without "going all the way down" to an argument using elements. We discuss this second procedure next.

Deductive Derivation Using Set Theorems

We have a list of 33 theorems involving sets. This list was given to illustrate some of the basic properties of unions, intersections, complements, and the subset relation. It was not intended to be a minimal list of theorems. That is, some of the theorems in the list are derivable from others. Also, it certainly is not a comprehensive list of all the major set theorems. There are many theorems about sets that do not appear in the list. To demonstrate how certain set theorems can be used to derive other set theorems, consider the following examples.

Example 2   Prove the absorption laws S13 and S14.

$$\text{S13} \quad (A \cup B) \cap B = B$$

$$\text{S14} \quad (A \cap B) \cup B = B$$

We shall prove both simultaneously.

Proof $(A \cup B) \cap B = B \cap (A \cup B)$ (commutative S2)

$$= (B \cap A) \cup (B \cap B) \quad \text{(distributive S5)}$$

$$= (A \cap B) \cup (B \cap B) \quad \text{(commutative S2)}$$

$$= (A \cap B) \cup B \quad \text{(idempotent S10)}$$

Thus, we have so far

$$(A \cup B) \cap B = (A \cap B) \cup B$$

Note

$$(A \cup B) \cap B \subseteq B \quad \text{(intersection, subset S24)}$$

and

$$B \subseteq (A \cap B) \cup B \quad \text{(subset, union S25)}$$

Hence,

$$(A \cup B) \cap B = B = (A \cap B) \cup B \qquad \text{(double inclusion S23)}$$

The above proof used no definitions; instead, moves from step to step were justified (and motivated) by known set theorems.

Example 3   Prove the theorem, for all sets A, B, and C:

$$A \cap (B - C) = (A \cap B) - C$$

Proof $\begin{aligned} A \cap (B - C) &= A \cap (B \cap \bar{C}) \qquad \text{(relative complement S15)} \\ &= (A \cap B) \cap \bar{C} \qquad \text{(associativity S4)} \\ &= (A \cap B) - C \qquad \text{(relative complement S15)} \end{aligned}$

$$\therefore \quad A \cap (B - C) = (A \cap B) - C$$

The basic idea in proving set equality by derivation from known set theorems is manipulation. That is, we start with one side of the equation, and by manipulating and replacing equal expressions in a series of steps, we try to develop an equation with the other side. Sometimes it helps to work on *both* sides, reducing each side to a common expression.

The reasoning process is difficult to describe for all cases since it varies with the nature and complexity of the theorem to prove as well as the recognition of key theorems which might be useful. This is unlike the pick-a-point process, where we know what we have to do, and the rest is "plugging" away. In Example 2, for instance, the theorem to prove involved operations of intersection and relative complement. Our theorem list contains lots of information about intersections and unions, but not much about relative complements. Therefore, it would be helpful to convert the relative complement statement into a statement about intersections or unions. The key theorem that does the job is S15. One should think of this theorem as "a way of changing from relative complement to intersection." In that way, it becomes a useful device for this and related proofs.

As a final example of worked-out proofs of set equality, the next will be done both ways—by the *pick-a-point* process, then by the *deductive derivation* method.

Example 4   Prove Theorem S32 (De Morgan's law for relative complements):

$$[A - (B \cup C)] = [(A - B) \cap (A - C)]$$

For the sake of simplicity, proof sketches (without complete justifications) will be given here. The reader should fill in the reasons for each step.

Method 1 (Pick-a-Point)

$(\subseteq)$ Let $x \in [A - (B \cup C)]$

then $x \in A$ and $x \notin (B \cup C)$

$\therefore \quad x \in A$ and $\sim [x \in B$ or $x \in C]$

$\therefore \quad x \in A$ and $(x \notin B$ and $x \notin C)$

$\therefore \quad (x \in A$ and $x \notin B)$ and $(x \in A$ and $x \notin C)$

$\therefore \quad x \in (A - B)$ and $x \in (A - C)$

$\therefore \quad x \in [(A - B) \cap (A - C)]$

Hence, $[A - (B \cup C)] \subseteq [(A - B) \cap (A - C)]$.

$(\supseteq)$ Let $x \in [(A - B) \cap (A - C)]$

then $x \in (A - B)$ and $x \in (A - C)$

$\therefore \quad (x \in A$ and $x \notin B)$ and $(x \in A$ and $x \notin C)$

$\therefore \quad x \in A$ and $(x \notin B$ and $x \notin C)$

$\therefore \quad x \in A$ and $\sim [x \in B$ or $x \in C]$

$\therefore \quad x \in A$ and $x \notin (B \cup C)$

$\therefore \quad x \in [A - (B \cup C)]$

Hence, $[(A - B) \cap (A - C)] \subseteq [A - (B \cup C)]$.

Therefore, by double inclusion,

$$[A - (B \cup C)] = [(A - B) \cap (A - C)]$$

Method 2 (Deductive Derivation)

$$
\begin{aligned}
A - (B \cup C) &= A \cap (\overline{B \cup C}) \\
&= A \cap (\overline{B} \cap \overline{C}) \\
&= (A \cap A) \cap (\overline{B} \cap \overline{C}) \\
&= (A \cap \overline{B}) \cap (A \cap \overline{C}) \\
&= (A - B) \cap (A - C)
\end{aligned}
$$

To conclude this section, we list some general heuristic strategies you should try when proving certain kinds of mathematical theorems.

1. To prove one set is a *subset* of another $(A \subseteq B)$,

   (a) pick any arbitrary element (point) in A and show it must also be an element (point) in B, or

   (b) use known theorems involving the subset relation, or

   (c) assume $A \not\subseteq B$ (there is an element in A which is not in B) and try to reach a contradiction.

2. To prove one set is *equal* to another ($A = B$),

(a) use the pick-a-point process and double inclusion ($A \subseteq B$ and $B \subseteq A$) (this is a two-part proof, ($\subseteq$) and ($\supseteq$)), or

(b) use known theorems involving set equalities (deductive derivation method).

    In both (1) and (2) above, remember the close analogy between logic and sets. Capitalize on that analogy, converting back-and-forth between the two systems of representation. Also, know your definitions well.

3. To prove a *conditional statement* ($P \rightarrow Q$),

(a) assume that P holds and try to deduce Q (conditional proof), or

(b) assume $\sim Q$ holds and try to deduce $\sim P$ (proof by contraposition or indirect proof).

4. To prove a *biconditional* statement ($P \leftrightarrow Q$), prove *two* conditionals $P \rightarrow Q$ *and* $Q \rightarrow P$ (this is a two-part proof, ($\rightarrow$) and ($\leftarrow$)).

5. To prove a series of *equivalences* ($P \leftrightarrow Q \leftrightarrow R \ldots$)

There are various ways of doing this. A common procedure is the *round-robin* proof where you start at any point and work through a series of conditionals that end at that same point. For example, $Q \rightarrow R \rightarrow P \rightarrow Q$. The latter would involve a three-part proof ($Q \rightarrow R$), ($R \rightarrow P$), and ($P \rightarrow Q$). In general, for n equivalences ($\leftrightarrow$), there will be $n + 1$ parts to the proof, one for each conditional ($\rightarrow$). The order is immaterial, but there must be a circuit through the statements involved.

6. To prove a statement holds *for all positive integers n*, try a proof by mathematical induction.

Assigned Problems

1. Prove Theorem S6, distributivity for union over intersection, using the *pick-a-point* process.

S6   For all sets $A, B, C, [A \cup (B \cap C)] = [(A \cup B) \cap (A \cup C)]$

2. Prove Theorem S33, De Morgan's law for relative complements, using the *deductive derivation* method and any set theorems you know.

S33   For all sets $A, B, C, [A - (B \cap C)] = [(A - B) \cup (A - C)]$

3. (a) Prove Theorem S26:

For all sets $A, B, (A \subseteq B) \leftrightarrow (\overline{B} \subseteq \overline{A})$

(b) For which tautology in logic is S26 the analogue?

4. Consider the following theorem of set theory:

$$\text{For any two sets } A, B, [A \cup (B - A)] = A \cup B$$

(a) Sketch a Venn diagram to illustrate this equality.

(b) Construct a formal proof using the pick-a-point process.

(c) Construct a formal proof using the deductive derivation method.

5. Is this statement true?

$$\text{For all sets } A, B, [A - (A - B)] = [B - (B - A)]$$

If true, verify it using any method you wish.

If false, produce a counterexample.

6. Generalize the distributive law for intersection of a given set A over the generalized union of any collection of n sets ($n \in I^+$). That is, prove

$$\forall_{n \in I^+} \ A \cap \left(\bigcup_{i=1}^{n} B_i \right) = \bigcup_{i=1}^{n} (A \cap B_i)$$

7. Recall the *symmetric difference* operation on sets, defined by

$$A * B = (A - B) \cup (B - A) \qquad \text{(see Problem 7, Section 3.1)}$$

Prove the operation of intersection distributes over symmetric difference. That is,

$$A \cap (B * C) = (A \cap B) * (A \cap C)$$

8. Prove

$$A * B = (A \cup B) - (A \cap B)$$

where "$*$" refers to the symmetric difference operation.

9. Specify which of the following statements about sets are true or false. If true, give a proof; if false, give a counterexample.

(a) $M * M = \emptyset$

(b) $M \cup (L - N) = (M \cup L) - (M \cup N)$ (i.e., is union distributive over relative complement?)

(c) $(M * L) * N = M * (L * N)$ (i.e., is symmetric difference an associative operation on sets?)

(d) $\left(\bigcap_{i=1}^{n} A_i \right) \cup \left(\bigcap_{i=1}^{n} B_i \right) = \bigcap_{i=1}^{n} (A_i \cup B_i)$ (i.e., is generalized intersection distributive over union of generalized intersection?)

(e) $A \cup (B * C) = (A \cup B)*(A \cup C)$ (i.e., is union distributive over symmetric difference?)

10. Let $A = \{x | x \in R \text{ and } x^2 - 5 = 0\}$
       $B = \{x | x \in R \text{ and } x^4 - 9x^2 + 20 = 0\}$

    Show that $A \subseteq B$.

4

Relations

4.1   Binary Relations

Intuitively, most of us already have a reasonable grasp of the general concept of *relationship*. We all have relatives—mothers, fathers, sisters, brothers, aunts, uncles, and so on. Motherhood, for example, could be abstractly defined as some huge collection of all ordered pairs of persons such that for every pair (x, y) in the set, person x is the mother of person y. We could symbolize this relationship by letting M represent the set of all (mother, offspring) pairs, and then

$(x, y) \in M$ means "x is the mother of y"

Another form of symbolization for the relation of motherhood might be to interpret $(x, y) \in M$ to mean "x stands in the relationship of *mother* to y," writing

xMy means "x is the mother of y"

Still another way of representing the motherhood relation would be to regard the statement $M(x, y)$ as a two-place predicate statement of logic, "x is the mother of y," where we can make a judgment True or False. This is exactly what we did with the *divides* relation: $D(x, y)$ means "x divides y" (where x and y were assumed to be integers). In this case,

$M(x, y)$ means "x is the mother of y"

Thus, there are different ways to represent the idea of two objects being related—as an ordered pair in a set, as one object standing in the relation M to the other, or as a two-place predicate logic statement. In each case, we are talking about *two* objects, and the relations involved are called *binary relations*.

For the previous discussion of motherhood, the *motherhood* relation among human beings could have been replaced by *brotherhood* (human beings), *less than* (real numbers), *is parallel to* (lines in a plane), *ends in the same digit as* (all three-digit numerals), *is married to* (human beings), *is a subset of* (all sets), *has the same domain as* (all functions), and numerous other possibilities. All of these are examples of *binary relations* defined on a particular set.

It is noteworthy that binary relations are not the only kinds of relations in mathematics, just as two-place predicate logic statements are not the only statements in quantificational logic. As an example of a *unary* relation, consider the following. Take as a reference set the set of persons in our class, and fix an integer—20, for example. Now, assuming that no person in class is exactly 20 years of age (today), we could define the relation Y to mean "is younger than 20 years of age." For each individual class member, we could make a judgment True or False with respect to the relation Y. Here again, we could think of Y as a set—the set of all members of our class who are younger than 20: $x \in Y$ means "x is younger than 20." Alternately we could think of Y as a single-place predicate statement: $Y(x)$ means "x is younger than 20." In any case, however we choose to represent it, Y is an example of a *unary* relation, since we are considering only one object at a time.

There are *n-ary* relations for any positive integer n. An example of a *ternary* (three-way) relation would be *betweenness* with respect to points on a line. If B represents this relation, $B(x, y, z)$ might mean "x is between y and z." If the reference set were persons in a theatre, $B(x, y, z)$ might mean "x sits between y and z." In either case, a relation involving *three* objects is being defined. For $n = 4$, a good example of a *quaternary* relation is *equidistance* involving ordered four-tuples of points in a plane. For example, we might select the symbol

$E(x, y, z, w)$ to mean "x is the same distance from y as z is from w"

The latter is an example of a relation involving *four* objects at a time.

Since *binary relations* occur most frequently in modern abstract mathematics, our discussion will be restricted to that class of relations. In fact, there are three major kinds of binary relations which are fundamental to mathematics. These are:

1. Equivalence relations

2. Order relations

3. Functions

After an introduction to the general characteristics of binary relations (the emphasis of this section), we shall turn our attention to those three fundamental relations.

The foregoing discussion was aimed at intuitive aspects of binary relations. We have not yet defined what exactly a *binary relation* is. To do this more formally, we need several intervening ideas.

Ordered Pair

From set theory, we know we can think of the set of two objects $\{a, b\}$ as an *unordered pair*. That is, the order of elements in the set makes no difference.

$$\{a, b\} = \{b, a\}$$

We could introduce the concept of *ordered set*, therefore, by requiring that two ordered sets are equal if and only if not only are their elements identical, but also the *order* of listing in the set must be the same. In the case of ordered sets, we shall drop the set braces notation and use parentheses to avoid confusion with standard sets. If the ordered set has two elements, a and b, we shall call it an *ordered pair* (a, b).

Equality of Ordered Pairs

$$(a, b) = (c, d) \text{ if and only if } a = c \text{ and } b = d$$

(Recall Problem 10, Section 3.2.) Note that this axiom on equality of ordered pairs extends to all ordered sets. For example, if $\{a_1, a_2, a_3, \ldots a_n\}$ is an ordered set, we denote it as an *ordered n-tuple*, and then equality of ordered *n*-tuples would require componentwise equality of elements. That is,

$$[(a_1, a_2, \ldots a_n) = (b_1, b_2, \ldots b_n)] \leftrightarrow (a_i = b_i) \ \forall_{i=1,2,\ldots n}$$

Cartesian Product of Sets

If A and B are sets, the *Cartesian product* of A and B, denoted $A \times B$ (read "A cross B") is the set of all ordered pairs (a, b) where the first component a is selected from set A and the second component b is selected from set B.

$$A \times B = \{(a, b) | a \in A \text{ and } b \in B\}$$

Note that the definition of a Cartesian product extends to any indexed collection of sets, here shown for integer indexing,

$$A_1 \times A_2 \times \cdots \times A_n = \{(a_1, a_2, \ldots a_n) | a_i \in A_i \ \forall_{i=1,2,\ldots n}\}$$

The members of the generalized Cartesian product of n sets are ordered *n*-tuples.

Similar to the way we use Venn diagrams to describe setwise relationships, there are suitable diagrams for visualizing Cartesian products (at least up to three sets). For example, for two sets we could think of one set A having its elements lined up along the lower border of a rectangular region, and the second

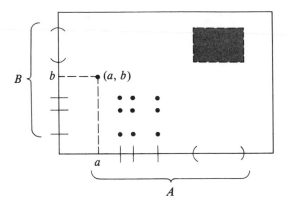

Figure 4.1

set B arranged along the left border. Then the Cartesian product $A \times B$ could be pictured in the interior as collections of points at the intersections of vertical lines through members of A and horizontal lines through members of B (see Figure 4.1). This is the basic idea of the well-known Cartesian (rectangular) coordinate system.

Example 1   Let R represent the set of real numbers. We can produce a visual display of R by using a device called a number line. Select any point as origin, decide on orientation (e.g., right of origin is $+$, left of origin is $-$), and decide on unit distance (distance from origin to point labeled "1"). These three characteristics define a one-to-one correspondence between real numbers and points on a line.

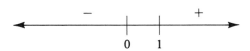

Now, if we take *two* number lines, one perpendicular to the other at their origins, we can produce a scaling device for the entire plane. Every point in the plane becomes representable by *two* coordinates—its (perpendicular) projected location on each of the respective number lines. This defines a one-to-one correspondence between points in the plane and ordered pairs of real numbers. The reference system so devised is well known as the Cartesian coordinate system, and the entire set being represented is $R \times R$ (or in simplified notation, R^2)—the Cartesian product of real numbers cross real numbers, or the set of all ordered pairs whose first component is any real number and whose second component is any real number.

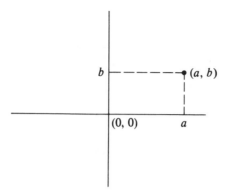

Finally, we can visually extend these ideas to three-dimensional space. Take three mutually perpendicular number lines intersecting at their origins (this establishes an origin we call $(0, 0, 0)$). Now there exists a one-to-one correspondence between points in space and all ordered triples of real numbers. The latter set is

$$R \times R \times R = \{(x, y, z) | x, y, z \text{ are real numbers}\}$$

which is sometimes abbreviated R^3.

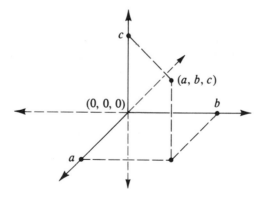

In general, R^n represents the n-fold Cartesian product of real numbers, $R \times R \times \cdots \times R$ (n times), which is the set of all ordered n-tuples of real numbers. While we cannot visually represent all such systems as we have above, they are nevertheless important in mathematics, since they are the fundamental idea behind the study of finite-dimensional vector spaces.

Now that certain preliminary ideas have been considered, we are able to return to *binary relations* in a more formal sense.

Binary Relation

> A *binary relation from set A to set B* is any subset of $A \times B$.
> A *binary relation on a set A* is any subset of $A \times A$.

Example 2   Suppose $A = \{1, 2, 3\}$ and $B = \{x, y\}$. Then

$$A \times B = \{(1, x), (1, y), (2, x), (2, y), (3, x), (3, y)\}$$

Now any subset of $A \times B$ will be a binary relation from A to B. Therefore the empty set and the entire set $A \times B$ qualify as binary relations. Also, the sets

$$R_1 = \{(1, x), (2, y)\}$$
$$R_2 = \{(1, y), (2, y), (3, y)\}$$
$$R_3 = \{(3, x)\}$$
$$R_4 = \{(1, x), (1, y), (2, y), (3, x)\}$$

are all examples of binary relations from A to B. How many different binary relations from A to B are there?

Notational Conventions Involving Binary Relations

We shall use the symbol R as a general symbol for abstract binary relations. Thus, to say "R is a binary relation from A to B" will mean "R is a subset of $(A \times B)$"; to say "R is a binary relation on A" will mean "R is a subset of $(A \times A)$." If $(x, y) \in R$, we shall usually denote this by

$$xRy \quad (x \text{ stands in the relation } R \text{ to } y)$$

The *domain* of a binary relation R (abbreviated dom R) is the set of all first components in the ordered pairs of the relation.

$$\text{dom } R = \{x | (\exists y)(xRy)\}$$

The *range* of a binary relation R (abbreviated ran R) is the set of all second components in the ordered pairs of the relation.

$$\text{ran } R = \{y | (\exists x)(xRy)\}$$

The *inverse* of a binary relation R (denoted R^{-1}) is the set of all "reverse-ordered" pairs in the relation R.

$$R^{-1} = \{(y, x) | (x, y) \in R\}$$

Another way of viewing the inverse relation is

$$xRy \leftrightarrow yR^{-1}x$$

Finally, the *graph* of a binary relation R is the set of all ordered pairs (x, y)—represented as points in a plane—for which the statement xRy is true.

Example 3   Using the sets A, B, R_1, R_2, R_3, and R_4 from Example 2, observe the following.

- Since $(2, y) \in R_1$, we denote this by $2R_1y$

- dom $R_2 = \{1, 2, 3\}$

- ran $R_3 = \{x\}$

- $R_1^{-1} = \{(x, 1), (y, 2)\}$

- graphs of R_4 and R_4^{-1} might look like:

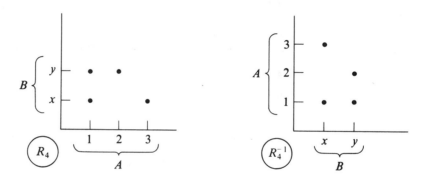

For the most part, in our current discussion of binary relations, which prefaces the study of equivalence and order relations, we shall be concerned only with *binary relations R defined on a set A*. Later, when we begin to study functions, we shall return to *binary relations R from A to B*. In each case where we discuss binary relations, it is important to know or have specified the reference set A (or sets A, B) upon which the binary relation is defined. Following are a series of examples of binary relations R on a given set A. Note in every case that R is a subset of $A \times A$.

Example 4   Here are fourteen different examples of binary relations on a set.

1. $A = \{$human beings$\}$
   R means "is the sister of"
   xRy means "x is the sister of y"

2. $A = [-1, 1]$ closed interval on the real line
$R = \{(x, y) | x^2 + y^2 = 1\}$
xRy means "$x^2 + y^2 = 1$"
Note: The graph of R is the unit circle.

3. $A = \{$lines in a plane$\}$
R means "is parallel to"   (Note: $l_1 \cap l_2 = \emptyset$, as point sets)
$l_1 R l_2$ means "l_1 is parallel to l_2".

4. $A = \{$lines in a plane$\}$
R means "is $\perp$ to"
$l_1 R l_2$ means "$l_1 \perp l_2$"

5. $A = \{$real numbers$\}$
R means "is less than or equal to"
xRy means "$x \leq y$"

6. $A = \{$positive integers$\}$
R means "is a divisor of"
mRn means "m is a divisor of n" (or "$m|n$")

7. $A = $ any set
R means "is equal to"
xRy means "$x = y$"
Note: This particular relation is called the *identity* relation.

8. $A = \left\{ \text{all fractions } \dfrac{a}{b} \text{ where } a, b \in I \text{ and } b \neq 0 \right\}$
R is defined by

$$\left(\frac{a}{b} R \frac{c}{d} \right) \quad \text{if and only if} \quad (ad = bc)$$

What are these kinds of fractions called?

9. $A = \{n | n \in I^+ \text{ and } 1 \leq n \leq 1000\}$
R means "ends in same digit as"
nRm means "n ends in the same digit as m"

10. $A = \{$human beings$\}$
R means "weighs within two pounds of"
xRy means "x weighs within two pounds of y"

11. $A = $ a collection of sets
R means "is a subset of"
MRN means "$M \subseteq N$"

12. A is any set
$R = A \times A$
xRy means "$(x, y) \in A \times A$"

13. $A = \{$all real numbers$\}$
xRy means "$\sin x = \sin y$"

14. $A = \{$triangles in a plane$\}$
R means "is similar to"
[(triangle T) R (triangle S)] means "triangle T is similar to triangle S"

There are certain properties or characteristics that some binary relations on a set possess and others do not. These properties will help us to classify particular kinds of binary relations that play significant roles in the development of mathematics. They are defined below.

For each of the following, A is a nonempty set and R is a binary relation on A.

Reflexive Property

R is said to be *reflexive on A* if and only if

$$\forall_{x \in A} \ xRx$$

(This says that *every* element in the set must be related to itself via the relation R in order that R have the reflexive property.)

Symmetric Property

R is said to be *symmetric on A* if and only if

$$\forall_{x, y \in A} \ (\text{if } xRy, \text{ then } yRx)$$

(This says that whenever a pair of elements in A is related via R, then it must follow that the reverse-order pair is also related via R in order that R have the symmetric property.)

Transitive Property

R is said to be *transitive on A* if and only if

$$\forall_{x, y, z \in A} \ (\text{if } xRy \text{ and } yRz, \text{ then } xRz)$$

(This says that whenever a first element is related to a second *and* the second is related to a third, then it must follow that the first and third are related—if this holds for all elements of A, then R has the transitive property.)

> **Antisymmetric Property**
>
> R is said to be *antisymmetric on A* if and only if
>
> $$\forall_{x,\,y\,\in\,A} \;(\text{if } xRy \text{ and } yRx, \text{ then } x = y)$$
>
> (This says that if the only cases in which pairs are related both ways occur when the elements are equal, then the relation R has the characteristic of being antisymmetric.)

Look back at the 14 examples of binary relations in Example 4 and see if you agree with the following analyses of properties.

Example 5   Consider the "sisterhood" relation defined on the set of all human beings. (See Example 4, relation 1.) Is this a reflexive relation? That is, is the statement

$$\forall_{x\,\in\,A} \;(x \text{ is the sister of } x)$$

true? This would mean every human being would be his/her own sister. R (sisterhood) is clearly *not reflexive*.

How about symmetric? We test the statement.

$$\forall_{x,\,y\,\in\,A} \;(\text{if } x \text{ is the sister of } y, \text{ then } y \text{ is the sister of } x)$$

Momentarily, this may at first appear to be true. But considering Mary Smith and her brother Jim, we have a counterexample. Therefore, sisterhood is *not symmetric*. To be a transitive relation, we would have

$$\forall_{x,\,y,\,z\,\in\,A} \;(\text{if } x \text{ is the sister of } y \text{ and } y \text{ is the sister of } z, \text{ then } x \text{ is the sister of } z)$$

This also may appear to be true, since the antecedent of the conditional forces both x and y to be females and z can be either female or male. But it is possible that x and z may be the same person. Thus the sisterhood relation is not a *transitive relation*. Finally, suppose we assume x is the sister of y and y is the sister of x; does it necessarily follow that x and y are the same person? No, so this relation is *not antisymmetric*.

Example 6   See Example 4, relation 2—the unit circle relation. Reflexive? Is xRx for every $x \in [-1,1]$? That is, is the equation $x^2 + x^2 = 1$ true for all x in the closed interval from -1 to 1? No. It *is* true for two points, however:

$$\left(\sqrt{\frac{1}{2}}, \sqrt{\frac{1}{2}}\right) \quad \text{and} \quad \left(-\sqrt{\frac{1}{2}}, -\sqrt{\frac{1}{2}}\right)$$

Symmetric? Assume xRy; $x^2 + y^2 = 1$. Does it follow that yRx? Yes, since if $x^2 + y^2 = 1$, then $y^2 + x^2 = 1$. Note that the unit circle has graph symmetry with respect to the line $y = x$.

Transitive? Assume xRy and yRz; that is, $x^2 + y^2 = 1$ and $y^2 + z^2 = 1$. Does it necessarily follow that $x^2 + z^2 = 1$? No, take for a counterexample $x = 1$, $y = 0$, and $z = 1$. Therefore this relation is not transitive.

Antisymmetric? No, because if $x^2 + y^2 = 1$ and $y^2 + x^2 = 1$, this does not force $x = y$. Take $x = 1$ and $y = 0$.

Example 7   As a final example, consider the relation of Example 4, relation 14—triangle similarity. Any triangle in the plane is similar to itself; if a given triangle is similar to another, then the second is similar to the first; and if a first triangle is similar to a second, and the second is similar to a third, then it must follow that the first is similar to the third. Thus the relation of similarity among triangles is reflexive, symmetric, and transitive. It is not, however, antisymmetric since two triangles that are similar to each other need not be the same triangle.

**Assigned
Problems**

1. Study each of the examples of binary relations in Example 4 that were not discussed above. Using the symbols r for reflexive, s for symmetric, t for transitive, and a for antisymmetric, decide which combinations of these four properties each of the eleven remaining binary relations possesses.

2. Prove that relation 8 of Example 4,

$$\left(\frac{a}{b} R \frac{c}{d}\right) \quad \text{if and only if} \quad (ad = bc)$$

where a, b, c, d are integers, $b \neq 0$, and $d \neq 0$, is reflexive, symmetric, and transitive. (Such a relation is an example of an equivalence relation; see section 4.2).

3. Sketch a graph of the following binary relation on the system of real numbers, and give its domain and range.

$$R = \{(x, y) | x^2 + y^2 \leq 2\}$$

4. Consider the following two problems:

   (a) There are k roads connecting the towns of Smithville and Grant, and l roads connecting Grant and Winchester. In how many different ways can one travel from Smithville to Winchester via Grant?

   (b) A room has m doors. In how many ways can an individual make a trip into this room and out again if she must enter and leave only by means of the doors?

   Translate these problems into statements involving Cartesian products of sets. What general conclusion can you make about the number of elements (ordered pairs) in a Cartesian product?

5. Prove that $[(A \cup B) \times C] = [(A \times C) \cup (B \times C)]$ for any sets A, B, C. In other words, Cartesian product distributes over set union.

6. Let A be a set with $n(A) = 10$. Show that there are over 10^{30} different binary relations on A.

7. (a) Generalize Problem 6 to the case where $n(A) = k$ (k some positive integer). How many different binary relations on A can there be?

(b) Show that of all the binary relations on A where $n(A) = k$, exactly $(2^k - 1)^k$ of them have all of A for their domain.

(c) Show that of all the $(2^k - 1)^k$ binary relations in (b), exactly $2^{(k^2 - k)}$ of them are reflexive on A.

8. Suppose $A = \{a, b, c, d\}$

Judge the following relations on A with respect to reflexive, symmetric, and transitive properties.

(a) $R_1 = \{(a, a), (a, b), (b, a), (c, c), (b, b), (d, d), (b, c), (c, b)\}$

(b) $R_2 = \{(a, b), (b, c), (c, d), (d, a)\}$

(c) $R_3 = \{(a, a), (b, b), (c, c), (d, d)\}$

(d) $R_4 = \{(a, a), (b, b), (c, c), (d, d), (a, b), (b, a), (a, c), (c, a), (a, d),$
$\quad\quad (d, a), (b, c), (c, b), (b, d), (d, b), (c, d), (d, c)\}$

9. Is the statement below true for all sets A, B, C, and D? If so, prove it; if not, provide a counterexample.

$$[(A \times B) \cup (C \times D)] = [(A \cup C) \times (B \cup D)]$$

10. Consider the set of real numbers R and the relation L defined on R by

$$L = \{(x, y) | y^2 = x^2\}$$

(a) Determine which properties (reflexive, symmetric, transitive, antisymmetric) L has and verify or provide a counterexample in each case.

(b) What is the domain and range of L?

(c) Sketch a graph of L.

4.2   Equivalence Relations

An *equivalence relation* E on a set S is a relation for which all three of the reflexive, symmetric, and transitive properties hold. That is, *E is an equivalence relation on S if and only if*

(a) (reflexive) $\forall_{x \in S}$ (xEx).

(b) (symmetric) $\forall_{x, y \in S}$ (if xEy, then yEx).

(c) (transitive) $\forall_{x, y, z \in S}$ (if xEy and yEz, then xEz)

We shall consider only situations where the set S is a nonempty set (and the relation R is a nonempty relation). Before we consider examples, it would be wise to think about what it really means for a relation R on a set S to *fail to have* any or all of these three properties. By definition (and a little logic), a relation R on S is *not* an equivalence relation if at least one of the properties of reflexivity, symmetry, or transitivity fails. Applying the logic of negating quantified sentences, a relation R on S is

- *not* reflexive if and only if $\exists_{x \in S}(x\cancel{R}x)$
- *not* symmetric if and only if $\exists_{x, y \in S}(xRy \text{ but } y\cancel{R}x)$
- *not* transitive if and only if $\exists_{x, y, z \in S}(xRy \text{ and } yRz, \text{ but } x\cancel{R}z)$

Example 1 Let $S = R = \{\text{real numbers}\}$ and E represent the equality relation on R. That is,

$$\forall_{x, y \in R} \quad (xEy \text{ if and only if } x = y)$$

(a) E is reflexive on R since $\forall_{x \in R} (x = x)$

(b) E is symmetric on R since $\forall_{x, y \in R}$ (if $x = y$, then $y = x$).

(c) E is transitive on R since $\forall_{x, y, z \in R}$ (if $x = y$ and $y = z$, then $x = z$).

Therefore, equality is an equivalence relation on the set R of real numbers.

Example 2 Let S be the set of integers I. Define a relation E on I by

$$\forall_{x, y \in I} \; xEy \text{ if and only if } (x - y = 5k \text{ for some integer } k)$$

This means that two integers are related (via E) if and only if their difference is a multiple of five. We claim E as defined is an equivalence relation on the set I of all integers. The proof of this assertion is as follows:

(a) E is reflexive on I, because $\forall_{x \in I} \; xEx$.

That is true since $x - x = 0$ and $0 = 5(0)$. Therefore, $x - x = 5(0)$ (hence 0 is the k we need).

(b) E is symmetric on I.

We need to prove: for all integers x and y, if xEy, then yEx. So, let x, y be any integers and assume xEy (we are applying conditional proof). That is, assume $x - y = 5k$ for some integer k (this is the definition of E). Now consider $y - x$. Is this expressible as five times (some integer)? Yes, since

$$y - x = -5k$$
$$= 5(-k)$$

and $-k$ is an integer if k is an integer. Therefore, we have yEx. Hence E is symmetric on I.

(c) E is transitive on I.

We need to prove here: for all integers x, y, and z, if xEy and yEz, then xEz. Thus, we start by assuming that x, y, and z represent any integers, and xEy and yEz. This means

$$\exists_{k_1 \in I} \text{ such that } x - y = 5k_1$$

and

$$\exists_{k_2 \in I} \text{ such that } y - z = 5k_2$$

Now we *want* $x - z = 5$ times (some integer), and we *can get* the form $x - z$ by adding the left members of the last two equations. Therefore,

$$(x - y) + (y - z) = 5k_1 + 5k_2$$

and so

$$x - z = 5(k_1 + k_2)$$

and $k_1 + k_2$ is an integer since k_1, k_2 are both integers. So $k_1 + k_2$ is the integer we desire.

Therefore, we have shown xEz. Hence E is transitive on I.

The foregoing steps (a), (b), and (c) taken together verify that E is an equivalence relation on I. Note that the proof that a relation E is an *equivalence relation* on a set S really involves three subproofs—one each for reflexive, symmetric, and transitive properties. Note also that the proofs for symmetric and transitive properties are *conditional* proofs. In each case we make an assumption (assume xEy for symmetric property; assume xEy and yEz for transitive property). Using these assumptions, we try to deduce the appropriate conclusion (yEx for symmetric property; xEz for transitive property). It is important to see the distinction here. For example, in verifying that the symmetric property holds, we are *not* trying to show yEx for all x, $y \in S$; we *are* trying to show, *if xEy, then yEx* for all x, $y \in S$.

One further reflection on the particular proof given in Example 2—each part (reflexive, symmetric, transitive) is really an *existence proof*. In each case we are required to come up with an integer k such that the difference of two integers is five times this k. For the reflexive proof, the k is 0; for the symmetric proof, the k is $-k$, the additive inverse of the k from the assumption; for the transitive proof, the k is $k_1 + k_2$, where k_1 and k_2 are the respective k's from each part of the assumption. Note especially that it is incorrect to use the *same k* (instead of possibly different k_1, k_2) in the transitive part above. Doing so would imply that $x - y$ and $y - z$ are the same number, which is a special case of the choice of x, y, and z.

It is interesting to note that there is an equivalent way of defining the equivalence relation in Example 2. To say "the difference between two integers is a multiple of 5" is equivalent to saying "two integers leave the same remainder

upon division by 5." This equivalence of statements is verified in the following •
theorem:

Theorem   (On Divisibility by 5)

Let x and y be integers. Then $x - y = 5k$ for some integer k if and only if x and y leave the same remainder upon division by 5.

Proof: ($\rightarrow$) Assume $x - y = 5k$ for some integer k. We try to show that x and y leave the same remainder upon division by 5. The *Division Algorithm* for integers states that given any two integers a and b ($b \neq 0$), there exist unique integers q (quotient) and r (remainder) such that

$$a = bq + r \qquad \text{where } 0 \leq r < |b|$$

Applying the Division Algorithm to x and y, individually divided by 5, we have the following two equations:

$$x = 5q_1 + r_1 \qquad \text{where } 0 \leq r_1 < 5$$
$$y = 5q_2 + r_2 \qquad \text{where } 0 \leq r_2 < 5$$

Now consider the integer $x - y$. From the foregoing two equations we have

$$x - y = 5(q_1 - q_2) + (r_1 - r_2) \text{ where } 0 \leq |r_1 - r_2| < 5$$

Now, either $r_1 - r_2 \geq 0$ or $r_2 - r_1 \geq 0$. In either case we know that there is some integer k such that

$$x - y = 5k \qquad \text{(assumption)}$$

And since this latter equation says that the remainder $x - y$ leaves upon division by 5 is 0, and the former says that the unique remainder left by $x - y$ upon division by 5 is

$$|r_1 - r_2|$$

it must follow that $|r_1 - r_2| = 0$. Therefore, $r_1 = r_2$. That is, x and y leave the same remainder upon division by 5. So we have proved: *if* $x - y = 5k$ (for some integer k), *then* x and y leave the same remainder upon division by 5.

($\leftarrow$) Assume, next, that x and y leave the same remainder upon division by 5. We now try to show that $x - y = 5k$ for some integer k. (This is an existence proof, and we are required to produce the integer k.) Since x, y leave the same remainder upon division by 5, suppose we call it r. Then, by the Division Algorithm,

$$x = 5q_1 + r \qquad \text{where } 0 \leq r < 5$$
$$y = 5q_2 + r \qquad \text{where } 0 \leq r < 5$$

Thus, $x - y = 5(q_1 - q_2)$. Since $q_1 - q_2$ is an integer (since q_1, q_2 are integers), then this must be the integer k we want. We have now shown that *if* x and y

leave the same remainder upon division by 5, *then x − y = 5k* for some integer *k*.

Putting both parts of the proof together [(→) and (←)], we have proved the desired equivalence (↔).

The above proof can be generalized by replacing each instance of "5" with "*n*," where *n* represents any positive integer. That is, we have essentially proved the following:

Theorem   (On Divisibility by n)

Let *x*, *y*, and *n* be integers, *n* > 0; then *x* − *y* = *nk* for some integer *k* if and only if *x* and *y* leave the same remainder upon division by *n*.

The notion of two integers being related if and only if they leave the same remainder upon division by *n* is known as the *congruence relation modulo n* on the integers. It is denoted by the symbol

$x \equiv y(\text{mod } n)$, which is read, "*x* is congruent to *y*, modulo *n*."

We have seen that the following three statements are equivalent:

- $x \equiv y(\text{mod } n)$

- *x* and *y* leave the same remainder upon division by *n*

- *x* − *y* = *nk* for some integer *k*

Using the logically equivalent form of the statement in Example 2, it is easy to see why the relation *E* is indeed an equivalence relation on *I*. All we need to test is statements such as the transitive property: Assume *x* and *y* leave the same remainder upon division by 5 and that *y* and *z* leave the same remainder upon division by 5; do *x* and *z* leave the same remainder upon division by 5?

Example 3   Let *S* be the set *I* of all integers. Define a relation *R* on *I* by

$$\forall_{x,\,y \in I} \, [\, xRy \text{ if and only if } (\, x + y = 5k, \text{ for some integer } k)\,]$$

Is *R* an equivalence relation on *I*? Show why or why not. To show *R is* an equivalence relation on *I* requires showing three things—that *R* is reflexive, symmetric, and transitive on *I*. To show *R* is *not* an equivalence relation on *I* requires a *counterexample* to *any one* of these three properties. The first property we test is the *reflexive property*. Is it true that for every integer *x*, *xRx*? That is (applying the definition of *R*), is *x* + *x* = 5*k* for some integer *k*? This would require *every* integer *x* to have the property that

$$2x = 5k \text{ for some integer } k$$

As a counterexample, choose *x* = 1. There is no integer *k* such that

$$2 = 5k$$

Therefore, the given relation R is *not* reflexive on I. Hence, R is *not* an equivalence relation on I.

Example 4 The following are examples of equivalence relations defined on the respective sets given. Mentally check the (r, s, t) properties for yourself in each case to see that each holds.

(a) "sits in the same row as" on the set of all students in class

(b) "ends in the same digit as" on $\{1, 2, 3, \ldots 100\}$

(c) "is congruent to" on the set of all triangles in a plane

(d) $\{(x, y)|\ x$ and y are real numbers, and $x^2 = y^2\}$

(e) "$A \times A$ for any set A" (that is, the entire Cartesian product of a set cross itself designates a binary relation on A that is an equivalence relation on A).

(f) "xEy if and only if $[\![x]\!] = [\![y]\!]$" on the set of real numbers. (*Note:* $[\![x]\!]$ means greatest integer in x. $[\![x]\!] = n$ if and only if $n \le x < n + 1$, where n is an integer.)

Example 5 This is an example showing how we might construct an abstract equivalence relation E given its reference set A. Suppose

$$A = \{p, q, r, s\}$$

Since an equivalence relation on A is a binary relation on A, and a binary relation on A is a subset of $A \times A$, we shall form a subset of $A \times A$. However, it is necessary to keep in mind that the reflexive, symmetric, and transitive properties must hold. Since we are working with ordered pairs, we should recall also the notational convention,

$$xEy \quad \text{means the same as} \quad (x, y) \in E$$

Now, if the reflexive property is to hold, we must have

$$\forall_{x \in A}\ xEx, \text{ or equivalently } (x, x) \in E$$

Thus, any equivalence relation E on A *must* have at least four ordered pairs in it:

$$E = \{(p, p), (q, q), (r, r), (s, s)\}$$

Is this an equivalence relation on A as it stands? A little reflection would show that it is (in fact it is the *equality* relation on A) since the reflexive property holds and both the symmetric and transitive properties hold trivially. That is to say, we never meet the condition "*if xEy ...*" when x is not equal to y, and it is always true that *if xEx*, then xEx, so E is clearly symmetric. By similar reasoning, E is also transitive.

Can we define any other equivalence relations on A besides the equality relation? Suppose we toss in (with the elements of E above) the ordered pair (p, q). Then symmetry demands that the ordered pair (q, p) also appear. It turns out that the relation

$$E_2 = \{(p, p), (q, q), (r, r), (s, s), (p, q), (q, p)\}$$

is also an equivalence relation on A.

By reasoning identical to that used for constructing E_2, we can produce two more equivalence relations on A:

$$E_3 = \{(p,p),(q,q),(r,r),(s,s),(p,r),(r,p)\}$$
$$E_4 = \{(p,p),(q,q),(r,r),(s,s),(p,s),(s,p)\}$$

Suppose we toss into E *two* ordered pairs (p, q) and (p, r). Then symmetry will require (q, p) and (r, p) to also appear. Now, however, we would have (r, p) and (p, q) in the set, so *transitivity* would require (r, q) and symmetry would require (q, r). Thus, we would have

$$E_5 = \{(p,p),(q,q),(r,r),(s,s),(p,q),(p,r),(q,p),(r,p),(r,q),(q,r)\}$$

By similar reasoning to that used in building E_5, we could produce other equivalence relations. For example, toss in (p, q) and (p, s), or (p, r) and (p, s), or (q, r) and (q, s). Additionally, we could add pairs that do not share a component—e.g., (p, q) and (r, s), or (p, r) and (q, s), or (p, s) and (q, r).

Finally we could take the entire Cartesian product $A \times A$ and this would define still another equivalence relation on A. It should be clear from this discussion that given a set S, there can be more than one equivalence relation defined on S.

Following are some definitions and notational conventions about equivalence relations and their associated sets. Suppose S is a set, and E is an *equivalence relation* defined on S, then

1. Any two elements in S that are related via E are said to be *equivalent* elements (with respect to the relation E). That is, if xEy, then we say "x is equivalent to y (with respect to E)."

2. Take any element x in S. There is at least one element (possibly more) in S to which it is equivalent (via E), namely itself (since E has the reflexive property). The set of all elements in S that are equivalent to x (related to x via E) is called the *equivalence class with representative x*, and denoted $\bar{x}$. In other words,

$$\forall_{x \in S} \ \bar{x} = \{y|y \in S \text{ and } xEy\}$$

3. The collection of all equivalence classes (subsets containing equivalent elements) of a given set S with respect to an equivalence relation E is denoted S/E.

$$S/E = \{\bar{x}|x \in S\}$$

where $\bar{x}$ is the equivalence class of all elements in S that are equivalent to x via the relation E.

Note that S/E is a set whose "elements" are sets (equivalence classes).

This set of sets is sometimes called "S modulo E" or "S mod E."

Example 6   Consult the equivalence relations (a) and (b) defined in Example 4.

(a) if $C = \{\text{students in class}\}$

$$E = \text{"sits in same row as"}$$

Then suppose Jim is a member of the class and Jim sits in the same row as Sue. We indicate this mathematically by

$$(\text{Jim}) \; E \; (\text{Sue})$$

Now, if we choose Sue as the representative of the equivalence class of all students sitting in her row, then

$$\overline{\text{Sue}} = \{\text{Sue, Jim, any other students sitting in Sue's row}\}$$

Thus, Sue is a person (an element in C), but $\overline{\text{Sue}}$ is a set (set of persons sitting in Sue's row). The complete set of equivalence classes would correspond to the set whose elements would be sets of persons sitting in each row of the class. It should be carefully noted that each equivalence class is independent of how we choose to name it, and we can name an equivalence class by any representative in the class. Thus, $\overline{\text{Jim}}$ is another name for $\overline{\text{Sue}}$, and if Chris is a student sitting in the same row as Sue and Jim, we might choose to name the class $\overline{\text{Chris}}$.

(b) If $H = \{1, 2, 3, \ldots 100\}$ and E means "ends in same digit as," then all integers in H that end in "1" are equivalent (via E), and all integers in H that end in "2" are equivalent, and so on.

$$\overline{1} = \{1, 11, 21, 31, 41, 51, 61, 71, 81, 91\}$$
$$\overline{2} = \{2, 12, 22, 32, 42, 52, 62, 72, 82, 92\}$$
$$\vdots$$
$$\overline{9} = \{9, 19, 29, 39, 49, 59, 69, 79, 89, 99\}$$
$$\overline{10} = \{10, 20, 30, 40, 50, 60, 70, 80, 90, 100\}$$

Note that $\overline{1} = \overline{11} = \overline{71}$ and so on, because the numerals under the bar are just different representatives of the same equivalence class. Similarly $\overline{2} = \overline{12} = \overline{22} = \cdots = \overline{92}$, but $\overline{2} \neq \overline{9}$ since 2 and 9 are *not* equivalent (via E); that is, 2 and 9 are not in the same class—they do not end in the same digit. The collection of all equivalence classes is H/E.

$$H/E = \{\overline{1}, \overline{2}, \overline{3}, \ldots \overline{9}, \overline{10}\}$$

Example 7   Consult the equivalence relations E_3 and E_5 constructed in Example 5.

$$A = \{p, q, r, s\}$$
$$E_3 = \{(p, p), (q, q), (r, r), (s, s), (p, r), (r, p)\}$$

This equivalence relation has p related to itself and r, while q and s are related only to themselves. Thus, the equivalence classes of A produced by E_3 are

$$\bar{p} = \{p, r\}$$
$$\bar{q} = \{q\}$$
$$\bar{s} = \{s\}$$

Note that $\bar{p} = \bar{r}$. The two equivalence classes $\bar{p}$ and $\bar{r}$ are *equal*; the representative elements p and r are *equivalent*.

On the other hand,

$$E_5 = \{(p, p), (q, q), (r, r), (s, s), (p, q), (p, r), (q, p), (r, p), (r, q), (q, r)\}$$

In E_5, p is related to itself, q, and r, and s is the only element related only to itself. Thus, the equivalence classes of A produced by E_5 are

$$\bar{p} = \{p, q, r\}$$
$$\bar{s} = \{s\}$$

Note that $\bar{p} = \bar{q} = \bar{r}$.

From the foregoing examples, there are some general properties of equivalence classes that we consolidate and prove in the next theorem.

Theorem (EC)   (Properties of Equivalence Classes)

Let S be a nonempty set and E be an equivalence relation on S. Let $x, y \in S$; then

(1) $y \in \bar{x} \leftrightarrow xEy$
(an element is a member of an equivalence class if and only if a representative of the equivalence class is related [equivalent] to it)

(2) $x \in \bar{x}$
(every element is a member of the equivalence class named by itself as representative)

(3) $\bar{x} \neq \varnothing$
(every equivalence class is nonempty)

(4) $\bar{x} \cap \bar{y} \neq \varnothing \leftrightarrow xEy$
(the intersection of any two equivalence classes is nonempty if and only if the representatives are related [equivalent])

(5) $\bar{x} \cap \bar{y} \neq \varnothing \leftrightarrow \bar{x} = \bar{y}$
(the intersection of any two equivalence classes is nonempty if and only if the equivalence classes are equal)

Proof: (1) ($\rightarrow$) Assume $y \in \bar{x}$. Then, by definition of $\bar{x}$, xEy.
          ($\leftarrow$) Assume xEy. Then, by definition of $\bar{x}$, $y \in \bar{x}$.

(2) We know $x \in \bar{x}$ since E is an equivalence relation; hence E has the reflexive property, which says

$$\forall_{x \in S} \; xEx$$

But this means $x \in \bar{x}$ by definition of $\bar{x}$.

(3) $\bar{x} \neq \varnothing$ follows immediately from (2), since $\bar{x}$ has an element in it, namely x.

(4) ($\rightarrow$) Assume $\bar{x} \cap \bar{y} \neq \varnothing$. We show xEy. Since the intersection of $\bar{x}$ and $\bar{y}$ is not empty, there must be some element in that set. Call the element z. Then z is an element of $\bar{x}$ and z is an element of $\bar{y}$. But then we have xEz and yEz from property (1). Since E has the symmetric property and yEz, then zEy. Thus we have

$$xEz \quad \text{and} \quad zEy$$

from which the desired conclusion (xEy) follows by the transitive property of E.

($\leftarrow$) Assume xEy. We show $\bar{x} \cap \bar{y} \neq \varnothing$. To do this, all we need to show is that there is some element in *both* $\bar{x}$ and $\bar{y}$. We know y itself is an element in $\bar{y}$. Also, since xEy by assumption, then we know y must also be an element in $\bar{x}$. Thus y is the element we desire. Summarizing, we have

$$y \in \bar{x} \quad \text{and} \quad y \in \bar{y}$$

Therefore,

$$y \in \bar{x} \cap \bar{y}$$

and

$$\bar{x} \cap \bar{y} \neq \varnothing$$

Part (4) follows as a biconditional from the two subproofs ($\rightarrow$) and ($\leftarrow$).

(5) ($\rightarrow$) Assume $\bar{x} \cap \bar{y} \neq \varnothing$. We show $\bar{x} = \bar{y}$. This is the same as showing two sets are equal. We do this by double inclusion.

($\subseteq$) Let z be any element in $\bar{x}$. Then, by property (1), we have xEz. Now since the intersection of $\bar{x}$ and $\bar{y}$ is not empty, then by property (4) we know xEy. Thus, by symmetry, yEx. Hence, we have yEx and xEz, from which we deduce yEz by transitivity. By (1) it follows that z is an element of $\bar{y}$. Therefore, by definition of subset, $\bar{x}$ must be a subset of $\bar{y}$.

($\supseteq$) Let z be any element in $\bar{y}$. Then, by property (1), we have yEz. Again, since the intersection of $\bar{x}$ and $\bar{y}$ is not empty, property (4) gives us xEy. Thus, we have xEy and yEz, from which xEz is true by transitivity. From this we see that z is an element of $\bar{x}$ (property (1)). Therefore, by definition of subset, $\bar{y}$ must be a subset of $\bar{x}$.

By the double inclusion argument, we now have $\bar{x} = \bar{y}$. So far we have proved: *if $\bar{x} \cap \bar{y} \neq \varnothing$, then $\bar{x} = \bar{y}$.*

($\leftarrow$) Assume $\bar{x} = \bar{y}$. We show $\bar{x} \cap \bar{y} \neq \varnothing$. We must show there is some element in *both* $\bar{x}$ and $\bar{y}$. We know x is an element in $\bar{x}$. But then x must also be an element in $\bar{y}$, by our assumption that $\bar{x}$ and $\bar{y}$ are equal sets. Therefore, we have an element that is in both $\bar{x}$ and $\bar{y}$, and their intersection is not empty.

Part (5) follows as a biconditional from the two subproofs ($\rightarrow$) and ($\leftarrow$). Thus the complete theorem on elementary properties of equivalence classes has been verified. Note that the effect of proving parts (4) and (5) in this theorem is to say that the following three statements are logically equivalent:

1. xEy,

2. $\bar{x} = \bar{y}$, and

3. $\bar{x} \cap \bar{y} \neq \varnothing$

Equivalence relations are extremely important in many branches of mathematics. They are used, for example, in the formal construction of one number system from another. The integers can be constructed from the natural numbers by the following development:

1. Let $N = \{1, 2, 3, 4, \ldots\}$ and consider $N \times N$

2. Define a relation E on $N \times N$ by

$$(a, b)E(c, d) \leftrightarrow a + d = b + c$$

E is an equivalence relation on $N \times N$.

3. The resulting equivalence classes define what we mean by "integer." That is, each equivalence class is an integer.

4. Operations $+$ and $\cdot$ are defined on the integers in certain ways, and properties are studied. This produces a structure on the set of integers leading to a complete description of the *system of integers.*

As another example, the system of rational numbers can be constructed formally from the system of integers by a similar development.

1. Let $I = \{\ldots -3, -2, -1, 0, 1, 2, 3, \ldots\}$
   Let $I_0 = $ Integers $- \{0\}$
   Consider $I \times I_0$.

2. Define a relation E on $I \times I_0$ by

$$(a, b)E(c, d) \leftrightarrow ad = bc$$

E is an equivalence relation on $I \times I_0$.

3. The resulting equivalence classes define what we mean by "rational number." That is, each equivalence class is a rational number. We might

think of the ordered pairs within each class as fractions (i.e., *equivalent fractions*) and the entire class as a rational number. Thus fractions are said to be "equivalent" if they are related via E (in the same class), and any representative of the class could be used to name the rational number. That is,

$$\frac{1}{2} = \frac{2}{4} = \frac{-1}{-2} = \frac{3}{6} = \text{etc.} \ldots$$

4. Operations $+$ and $\cdot$ are defined on the rational numbers in certain ways, and properties are studied. This produces a structure on the set of rational numbers leading to a complete description of the *system of rational numbers*.

You may have noticed in all the discussion above that whenever an equivalence relation E is defined on a set S, the set of equivalence classes "chops up" S into a collection of disjoint subsets. This kind of subdivision S/E is an example of a *partition* of the set S, which we shall study next.

Assigned Problems

1. Classify each of the relations defined on the given sets below as reflexive (r), symmetric (s), and transitive (t). Pick out the equivalence relations.

   (a) *PEQ* means "$P \to Q$ is true" on the set of all sentences in logic.

   (b) *PEQ* means "$P \wedge Q$ is true" on the set of all sentences in logic.

   (c) *PEQ* means "$P \vee Q$ is true" on the set of all sentences in logic.

   (d) *PEQ* means "$P \leftrightarrow Q$ is true" on the set of all sentences in logic.

2. Do the same as Problem 1 for the following relations defined on sets.

   (a) $\leq$ on the set of all real numbers

   (b) "is the grandmother of" on the set of all persons

   (c) "lives less than a mile from" on the set of all persons

   (d) "has the same length as" on the set of all line segments in a plane

   (e) "is parallel to" on the set of all lines in a plane. (Assume $l_1 \| l_2 \leftrightarrow l_1 \cap l_2 = \varnothing$.)

   (f) "is perpendicular to" on the set of all lines in a plane

   (g) "is similar to" on the set of all triangles in a plane

   (h) "is a divisor of" on the set of all positive integers

(handwritten margin notes at top: "Turn-in" / p.173 4.1 - 5,8,9 / 4.2 - 5,6,9)

(i)   "precedes" on the set of all letters of the alphabet

(j)   "has as many digits as" on the set $\{1, 2, 3, \ldots 1000\}$

(k)   "loves" on the set of all egotists

(l) $\{(1, 1), (2, 2), (3, 3), (1, 2), (1, 3)\}$ on $\{1, 2, 3\}$

(m)   "is in the same subset as" on the elements of any set *(handwritten: no claro)*

(n) fEg means "$\int_0^1 f(x)\,dx \leq \int_0^1 g(x)\,dx$" when f, g are continuous functions on $[0, 1]$

(o)   "x is congruent to y (modulo 2)" on the set of integers

3. Prove that the two relations cited in this section for the construction of integers from natural numbers and for the construction of rational numbers from integers are really equivalence relations:

   (a) On set $N \times N$,

   $$(a, b)E(c, d) \leftrightarrow a + d = b + c$$

   (b) On set $I \times I_0$,

   $$(a, b)E(c, d) \leftrightarrow ad = bc$$

   List some members of various equivalence classes for each of these equivalence relations on their respective sets.

4. Complete Example 5 by defining each equivalence relation possible on the set $A = \{p, q, r, s\}$. How many equivalence relations on A are there?

5. Where is the flaw in the following reasoning? Let R be a symmetric and transitive relation on a set A. Let $x, y \in A$ and xRy. Then by symmetry, yRx. Since xRy and yRx, then by transitivity, xRx. Therefore, R is reflexive, and we have shown that all symmetric and transitive relations are also reflexive. Thus, reflexivity is not an independent property.

6. Consider the congruence mod 5 relation on the set of integers

   $$\forall_{x, y \in I} ([x \equiv y(\mathrm{mod}\,5)]$$

   $$\leftrightarrow [x \text{ leaves the same remainder as } y \text{ upon division by } 5])$$

   or, alternately,

   $$\forall_{x, y \in I} ([x \equiv y(\mathrm{mod}\,5)] \leftrightarrow [x - y = 5k, \text{ for some integer } k])$$

   (a) How many equivalence classes are there in $I/\equiv$ (mod 5)? *(handwritten: [0],[1],[2],[3],[4])*

   (b) List six elements in each equivalence class.

   (handwritten:
   $[0, 5, 10, 15, 20, 25]$ $[3, 8, 13, 18, 23, 28]$
   $[1, 6, 11, 16, 21, 26]$ $[4, 9, 14, 19, 24, 29]$
   $[2, 7, 12, 17, 22, 26]$ *)*

7. Discuss, intuitively, the nature of the equivalence classes for each of the equivalence relations and associated sets given in Example 4. Note that the first two ((1) and (2)) were discussed in Example 6.

8. Let L be any line in a plane, and A, B be half-lines. Define a relation R on the set of all half-lines in L by

$$ARB \leftrightarrow [A \subseteq B \text{ or } B \subseteq A]$$

(a) Prove R is an equivalence relation on the set of all half-lines in L. In geometry this is called the "orientation" relation on a line.

(b) How many equivalence classes are there? What are they?

9. Let R be the set of real numbers. Prove that the relation E described by the set of ordered pairs (in which each component is itself an ordered pair) is an equivalence relation on $R \times R$.

$$E = \{((a, b), (c, d)) | a^2 + b^2 = c^2 + d^2\}$$

10. Describe *geometrically* the equivalence classes in $R \times R$ associated with the equivalence relation E given in Problem 9.

4.3 *Partitions*

The reader may recall the development of the Riemann definite integral from calculus. In this development, we start with a continuous function f defined on a closed interval $[a, b]$. Next, we subdivide $[a, b]$ into a collection of non-empty, nonoverlapping subintervals having endpoints $x_0 = a, x_1, x_2, \ldots x_{n-1}, x_n = b$ (see Figure 4.2).

This collection of subintervals $\{[x_{i-1}, x_i] | x_i \in [a, b] \text{ and } i = 1, 2, \ldots n\}$ was called a *partition* of the original interval $[a, b]$. The partition had two major characteristics:

1. The subintervals are disjoint (except at endpoints).

2. The union of all subintervals is $[a, b]$.

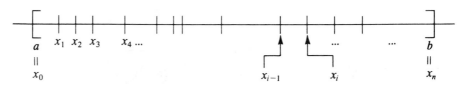

Figure 4.2

The rest of the development involved selecting a point w_i in each subinterval, forming products $f(w_i)(x_i - x_{i-1})$ that corresponded with areas of rectangles (if $f(w_i) > 0$), summing up all such products to obtain one approximation to the definite integral (a Riemann sum), and defining the definite integral as the limit of collections of Riemann sums as the norm of the partition (length of longest subinterval) approaches zero.

In this section we shall be concerned primarily with the notion of *partition of a set* and how partitions of sets and equivalence relations on sets are related.

Partition of a Set

Let A be any nonempty set. A *partition of A* is a collection $\{A_\delta | \delta \in \Delta\}$ of subsets of A with the following three properties:

1. $\forall_{\delta \in \Delta}\ A_\delta \neq \emptyset$ (every subset in the collection has at least one element),

2. $\forall_{\delta, \gamma \in \Delta}$ if $A_\delta \neq A_\gamma$, then $A_\delta \cap A_\gamma = \emptyset$ (the subsets are pairwise disjoint), and

3. $\bigcup_{\delta \in \Delta} A_\delta = A$ (the union of all the subsets is A).

Simply put, a *partition of A* is a collection of nonempty, mutually disjoint subsets of A whose union exhausts all of A. Note that the *elements* of the partition are certain *subsets* of A. A simple example of a partition of the set of all living humans might be $\{$males, females$\}$. Each set $\{$males$\}, \{$females$\}$ is nonempty (condition 1), every human is in one of these two sets (condition 3), and no human is in both sets (condition 2). There are numerous ways of restating the last two conditions for a partition. Condition 2 could be restated as

$$\forall_{\delta, \gamma \in \Delta}\ A_\delta = A_\gamma \quad \text{or} \quad A_\delta \cap A_\gamma = \emptyset$$

(Either two subsets in the partition are the same or they share no elements.)

or

$$\text{If } \exists_{x \in A} \text{ such that } x \in A_\delta \cap A_\gamma, \text{ then } A_\delta = A_\gamma.$$

(If even one element is shared by two subsets, the subsets are equal.)

Similarly, condition (3) could be restated as

$$A \subseteq \bigcup_{\delta \in \Delta} A_\delta$$

(*A* is contained in the union of all the subsets, since clearly the general union is contained in *A*.)

or

$$\forall_{x \in A} \ \exists_{\delta \in \Delta} \text{ such that } x \in A_\delta$$

(Every element in *A* is an element of some subset in the partition.)

A partition of a set *A* might be visualized as follows—a breakup of a set into a family of disjoint classes. In Figure 4.3, the classes are indexed by the first seven positive integers (of course, any indexing set might be used).

We shall generally use the symbol $\mathscr{P}$ to denote a partition of a given set. The subsets of *A* that are members of the partition will be referred to as *cells of the partition*.

Several cautionary remarks are in order here. First, one of several colloquial English usages of the word *partition* is that of a verb. That is, we *partition* something; we chop it up or subdivide it. Sometimes, we use *partition* as a noun (e.g., "my office *partition* is not soundproof"). In contrast, the mathematical usage of *partition* is always that of a noun. A mathematical partition is a thing —it is not something one does; it is rather the *consequence* of subdividing. The partition *is* the collection of cells; it *is* a set of sets. Secondly, it is required that the cells are mutually disjoint. You may have noticed in the partition leading to Riemann integrals that the cells (subintervals, in that case) shared endpoints, hence were not disjoint. Strictly speaking, that was *not* a formal partition of [*a*, *b*]. The endpoints were needed, however, to satisfy a theorem requiring continuity of the function over closed intervals, and the term *partition* was used as a matter of convenience in description.

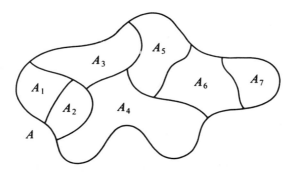

Figure 4.3

Example 1   Let W be the set of whole numbers $\{0, 1, 2, 3, \ldots\}$. Consider the following three subsets of W:

$$S_0 = \{3n \mid n \in W\} = \{0, 3, 6, 9, 12, \ldots\}$$
$$S_1 = \{3n + 1 \mid n \in W\} = \{1, 4, 7, 10, 13, \ldots\}$$
$$S_2 = \{3n + 2 \mid n \in W\} = \{2, 5, 8, 11, 14, \ldots\}$$

The collection $\{S_0, S_1, S_2\}$ is a partition of W. Note that each $S_i \neq \varnothing$. Also,

(a) The cells are mutually disjoint; that is,

$$S_0 \cap S_1 = \varnothing, \quad S_0 \cap S_2 = \varnothing, \quad S_1 \cap S_2 = \varnothing$$

(b) The union of the cells exhausts all of W:

$$S_0 \cup S_1 \cup S_2 = W$$

These cells can be characterized in a different way—as remainder classes for division by 3:

$S_0 = \{$whole numbers that leave a remainder of 0 upon division by 3$\}$

$S_1 = \{$whole numbers that leave a remainder of 1 upon division by 3$\}$

$S_2 = \{$whole numbers that leave a remainder of 2 upon division by 3$\}$

Since, by the division algorithm, there are only three possible remainders a whole number can have upon division by 3, namely 0, 1, or 2, then every whole number is in one of these classes. Since the remainder is unique (a given whole number cannot have two different remainders), then no whole number lies in more than one class.

You might have noticed that these three *cells* of the partition are precisely the same as the three *equivalence classes* associated with the congruence (modulo 3) equivalence relation on W. This is not a coincidence. There is a very close relationship between equivalence relations and partitions, in fact a one-to-one correspondence, as the next theorem points out.

Theorem   (Equivalence Relations and Partitions)

1. Existence

   (a) With every equivalence relation E defined on a set A, there is associated a partition $\mathscr{P}$ of A.

   (b) With every partition $\mathscr{P}$ of a set A, there is associated an equivalence relation E defined on the set A.

2. Uniqueness

   (a) A partition $\mathscr{P}$ associated with a given equivalence relation E defined on a set A is unique.

   (b) An equivalence relation E on A associated with a given partition $\mathscr{P}$ of A is unique.

We shall prove this theorem in a series of four stages, each with some commentary on the nature of the proof.

Proof of 1(a):

Assume A is a nonempty set and E is any equivalence relation defined on A. We shall show there exists a partition of A "induced" by E. This is an existence proof, so we need to produce the collection of sets that is the partition, and then use the definition of partition to prove it really is what we claim it is—a partition of A.

An excellent candidate for the partition is A/E, the collection of all *equivalence classes* of A associated with the equivalence relation E.

$$A/E = \{\, \bar{x} \mid x \in A \,\} \qquad \text{where } \bar{x} = \{\, y \mid y \in A \text{ and } xEy \,\}$$

Claim: A/E is a partition of A. By the definition of partition, we need to show three things: each "cell" is nonempty, they are mutually disjoint, and their union is all of A. Moreover, we know from a prior theorem on equivalence classes (see Section 4.2, Theorem EC) that every element is a member of the equivalence class named by itself. Therefore, every equivalence class is nonempty.

Next, suppose the intersection of any two equivalence classes $\bar{x}$ and $\bar{y}$ is nonempty. Then by Theorem EC (5) we have $\bar{x} = \bar{y}$. This says the classes are all mutually disjoint.

Finally, let z be any element of A. Then by Theorem EC (2), z is an element in $\bar{z}$. Thus, there is a cell that contains z. Hence, z is an element of the union of all equivalence classes. Every element of A is in some class. This says that A is a subset of the union of all classes, and since it is obvious that the union of all classes is a subset of A, then we have equality, and condition 3 is met. By the definition of partition and with the help of Theorem EC, we have proved

$$A/E \text{ is a partition of } A$$

We call it the *partition of A induced by E*.

Proof of 1(b):

Assume $\mathscr{P}$ is any partition of A; that is, $\mathscr{P}$ is a collection of nonempty, mutually disjoint subsets of A, whose union is all of A. Suppose we index the elements (cells) in $\mathscr{P}$ as follows:

$$\mathscr{P} = \{\, A_\delta \mid \delta \in \Delta, \text{ for some indexing set } \Delta \,\}$$

We will produce an equivalence relation on A associated with this partition (again, we have an existence proof to construct here). In this case, once we have defined an equivalence relation on A, we must use the definition of equivalence relation (reflexive, symmetric, and transitive) to prove it really is what we claim it to be—an equivalence relation. Our first task is to come up with a suitable definition, and all we have to go on is that we're given a partition of A. If we *had* an equivalence relation on A, its equivalence classes should be the same as the cells of $\mathcal{P}$. This motivates the following definition, which we invent to help us with the proof.

Define a relation E on A by:

$$\forall_{x,\,y \in A} \left[xEy \text{ if and only if } \left(\exists_\delta (x \in A_\delta \text{ and } y \in A_\delta) \right) \right]$$

This says, in effect, that x is related E-wise to y if and only if x and y are in the same cell of the partition $\mathcal{P}$.

Claim: E is an equivalence relation on A.

1. Reflexive? xEx is true for every x that is an element of A, since every element is in the same cell as itself. That is,

$$\exists_\delta (x \in A_\delta \text{ and } x \in A_\delta)$$

2. Symmetric? Suppose xEy; is it true that yEx? Yes, since

$$xEy \rightarrow \exists_\delta (x \in A_\delta \text{ and } y \in A_\delta)$$

Therefore, $\exists_\delta (y \in A_\delta \text{ and } x \in A_\delta)$. So yEx.

3. Transitive? Suppose xEy and yEz; must it follow that xEz?

$$xEy \rightarrow \exists_{\text{a cell } A_\delta \in \mathcal{P}} \text{ such that } x, y \in A_\delta$$

$$yEz \rightarrow \exists_{\text{a cell } A_\gamma \in \mathcal{P}} \text{ such that } y, z \in A_\gamma$$

since y is an element of both A_δ and A_γ, then A_δ and A_γ have a nonempty intersection. But $\mathcal{P}$ is a partition of A, and this forces $A_\delta = A_\gamma$. Thus, x, y, and z (particularly x, z) are all in the same cell of $\mathcal{P}$. Therefore, xEz.

Hence, we have E is an equivalence relation on A. We call it the *equivalence relation on A associated with partition $\mathcal{P}$*. This completes the existence part of the proof. We now turn to uniqueness arguments.

Proof of 2(a):

To show that the partition $\mathcal{P}$ induced by a given equivalence relation on a set A is unique, we adopt the following strategy. We want to show that whenever we have an equivalence relation E defined on a set, the "natural" partition $\mathcal{P}$ whose existence was proved in part 1(a) is the *only* partition associated with E. To do this, we start with a partition $\mathcal{P}$ of A. By part 1(b) there is associated with $\mathcal{P}$ an equivalence relation E on A. In turn, this equivalence relation induced a partition $\mathcal{P}'$ of A, by 1(a). We shall show $\mathcal{P} = \mathcal{P}'$. This amounts to showing, via double inclusion, that every cell A_δ in $\mathcal{P}$ is also a cell in $\mathcal{P}'$, and conversely.

($\subseteq$) Assume A_δ is a cell in partition $\mathcal{P}$. We show A_δ is a cell in partition $\mathcal{P}'$. Since A_δ is in $\mathcal{P}$ and $\mathcal{P}$ is a partition of A, A_δ is nonempty and there is some element x in A_δ. Thus, since E is the equivalence relation on A associated with $\mathcal{P}$,

$$A_\delta = \{ y \,|\, y \in A \text{ and } xEy \}$$

Thus A_δ is an equivalence class of E-related elements, and since $\mathcal{P}'$ is a partition of A induced by E, then A_δ must be a cell in $\mathcal{P}'$. Therefore, partition $\mathcal{P}$ is a subset of partition $\mathcal{P}'$.

($\supseteq$) Assume A_δ is a cell in partition $\mathcal{P}'$. We show A_δ is a cell in partition $\mathcal{P}$. Since $\mathcal{P}'$ is a partition of A induced by E, then A_δ is some equivalence class of E-related elements. But all such equivalence classes are cells of partition $\mathcal{P}$. Thus, A_δ is a cell in $\mathcal{P}$. Hence partition $\mathcal{P}'$ is a subset of partition $\mathcal{P}$. Therefore, the two partitions $\mathcal{P}$ and $\mathcal{P}'$ are equal (really the same partition).

Proof of 2(b):

Analogous to the proof technique of 2(a), here we shall start with an equivalence relation E defined on A. By part 1(a) there is a partition $\mathcal{P}$ induced on A. By part 1(b), since $\mathcal{P}$ is some partition of A, there is associated with $\mathcal{P}$ some equivalence relation E' on A. We shall show $E = E'$. To accomplish that, it suffices to show

$$\forall_{x, y \in A}\ xEy \leftrightarrow xE'y$$

which is the same as showing

$$\forall_{(x, y) \in A \times A}\ (x, y) \in E \leftrightarrow (x, y) \in E'$$

Therefore, as sets of ordered pairs, the binary relations E and E' would be equal (really the same set). Thus, we have a two-part biconditional proof to construct.

($\rightarrow$) Assume xEy. Then, since $\mathcal{P}$ is a partition of A induced by E, x and y belong to the same cell of this partition. But if E' is an equivalence relation on A associated with the partition $\mathcal{P}$, then x and y must be related via E' also. Therefore, $xE'y$.

($\leftarrow$) Assume $xE'y$. Then since E' is an equivalence relation associated with the partition $\mathcal{P}$ of A, there exists a cell C of which x and y are members. But the cells of $\mathcal{P}$ are equivalence classes associated with the equivalence relation E. Therefore, xEy.

This concludes the entire proof of the theorem. The theorem says essentially that for any given equivalence relation defined on a set, there is exactly one partition of the set associated with it; and for any given partition of a set, there is exactly one equivalence relation associated with it. In short, *there is a one-to-one correspondence between equivalence relations and partitions of any non-empty set*. This correspondence is illustrated in Figure 4.4.

Example 2 Let I be the set of integers $\{\ldots -3, -2, -1, 0, 1, 2, 3, \ldots\}$. Consider the congruence (modulo 5) equivalence relation on I:

$$\forall_{x, y \in I}\ ([x \equiv y \,(\text{mod}\, 5)] \leftrightarrow [x \text{ and } y \text{ leave the same remainder upon division by } 5])$$

The partition $\mathcal{P}$ of I induced by $\equiv$ (mod 5) is the set of equivalence classes each of whose elements leave the same remainder upon division by 5. Since there are only five possible remainders 0, 1, 2, 3, and 4, there are only five equivalence classes:

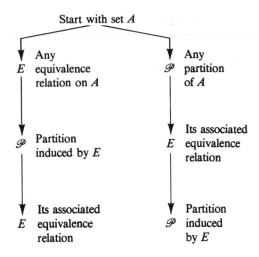

Start with set A

| Any E equivalence relation on A | Any $\mathscr{P}$ partition of A |

| $\mathscr{P}$ Partition induced by E | Its associated E equivalence relation |

| Its associated E equivalence relation | Partition $\mathscr{P}$ induced by E |

Figure 4.4

$$\bar{0} = \{\ldots -15, -10, -5, 0, 5, 10, 15, \ldots\}$$

$$\bar{1} = \{\ldots -14, -9, -4, 1, 6, 11, 16, \ldots\}$$

$$\bar{2} = \{\ldots -13, -8, -3, 2, 7, 12, 17, \ldots\}$$

$$\bar{3} = \{\ldots -12, -7, -2, 3, 8, 13, 18, \ldots\}$$

$$\bar{4} = \{\ldots -11, -6, -1, 4, 9, 14, 19, \ldots\}$$

Thus $\mathscr{P} = \{\bar{0}, \bar{1}, \bar{2}, \bar{3}, \bar{4}\}$. Note that the members of $\mathscr{P}$ are *sets*, and those sets are nonempty, mutually disjoint, and their union exhausts all of I. Therefore, $\mathscr{P}$ is a *partition* of I.

Example 3   Let I be the set of integers $\{\ldots -3, -2, -1, 0, 1, 2, 3, \ldots\}$. Let $\mathscr{P}$ be the partition of I consisting of two classes

$$E = \{\text{even integers}\}$$

$$O = \{\text{odd integers}\}$$

That is, $\mathscr{P} = \{E, O\}$. What equivalence relation is associated with the partition $\mathscr{P}$? Define R on I by

$$\forall_{x, y \in I} \, [xRy \leftrightarrow (x \text{ and } y \text{ have the same parity})]; \text{ (that is, } x \text{ and } y \text{ are both even}$$
or x and y are both odd)

This relation R is an equivalence relation on I (verification left to the reader), and its equivalence classes are E and O. Hence, R is the equivalence relation associated with partition $\mathcal{P}$. Note that "having the same parity" is the same relation as "congruent (mod 2)," or "leaves the same remainder upon division by 2," or "difference is a multiple of 2."

Example 4   Recall the equivalence relation E on the set of real numbers defined by

$$\forall_{x, y \in R} \; (xEy \leftrightarrow [\![x]\!] = [\![y]\!])$$

What is the partition of R associated with E? By the definition of the greatest integer function, given any real number x, there exists an integer n such that

$$n \leq x < n + 1$$

or, equivalently, $x \in [n, n + 1)$. This n is the "greatest integer in x"; namely, $[\![x]\!]$. Now, two real numbers x, y have the same greatest integer; that is, $[\![x]\!] = [\![y]\!]$ if and only if both real numbers lie in the same half-open interval

$$[n, n + 1)$$

Therefore, the partition $\mathcal{P}$ of the set of real numbers is the collection of subintervals

$$\mathcal{P} = \{ [n, n + 1) | n \in I \}$$

Visually, this looks like

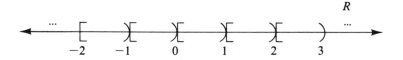

Note that this collection of nonempty subintervals has mutually disjoint members whose union is the entire set of real numbers.

Example 5   This is a situation analogous to the greatest integer example above, only extended to two-dimensional space. Consider the Cartesian plane R^2 and define a sequence of squares as follows:

$$S_{00} = \{ (x, y) | 0 \leq x < 1, 0 \leq y < 1 \}$$

$$S_{10} = \{ (x, y) | 1 \leq x < 2, 0 \leq y < 1 \}$$

$$S_{01} = \{ (x, y) | 0 \leq x < 1, 1 \leq y < 2 \}$$

$$\vdots$$

$$S_{ij} = \{ (x, y) | i \leq x < i + 1, j \leq y < j + 1 \} \qquad \text{where } i, j \text{ are integers.}$$

Two of these squares, S_{00} and S_{ij} $(i, j > 0)$, are shown in the coordinate plane below:

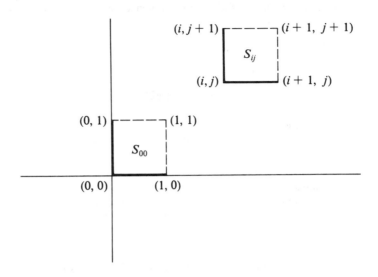

Note that in every square, two boundaries are excluded (dotted) and two boundaries are included (solid). The collection of all these squares is a partition of R^2:

$$\mathscr{P} = \{ S_{ij} | i, j \in I \}$$

Every ordered pair (x, y) of real numbers defines a point that lies in exactly one of these squares, since each component has a greatest integer, and the pair $(\llbracket x \rrbracket, \llbracket y \rrbracket)$ defines the southwest corner of the square containing (x, y). Also, no two squares overlap.

Furthermore, the equivalence relation defined on R^2 that is associated with this partition is

$$\forall_{(x, y),(z, w) \in R^2} ([(x, y) E(z, w)] \leftrightarrow (\llbracket x \rrbracket = \llbracket z \rrbracket \text{ and } \llbracket y \rrbracket = \llbracket w \rrbracket))$$

Example 6   This is a somewhat abstract example of the equivalence relation/partition connection. Let A be the set

$$A = \{ a, b, c, d \}$$

We know there are many different equivalence relations possible for a set with four elements. Moreover, there are also many different ways of subdividing the elements in A to form various partitions of A. The major theorem we proved in this section indicated that every partition has exactly one associated equivalence relation, and vice-versa. So, suppose we take any partition of A; call it $\mathscr{P}_1$:

$$\mathscr{P}_1 = \{ \{ a, b \}, \{ c, d \} \}$$

$\mathscr{P}_1$ has two cells:

$$M_1 = \{ a, b \}$$
$$M_2 = \{ c, d \}$$

How can we construct the associated equivalence relation? If the cells M_1 and M_2 are to be equivalence classes, then M_1 says a is related to itself and to b, and b is related to a and to itself. Thus, four ordered pairs in the equivalence relation would have to be

$$(a, a), (a, b), (b, a), (b, b)$$

Similar reasoning produces four more ordered pairs from M_2. Thus, the equivalence relation E_1 associated with $\mathscr{P}_1$ must be

$$E_1 = \{(a, a), (a, b), (b, a), (b, b), (c, c), (c, d), (d, c), (d, d)\}$$

Note that this is really the same as $(M_1 \times M_1) \cup (M_2 \times M_2)$. What happens if we change the partition? Do we get a different equivalence relation? Let's see.

$$\mathscr{P}_2 = \{\{a, c, d\}, \{b\}\}$$

Thus, $\mathscr{P}_2$ has two cells:

$$N_1 = \{a, c, d\}$$
$$N_2 = \{b\}$$

N_1 says that a, c, and d are all related; N_2 says that b is related only to itself. Thus the equivalence relation E_2 associated with partition $\mathscr{P}_2$ is

$$E_2 = \{(a, a), (a, c), (a, d), (c, a), (c, d), (d, a), (d, c), (c, c), (d, d), (b, b)\}$$

Again, note that $E_2 = (N_1 \times N_1) \cup (N_2 \times N_2)$.

Suppose we start with an equivalence relation and try to build the induced partition. Let

$$E_3 = \{(a, a), (b, b), (c, c), (d, d), (b, c), (c, b)\}$$

In this equivalence relation, a and d are related only to themselves, and b and c are related to each other. Thus, the induced partition $\mathscr{P}_3$ is clearly

$$\mathscr{P}_3 = \{\{a\}, \{b, c\}, \{d\}\}$$

If
$$T_1 = \{a\}$$
$$T_2 = \{b, c\}$$
$$T_3 = \{d\}$$

we observe that $E_3 = (T_1 \times T_1) \cup (T_2 \times T_2) \cup (T_3 \times T_3)$.

Finally, suppose we let E_4 be the *equality* relation on A:

$$E_4 = \{(a, a), (b, b), (c, c), (d, d)\}$$

The induced partition is clearly

$$\mathscr{P}_4 = \{\{a\}, \{b\}, \{c\}, \{d\}\}$$

since each element in A is only related to itself. Once again we see that the equivalence relation E_4 can be regarded as the union of Cartesian products of cells cross themselves.

Some of the observations in Example 6 lead to the following theorem, the proof of which is assigned as a problem.

Theorem   *(Equivalence Relations / Partitions of Finite Sets)*

Let A be a finite set, and let $\mathscr{P}$ be a partition of A into k different cells. Suppose

$$\mathscr{P} = \{C_1, C_2, \ldots C_k\}$$

Suppose also that the ith cell C_i contains t_i elements; that is,

$$n(C_i) = t_i \qquad i = 1, 2, \ldots k$$

Then the equivalence relation E associated with $\mathscr{P}$ is defined (as a set of ordered pairs) by

$$E = \bigcup_{i=1}^{k} (C_i \times C_i)$$

and the number of ordered pairs in E is given by

$$n(E) = \sum_{i=1}^{k} t_i^2$$

Assigned Problems

1. Prove the theorem given above in this section on equivalence relations and partitions of finite sets.

2. Let $S = \{1, 2, 3, 4, 5\}$.

   (a) Construct two different equivalence relations on S and produce their respective induced partitions of S.

   (b) Construct two different partitions of S and produce their respective associated equivalence relations on S.

3. Following are some equivalence relations defined on various sets. In each case, describe the equivalence classes of the induced partition.

   (a) R is the set of real numbers
   $$E = \{(x, y) \mid x, y \in R \text{ and } x^2 = y^2\}$$

   (b) R is the set of real numbers
   $$E = \{(x, y) \mid x, y \in R \text{ and } e^x = e^y\}$$

   (c) S is the set of all statements in logic

   E is the equivalence relation defined by

   $PEQ \leftrightarrow (P \leftrightarrow Q \text{ is true})$

   (d) $N = \{1, 2, 3, \ldots 100\}$

   xEy means "x has the same number of digits as y"

   (e) $N = \{1, 2, 3, \ldots 100\}$

   xEy means "x ends in the same digit as y"

$1-9, 10-99, 100$

4. If S is any nonempty set, then $S \times S$ is a relation on S. In fact, $S \times S$ is an equivalence relation on S. What is its associated partition?

5. Suppose A is a set with $n(A) = 3$. A relation $R \subseteq A \times A$ is selected at random. What is the probability that R is an equivalence relation on A?

6. Recall the equivalence relation E defined on $I \times I_0$ by

$$(a, b) E(c, d) \leftrightarrow ad = bc$$

(See Problem 3, Section 4.2.) Describe the partition of $I \times I_0$ associated with E. Take five or six different equivalence classes and list five elements in each. Suppose, instead of using ordered pair (a, b) notation, you used fraction notation, a/b. Does the partition look familiar?

7. Recall the equivalence relation E defined on $N \times N$ by

$$(a, b) E(c, d) \leftrightarrow a + d = b + c$$

(See Problem 3, Section 4.2.) Describe the partition of $N \times N$ associated with E. Take five or six different equivalence classes and list five elements in each. Suppose, instead of using ordered pair (a, b) notation, you used subtraction notation, $a - b$. Does the partition look familiar?

8. Suppose A is a set and $\mathscr{P}_1 = \{A_i | i \in$ some indexing set $I\}$ is a partition of A. Suppose also that B is a set and $\mathscr{P}_2 = \{B_j | j \in$ some indexing set $J\}$ is a partition of B. Prove that $\{A_i \times B_j | (i, j) \in I \times J\}$ is a partition of $A \times B$.

9. Let R be the set of real numbers. Consider the sets

$$L_r = \{(x, y) | y = x + r, \text{ where } r \text{ is a specific real number}\},$$

for r varying throughout the set of real numbers.

(a) Prove that $\{L_r | r \in R\}$ is a partition of $R \times R$.

(b) Describe geometrically the equivalence classes in this partition.

(c) Find the associated equivalence relation.

10. Let R be the set of real numbers. Consider the sets

$$C_r = \{(x, y) | x^2 + y^2 = r, \text{ where } r \text{ is a specific nonnegative real number}\},$$

for r varying throughout the set of all nonnegative real numbers

(a) Prove $\{C_r | r \in R^+ \cup \{0\}\}$ is a partition of $R \times R$.

(b) Describe geometrically the equivalence classes in this partition.

(c) Find the associated equivalence relation.

4.4   *Order Relations*

In this section, we shall examine the idea of binary relation on a set in the context of *order relations*. We will introduce two kinds of order relations that can be defined on a set: (a) a *partial order*, and (b) a *total* (or *linear*) *order*.

Partial Order

Let S be a set and R be a binary relation on S. R is a *partial order* on S if and only if:

1. R is reflexive: $\forall_{a \in S}\ aRa$

2. R is antisymmetric: $\forall_{a, b \in S}$ (if aRb and bRa, then $a = b$)

3. R is transitive: $\forall_{a, b, c \in S}$ (if aRb and bRc, then aRc)

There are three classic examples of partial orders on sets—the relation $\leq$ on the set of real numbers, the subset relation ($\subseteq$) on the power set $\mathscr{P}(A)$ of a given set A, and the divisibility relation ($a \mid b$) on the set of positive integers. The following three examples illustrate these partial orders.

Example 1   Let R be the set of real numbers and consider the relation $\leq$ ("is less than or equal to") on R. This is a partial order on R because

(a) $\leq$ is reflexive: $\forall_{x \in R}\ (x \leq x)$

(b) $\leq$ is antisymmetric: $\forall_{x, y \in R}$ (if $x \leq y$ and $y \leq x$, then $x = y$)

(c) $\leq$ is transitive: $\forall_{x, y, z \in R}$ (if $x \leq y$ and $y \leq z$, then $x \leq z$)

Example 2   Suppose A is a nonempty set, $\mathscr{P}(A)$ is its power set, and the relation under consideration is the subset ($\subseteq$) relation. This is a partial order on $\mathscr{P}(A)$:

(a) $\subseteq$ is reflexive: $\forall_{M \in \mathscr{P}(A)}\ M \subseteq M$
(Every set is a subset of itself.)

(b) $\subseteq$ is antisymmetric: $\forall_{M, N \in \mathscr{P}(A)}$ (if $M \subseteq N$ and $N \subseteq M$, then $M = N$)
(This is the double inclusion principle.)

(c) $\subseteq$ is transitive: $\forall_{M, N, T \in \mathscr{P}(A)}$ (if $M \subseteq N$ and $N \subseteq T$, then $M \subseteq T$)

Example 3   Let I^+ be the set of positive integers and consider the divisibility relation ($\mid$) defined by

$$\forall_{a, b \in I^+}\ (a \mid b \text{ means } \exists_{k \in I^+} \text{ such that } ak = b)$$

The divisibility relation is a partial order on I^+.

(a) | is reflexive: $\forall_{a \in I^+}$ $(a|a)$ since $a \cdot 1 = a$
(Every positive integer divides itself.)

(b) | is antisymmetric: $\forall_{a, b \in I^+}$ (if $a|b$ and $b|a$, then $a = b$). This is true, since

$$a \mid b \rightarrow ak = b \qquad \text{for some } k \in I^+$$

$$b \mid a \rightarrow bj = a \qquad \text{for some } j \in I^+ \qquad (\text{note the necessity to allow } k \text{ and } j \text{ to}$$
$$\text{be possibly different positive integers})$$

Therefore,

$$(ak)j = a$$

so

$$a(kj) = a$$

Thus,

$$kj = 1$$

Therefore,

$$k = j = 1 \text{ since } k, j \in I^+$$

Hence, $a = b$.

(c) | is transitive: $\forall_{a, b, c \in I^+}$ (if $a|b$ and $b|c$, then $a|c$). This is true, since

$$a|b \rightarrow ak = b \qquad \text{for some } k \in I^+$$
$$b|c \rightarrow bj = c \qquad \text{for some } j \in I^+$$

Now we *want* to show $a|c$, which means we want $al = c$ for some positive integer l. Going back to the two equations above, starting with the one involving c, we have

$$bj = c$$

Therefore,

$$(ak)j = c$$

Therefore,

$$a(kj) = c$$

Thus, kj is the "l" we want:

$$\therefore \quad a|c$$

When a nonempty set S has a partial order relation R defined on it, we call the ordered pair (S, R) a *partially ordered set*, abbreviated *POSET*. From the examples above, we see that $(R, \leq)$, $(\mathscr{P}(A), \subseteq)$, and $(I^+, |)$ are all partially ordered sets, or *posets*. Quite frequently in mathematics, we drop the ordered pair designation for partially ordered sets and use only the set symbol if the order relation is clearly understood. For example, when we say "the set of real numbers R is an ordered set," we generally are referring to the relation $\leq$.

If (S, R) is a poset, we say two elements a and b are *comparable* if aRb or bRa (or both, in which case $a = b$ by the antisymmetric property). Note that in a partially ordered set, it is not necessary that every pair of elements be comparable. For example, in $(I^+, |)$ 2 and 3 are not comparable. Also the subsets $\{a, b\}$, $\{a, c\}$ of $\{a, b, c\}$ are not comparable with respect to the $\subseteq$ relation. But in $(R, \leq)$, every pair of real numbers is comparable.

Linear (Total) Order

Let S be a set and R be a binary relation on S. R is a *linear* (*or total*) *order* on S if and only if:

1. *Trichotomy*: $\forall_{a, b \in S}$ (Exactly one of the following possibilities holds):

   $aRb \qquad a = b \qquad bRa$

2. *Transitive*: $\forall_{a, b, c \in S}$ (if aRb and bRc, then aRc).

The main distinction between linear order and partial order on a set S as defined here lies in the *comparableness* of elements in S. Under a linear ordering, *every* pair of elements is comparable; but this is not generally true of partially ordered sets. Note that this does *not* mean that if every pair of elements in a set S is comparable with respect to a partial order, then the partial order is a linear order. Rather, the converse is true: if a set S has a linear order defined on it, then every pair of elements in the set is comparable. A good example of this is the $\leq$ relation on the set of real numbers R. This is *not* a linear order on R since the trichotomy property does *not* hold. Yet every pair of real numbers is comparable with respect to $\leq$. The $\leq$ relation is a partial order on R, but not a total (linear) order on R.

Some mathematicians define a total order to be a partial order in which any two elements are comparable. Note that this is a *different* view than the one taken here. You may note, however, that strict inequality $(<)$ or $(>)$ *is* a linear order on the set of real numbers R.

Example 4 Following are some examples of linear orders on various sets. The reader should check to see in each case that the trichotomy and transitivity properties are holding.

- $<$ on the set of real numbers R

- $>$ on the set of real numbers R

- "is taller than" on the set of all humans

- "is responsible to" in an Army chain of command where each person is of different rank (e.g., private, corporal, sergeant, etc.)

- "is an integral power of" on the set $\{5^n | n \in I^+\}$

There is a convenient visual device for studying finite ordered sets—either partially ordered sets or linearly ordered sets. This device is the *lattice* or *nodal* diagram. In a lattice diagram the following conventions are used:

1. Every vertex corresponds to an element of the set, and vice-versa.

2. Two elements are comparable if the first lies at or above the second *and* if they are connected by line segments that ascend through a sequence of vertices only. That is, xRy if and only if one can ascend from x to y via a connected path of line segments.

Example 5   Let $S = \{a, b, c, d, e\}$.

Consider the lattice diagram below.

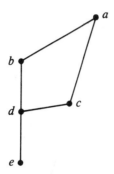

Using the conventions above, this diagram represents the relation R, given as a set of ordered pairs by

$$R = \{(b,a),(d,a),(e,a),(d,b),(e,b),(e,d),(d,c),(e,c),(c,a)\}$$

When the ordered pairs (a, a), (b, b), (c, c), (d, d), (ee) are added to the set R; R becomes a partial order on S. Note that R is not a linear order since b and c are incomparable. That is, $b\cancel{R}c$ and $c\cancel{R}b$ and $b \neq c$.

Example 6   Represent the power set $\mathscr{P}(A)$ of the set

$$A = \{a, b, c\}$$

in a lattice diagram, where the order relation is $\subseteq$.

Note that even though this might appear visually to be a space grid, it is intended to be a two-dimensional lattice diagram.

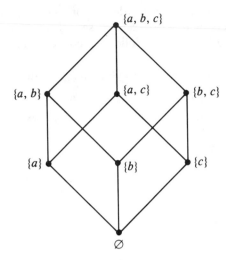

The vertices are marked by dots. Other intersections of line segments are not considered vertices. Thus, $\varnothing \subseteq \{c\} \subseteq \{a, c\} \subseteq \{a, b, c\}$. Also, the set $\{b\}$ is at the intersection of two segments coming down from $\{a, b\}$ and $\{b, c\}$; that is,

$$\{b\} \subseteq \{a, b\} \quad \text{and} \quad \{b\} \subseteq \{b, c\}$$

This lattice diagram conveys the partial order nature of $\subseteq$ since the reflexive, antisymmetric, and transitive properties hold. Moreover, it is clearly not a total (linear) order since there are various pairs of subsets that are not comparable (e.g., $\{a, b\}$ and $\{b, c\}$, $\{b\}$ and $\{c\}$, etc.). All eight subsets of $\{a, b, c\}$ appear in the diagram.

Example 7   Represent the partial order relation "divides" ($|$) with respect to the set

$$\{1, 2, 4, 8, 16\}$$

in a lattice diagram.

While this may look like a total (linear) order, it fails the test of trichotomy. For example, 2 divides 2 and 2 equals 2. However, using the same set with a different order relation ($<$, a *total* order), the same diagram represents the latter.

Following are a series of definitions for various terms that are used in mathematical discussions about partially ordered sets. In each case, it will be assumed that S is some nonempty set, M is a subset of S, and R is a partial order relation on S.

We are using the general symbol R for partial order relation because the definitions below hold when R is replaced by $|$, $\subseteq$, $\leq$, or any symbol for a known partial order. It may be convenient for the reader to think of the less than or equal to ($\leq$) relation on real numbers in each case.

1. *Upper bound of a set*: An element $u \in S$ is said to be an *upper bound* of M if and only if

$$\forall_{x \in M} \; xRu \quad (x \leq u)$$

2. *Lower bound of a set*: An element $l \in S$ is said to be a *lower bound* of M if and only if

$$\forall_{x \in M} \; lRx \quad (l \leq x)$$

3. *Bounded set*: The set M is said to be *bounded* if and only if M has both a lower bound and an upper bound.

Note that an upper bound u or a lower bound l of a set M *need not be a member of the set M*. If an upper bound u of M happens also to be a member of M, we call u a *greatest element of M*; if a lower bound l of M is also a member of M, then we call l a *least element of M*. For example, consider the set of real numbers with respect to the partial order ($\leq$). Let M be the subset containing all rational numbers whose squares are less than or equal to 2:

$$M = \{ x \mid x \in Q \text{ and } x^2 \leq 2 \}$$

M is a bounded set, since 5 is clearly an upper bound of M and -3 is a lower bound of M. [Note that, in this case, if there is one upper bound, there are infinitely many; and similarly a lower bound (if it exists) is not unique.] M does not have a greatest or least element since the square root of 2 is the smallest of all upper bounds, but it is not an element of M since it is irrational. Similarly, $-\sqrt{2}$ is the largest of all lower bounds, but it is not an element of M. This leads us to the concepts of least upper bound (lub) and greatest lower bound (glb) for a set.

4. *Least upper bound; supremum*: An element $U \in S$ is said to be a *least upper bound*, or *supremum* of M, if and only if

(a) U is an upper bound of M

(b) If u is any upper bound of M,

$$URu \qquad (U \le u)$$

(U is the "least" of all the upper bounds of M.)

5. *Greatest lower bound; infimum:* An element $L \in S$ is said to be a *greatest lower bound*, or *infimum* of M, if and only if

(a) L is a lower bound of M

(b) If l is any lower bound of M,

$$lRL \qquad (l \le L)$$

(L is the "greatest" of all lower bounds of M.)

The least upper bound of a set M (if it exists) is usually abbreviated "lub M" or "sup M"; the greatest lower bound of a set M (if it exists) is usually abbreviated "glb M" or "inf M." In the foregoing example involving the poset $(R, \le)$, where $M = \{x \mid x \in Q$ and $x^2 \le 2\}$,

$$\text{lub } M = \sqrt{2}$$
$$\text{glb } M = -\sqrt{2}$$

Note that, as with upper and lower bounds, the lub and glb of a set M (if they exist) need not be members of the set M.

When we are working with the real numbers and the partial order $\le$, the above terms involving "bounds" are somewhat natural because of our wide experience with this particular relation. However, the reader is cautioned to observe carefully the specific definitions of each idea, especially when working with somewhat unfamiliar order relations, as the following examples provide.

Example 8   Let I^+ be the positive integers and $\mid$ be the "divides" relation. Consider the subset

$$M = \{12, 30, 90, 210\}$$

- The integer 2 is a lower bound on M, since 2 divides every member of M

$$\forall_{x \in M} \; 2 \mid x$$

(3 is also a lower bound, so is 6)

- The integer 27720 is an upper bound on M since every member of M divides 27720

$$\forall_{x \in M} \; x \mid 27720$$

(2520 is also an upper bound, as is 1260; so is any multiple of 1260)

- M has *no* least element since no element *in the set* divides every element in the set; M has *no* greatest element since no element in the set is divided by every element in the set. (*Remember*: The relation in question here is $\mid$, not $\le$.)

- lub $M = 1260$
- glb $M = 6$

Note that when we are discussing the divisibility relation as an order relation, the lub and glb of a set are really the *least common multiple* and *greatest common divisor* for the set.

Example 9   Refer to the sets in Example 6.

$$A = \{a, b, c\}$$

$\mathcal{P}(A)$ is the power set of A (set of all subsets of A). The order relation is the subset ($\subseteq$) relation.

- $\varnothing$ is a lower bound (the only one) on $\mathcal{P}(A)$; it is also the least element of $\mathcal{P}(A)$, and the glb $\mathcal{P}(A)$, since $\varnothing$ is a subset of all elements (sets) in $\mathcal{P}(A)$; it is a member of $\mathcal{P}(A)$, and no member of $\mathcal{P}(A)$ is a subset of it (except itself).

- $A = \{a, b, c\}$ is an upper bound (the only one) on $\mathcal{P}(A)$; it is also the greatest element of $\mathcal{P}(A)$, and the lub $\mathcal{P}(A)$, since A contains all elements (sets) in $\mathcal{P}(A)$ as subsets; it is a member of $\mathcal{P}(A)$, and it is a subset of no member of $\mathcal{P}(A)$ except itself.

We shall conclude our discussion of order relations by stating and proving several major theorems about the order relation $\leq$ in the system of positive integers I^+. The reader should be cautioned that these proofs are given only to *demonstrate* mathematical reasoning that uses some of the tools we have developed so far. It is not our intention here to change gears to axiomatic mathematics.

There is a major theorem about the set of positive integers called the *Well-Ordering Principle*. It is stated as follows:

Well-Ordering Principle (WOP)
Every nonempty subset of the set of positive integers has a least element.

Taking the *Well-Ordering Principle* (*WOP*) as an axiom, we can prove the *Principle of Mathematical Induction* (*PMI*) as a theorem. We do this in two steps. First, we will state and prove a lemma. A lemma is a minor or "assist" theorem, used to help prove a more significant result.

Lemma   1 is the least positive integer.

Proof: Define a set M by

$$M = \{ m \mid m \in I^+ \text{ and } 0 < m < 1 \}$$

We argue by the indirect method. If $M = \varnothing$, then 1 is the smallest positive integer, and we're done. So suppose $M \neq \varnothing$. Then by the WOP, there is a least element in M; call it n. Since n is an element of M, then

$$0 < n < 1$$

Also, $0 < n^2 < n$ (multiplying through by n). But now we have $0 < n^2 < n < 1$, which puts n^2 in M, and n^2 is smaller than n. This contradicts that n is least element of M. Therefore we must conclude $M = \varnothing$; hence, 1 is the least positive integer.

We shall now use the Well-Ordering Principle and the lemma to deduce the Principle of Mathematical Induction as a theorem.

Theorem   (PMI: The Principle of Mathematical Induction)
If $M \subseteq I^+$ such that
      (a) $1 \in M$, and
      (b) $\forall_{n \in I^+}$ (if $n \in M$, then $n + 1 \in M$),
then $M = I^+$.

Proof: Our strategy here is by the indirect method of proof. We shall define a set

$$S = I^+ - M$$

If S is empty, then I^+ equals M and we're done. So we begin by supposing S is nonempty, and hope to establish a contradiction. Since S is nonempty, the WOP guarantees us that S must have a least element. Call the least element m. Now, m cannot be 1, since by hypothesis (a), 1 is an element of M. Since m is an element of S, m is not an element of M. Furthermore, since by the lemma, 1 is the least positive integer, then m must be strictly greater than 1. Now, consider the element $m - 1$. Since

$$m > 1 \quad \text{then} \quad m - 1 > 0$$

But m is the least element of S, so $m - 1$ must be an element of M (if $m - 1$ were an element of S, then it would be smaller than the least element of S). However, hypothesis (b) says that if $m - 1$ is an element of M, then so is $(m - 1) + 1$. This puts m as an element in M, contradicting that m is an element of S. Therefore, S must be empty, and M is the entire set of positive integers.

We have shown here that

> Well-Ordering Principle → Principle of Mathematical
> Induction

It can also be shown that if the PMI is taken as an axiom, then the Well-Ordering Principle can be deduced from it as a theorem. In short, the Well-Ordering Principle and the Principle of Mathematical Induction are mathematically equivalent (WOP ↔ PMI). The interested reader is advised to consult any good book on abstract algebra or number theory for the proof PMI → WOP.

Assigned   1. For each of the following relations defined on the respective sets, identify
Problems   which are partial orders, which are linear (total) orders, and which are neither.

     (a) $H = \{$humans, living or deceased$\}$

         xRy means "$x = y$ or x is an ancestor of y" R, A, T (not Total)

     (b) $L = \{$line segments in a plane$\}$

         lRm means "l is longer than m" neither

     (c) $A = \{$letters of the English alphabet$\}$

         aRb means "a precedes b"

     (d) $P = \{$products of prime numbers (including primes themselves)$\}$

         pRq means $p|q$.

     (e) $S = \{$rays in a line oriented in the same direction$\}$

         rRs means $r \subseteq s$

2. Let $S = \{a, b, c, d\}$. Consider the power set of S, $\mathscr{P}(S)$, and the subset relation on $\mathscr{P}(S)$. Draw a lattice diagram representing this relation.

3. Draw a lattice diagram for the relation "is responsible to" on the set of ranks for Army officers (use at least six ranks).

4. Let S be a nonempty set, and R be a partial order on S. Let $M \subseteq S$ and suppose lub M and glb M exist. Prove lub M and glb M are unique.

5. For the set $S = \{56, 84, 140, 308, 4004\}$ and divisibility relation ($|$), determine

     (a) a lower bound in I^+

     (b) an upper bound in I^+

     (c) Does S have a least element? A greatest element? Why or why not?

     (d) What is glb S? lub S?

⨯ 6. Let S denote the set of all nonnegative integers. Define a relation R on S by

$$\forall_{x,\,y \in S} \ (xRy \leftrightarrow \exists_{z \in S} \text{ such that } xz = y)$$

(a) Prove that R is a partial order relation on S.

(b) Prove that R is not a total order on S.

(c) For any x, $y \in S$, prove that lub$\{x, y\}$ and glb$\{y, x\}$ both exist in S with respect to the given order relation.

(d) Prove that $1Rx \ \forall_{x \in S}$.

(e) Prove that $\forall_{x \in S} \ xR0$.

7. Consider the set R^2 ($R \times R$), all ordered pairs of real numbers, and define a relation $\leq$ on R^2 by

$$\forall_{(a,\,b),\,(c,\,d) \in R^2}$$

$$(a, b) \leq (c, d) \leftrightarrow a \leq c \text{ and } b \leq d$$

Is $\leq$ defined this way a partial order on R^2? Show why or why not.

8. Suppose S is any set and M, N are subsets of S. Using the partial order $\subseteq$ on S, what are

(a) lub$\{M, N\}$?

(b) glb$\{M, N\}$?

9. How many different total order relations are there on a finite set S having n elements, where n is a positive integer? Explain.

10. Let $A = \left\{ \dfrac{x+1}{x} \,\middle|\, x \in R^+ \right\}$

Discuss the bounds on A. Does A have least and/or greatest elements? What about glb A? lub A? What process in calculus helps with your reasoning here?

4.5   Graphs

Mathematical modeling usually involves trying to solve (or approximate a solution to) a "real world" problem by abstracting some or all of its characteristics, representing them in some mathematical way (e.g., equations, inequalities, graphs, matrices), solving the problem in its mathematical context, and then translating that information back to the real situation. We also do considerable mathematical modeling within mathematics itself whenever we represent information in a different-looking mathematical medium that accommodates translation back and forth from the given situation to its model. For example, we are doing modeling within mathematics whenever we represent a function or relation by its Cartesian graph, a system of equations by a matrix equation, or the information in a problem by a diagram.

There is a compact mathematical modeling device for relations, also. It is called a *graph*, and there has evolved recently an entire branch of mathematics and computer science called "graph theory." We shall touch on a handful of the elements of graph theory so that we can use graphs to model relations, readily translate information about the properties of relations, help solve certain kinds of problems by visualization, and learn a little about this fascinating branch of mathematics. Generally, graphs will help us to aid, clarify, and visualize our thinking about certain relations and problems, as well as help us communicate our thinking to others.

Imagine a nonempty set V of points (called vertices) in a plane, and another set E (whose elements are called edges or arcs), where each edge connects at least one pair of vertices or a vertex with itself. The set V may be finite or infinite. For our work, we will restrict our attention to finite graphs. Also, the set E may be empty, in which case we have a graph with one or more *isolated* vertices. An *undirected graph* (or simply, *graph*) is a system G (sometimes denoted $G(V, E)$) consisting of these two sets.

Example 1 If $V = \{a, b, c, d, e\}$ and $E = \{e_1, e_2, e_3, e_4, e_5, e_6\}$, then the system $G(V, E)$ (Figure 4.5) is an undirected graph with 5 vertices and 6 edges.

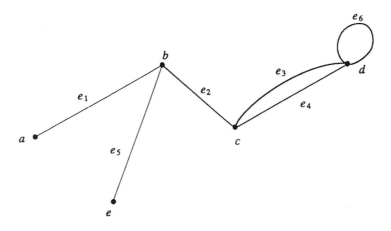

Figure 4.5

In a more formal sense, a graph consists of two sets: a nonempty set V of elements called vertices and a set E of elements called edges, where each edge is associated with a unique pair of vertices (not necessarily distinct). This definition of graph allows *isolated vertices*, *loops* (edges connecting a vertex with itself), and *parallel edges* (two or more edges associated with the same pair of vertices. Note that in the undirected graph of Example 1, any edge is associated with the unordered pair of vertices that it connects; here e_1 is associated with $\{a, b\}$.

If we associate edges with *ordered pairs* of vertices (so that (a, b) would symbolize a different edge than (b, a)), then a graph takes on the additional characteristic of direction. A *directed graph*, or *digraph G* is a graph in which each edge is associated with an *ordered pair* (v, w) of vertices. In a digraph, all edges are represented as arrows connecting the first component (vertex) of an ordered pair of vertices to the second component (vertex). We will use the notation $e = (v, w)$ to indicate that edge e connects vertex v with vertex w in a directed graph, and we draw an arrow with initial point at v and terminal point at w. In this case we say edge e is *incident* from vertex v to vertex w, and we call v and w *adjacent* vertices. The digraph in Figure 4.6 has vertices a, b, c, d, and e and edges (a, b), (b, b), (b, d), (b, c), (c, d), and (d, c).

One (of many) physical interpretations of a digraph is to think of it as a traffic network consisting of a system of one-way and two-way streets and traffic circles.

Some additional terminology about graphs which we will need later involves the ideas of *paths* and *circuits*.

A *path* of length n from vertex v to vertex w is a sequence of vertices $v_0, v_1, v_2, \ldots v_n$ $(n \geq 2)$ such that (v_0, v_1), $(v_1, v_2), \ldots (v_{n-1}, v_n)$ are distinct edges, and such that $v = v_0$ and $w = v_n$. We denote such a path by the ordered $(n + 1)$-tuple $(v_0, v_1, v_2, \ldots v_n)$.

A *circuit* is a path which begins and ends at the same vertex.

A *simple path* from v to w is a path $(v_0, v_1, \ldots v_n)$ where $v = v_0$, $w = v_n$, and all vertices v_i are distinct, except possibly v_0 and v_n $(i = 0, 1, \ldots n)$.

A *simple circuit* is a simple path that is a circuit.

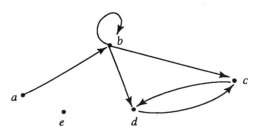

Figure 4.6

In the directed graph in Figure 4.6, there is a path of length 1—(b, d)—and a path of length 2—(b, c, d)—from vertex b to vertex d.

Since a digraph has edges associated with *ordered pairs* of vertices, it is natural to use digraphs to model binary relations. If R is a binary relation on set S, then the elements of R are ordered pairs from $S \times S$. Thus, we can think of S as supplying vertices, and each ordered pair in R as an edge, and we have ourselves a digraph which models the relation R. For example, if $S = \{a, b, c, d\}$, and $R = \{(a, b), (a, c), (a, d), (b, c), (d, c), (c, c)\}$, we obtain the directed graph in Figure 4.7.

Using a digraph to represent a relation affords us an easy visual interpretation of the reflexive, symmetric, transitive, and antisymmetric properties of nonempty binary relations. A nonempty binary relation R on S is:

(i) *reflexive* if and only if every element in S is a vertex in the digraph and every vertex has a loop.

(ii) *symmetric* if and only if every pair of distinct adjacent vertices has two parallel edges in opposite directions.

(iii) *transitive* if and only if for every path of length two from one vertex to another, there is also a path of length one from the first vertex to the second.

(iv) *antisymmetric* if and only if every pair of adjacent vertices has only a one-way connection (no parallel edges in opposite directions).

For example, if $S = \{1, 2, 3, 4\}$ and R is a binary relation on S having the digraph in Figure 4.8, then R is symmetric, but is neither transitive nor reflexive. As a set of ordered pairs,

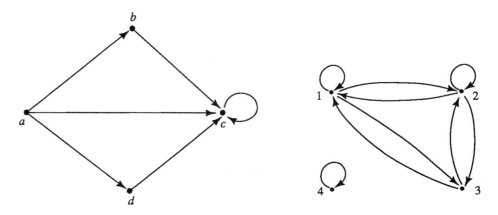

Figure 4.7   *Figure 4.8*

$$R = \{(1,1), (1,2), (2,1), (2,2), (2,3), (3,2), (1,3), (3,1), (4,4)\}$$

Additionally, the inverse of a relation R is easily pictured in a digraph. To see R^{-1}, all we need do is reverse the direction of all arrows.

Example 2   Let $S = \{a, b, c, d\}$ and

$$R = \{(a,b), (a,c), (b,b), (b,c), (c,a), (c,d), (d,a)\}$$

The following digraphs (Figure 4.9) illustrate the relationship between R, R^{-1}, $R \cup R^{-1}$, and $R \cap R^{-1}$. Note that

$$R^{-1} = \{(b,a), (c,a), (b,b), (c,b), (a,c), (d,c), (a,d)\}$$

$$R \cup R^{-1} = \{(a,b), (b,a), (a,c), (c,a), (b,b), (b,c), (c,b),$$

$$(c,d), (d,c), (a,d), (d,a)\}$$

$$R \cap R^{-1} = \{(a,c), (b,b), (c,a)\}$$

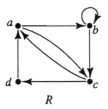

R

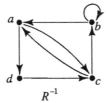

R^{-1}

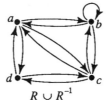

$R \cup R^{-1}$

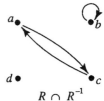

$R \cap R^{-1}$

Figure 4.9

We leave it as an exercise to show, by a set/element proof, that for any binary relation R on a set S, $R \cup R^{-1}$ and $R \cap R^{-1}$ are always symmetric relations.

The association between equivalence relations and partitions is also strengthened visually by modeling with directed graphs. We know that a digraph of an equivalence relation must exhibit a loop at every vertex, parallel edges in opposite directions must join any two distinct vertices that correspond to an ordered pair in the relation, and any two vertices joined by a path of length 2 must also be joined by a path of length 1 from the first to the last. Given that these conditions prevail, the equivalence classes in the associated partition are simply the disjoint (disconnected) subgraphs (subsets of the graph) that comprise the graph.

Example 3   Let $S = \{a, b, c, d\}$ and R be the equality relation. The associated partition of S consists of four disjoint cells:

$$\{a\}, \{b\}, \{c\}, \text{and } \{d\}$$

Likewise, the digraph in Figure 4.10 reflects this.

Figure 4.10

Example 4 If $S = \{a, b, c, d, e, f\}$, the following digraph in Figure 4.11 represents an equivalence relation on S having cells

$$\{a, b\}, \{c\}, \{d, e, f\}$$

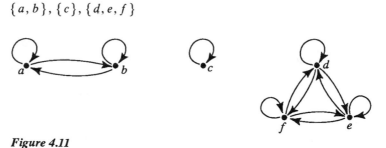

Figure 4.11

Graph theory has remarkably simple, yet widespread applications in modern discrete mathematics and in modeling. Computer scientists are using graphs whenever they construct flowcharts, chemists illustrate molecular structure with graphs in which the edges are chemical bonds, civil engineers study graphs as models of traffic-congestion networks, geneologists use special graphs to represent family lineage and relationships, geneticists model offspring production and crossbreeding with graphs, linguists parse sentences with graphs, statisticians create graphical representations of decisions or events and assign probabilities to edges, and electrical networks are real physical graphs. In fact, the most extensive current application of plane graphs is by electrical engineers, who use graphs in the design of circuits.

Historically, graph theory was invented in 1736 by L. Euler as a method of solving the now-famous "Seven Bridges of Königsberg" problem, which was posed as follows:

Two islands in the Pregel River at Königsberg (now Kalingrad in East Prussia) were connected to each other and to the regions along both sides of the river by 7 bridges, as shown in Figure 4.12.

The people of Königsberg used to take evening and Sunday strolls across these bridges. The question arose: Is it possible to start in some region (a, b, c, or d), walk over each bridge exactly once, and return to the point of origin?

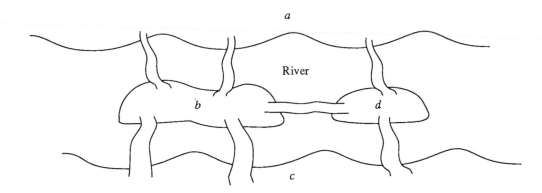

Figure 4.12

Euler's mathematical genius was evidenced when he created a simple abstract mathematical model of the situation. He reduced the four regions to points (vertices), and the seven bridges to arcs (edges) joining the regions, and thereby created an undirected graph which accurately represented the situation. (See Figure 4.13.)

To solve the Königsberg Bridge problem, we turn our attention to undirected graphs and use a proof employing modern terminology, but based on Euler's reasoning. We call the *order* of a graph the number of *vertices* in the graph; the *size* of a graph is the number of *edges*; and the *degree of a vertex* (denoted by $d(v)$) is the number of edges incident with v. (A loop would contribute 2 to the degree of its vertex.) For example, the graph in Figure 4.13 has order 4, size 7; and the degrees of vertices a, b, c, and d are 3, 5, 3, and 3, respectively, Also, a path "through a graph" which starts at a vertex, traverses each edge exactly once, visits each vertex at least once, and ends at the same vertex as it started is called an *Euler circuit*. Thus, the Königsberg problem reduces to finding whether or not there exists an Euler circuit through the graph. One solution proceeds as follows:

> We wish to start at any vertex, move along each edge exactly once, visit each vertex at least once, and return to the original vertex. Now, examine any vertex v other than the initial one. Each time we take an edge incoming to v, we must exit from v by a *different* edge. Thus, the edges incident with v must occur in pairs. That is, the degree of v must be even. This is also true of the starting vertex, since we must first exit it along one edge, and eventually return to it along another edge. Therefore, in order to have an Euler circuit in a graph, *every* vertex must have *even* degree. Since this is not true of the Königsberg model, the answer to the question is "No."

We have seen that *if* an Euler circuit exists in a graph, *then* every vertex in the graph must have even degree. Contrapositively, if any vertex has odd degree, then there is no Euler circuit in a graph. However, is the *converse* true? That is, if

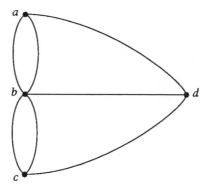

Figure 4.13

every vertex has even degree, then can we conclude that the graph has an Euler circuit? Unfortunately, the answer is "No." (Think about graphs with isolated vertices or disconnected graphs.) We need another property of graphs to hold—namely, "connectedness." We say a graph is *connected* if and only if there is a path joining every pair of distinct vertices.

Theorem   1

A graph has an Euler circuit if and only if it is connected and all vertices have even degree.

Clearly, if a graph has an Euler circuit, it must be connected. The fact that all vertices must have even degree was argued earlier in the Königsberg Bridge problem. To show that a connected graph with all vertices of even degree has an Euler circuit is really an existence proof. That is, we will start with a graph meeting the right conditions, and demonstrate precisely how to find an Euler circuit through the graph. We demonstrate here the reasoning for a specific graph, and leave the generalization of our proof as an exercise.

Example 5   Find an Euler circuit in the graph of Figure 4.14.

To begin, choose any vertex, arbitrarily say a, and exit by an edge, arriving at another vertex, say c. Now we must select an edge with which to exit c (one always will exist, since all vertices have even degree). Suppose we decide to travel along the circuit

$$(a, c, d, g, e, a)$$

Now, if we had traversed all edges exactly once, we'd have our Euler circuit and be done. But since we haven't, we will remove all edges traversed and isolated vertices. We are now left with the subgraph in Figure 4.15. Since we have removed an even number of edges from each vertex (e.g., (a, c) and (c, d) were removed from vertex c) and each vertex had even degree, there still remains an even number of edges incident with each remaining vertex. We repeat the procedure, this time in the subgraph. The original graph was

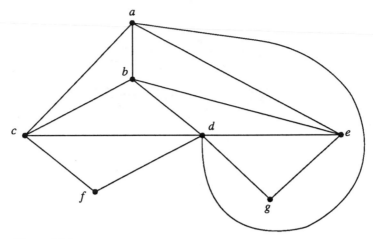

Figure 4.14

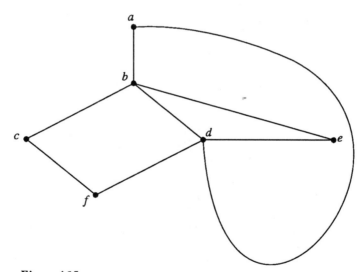

Figure 4.15

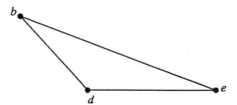

Figure 4.16

connected; thus the first circuit must have a vertex shared with the remaining subgraph. Select such a vertex, say d, and traverse another circuit, say

$$(d, a, b, c, f, d)$$

Repeat the procedure. Only the graph in Figure 4.16 remains. Start at a shared vertex (b) and traverse the circuit (b, d, e, b). Finally, to obtain our Euler circuit, we start with the first circuit until we encounter the first shared vertex (it was d), then we take a detour along the second circuit until we arrive at the next shared vertex (b), taking its circuit, returning to the second circuit, and finally completing the remainder of the first circuit. We have

$$\left(a, c, \underbrace{d, a, \underbrace{b, d, e, b,}_{\text{circuit 3}} c, f, d, g, e, a}_{\text{circuit 2}} \right)$$

To get a feeling for this procedure, it is helpful to construct several different connected graphs in which all vertices are of even degree, and apply the reasoning in several different ways to each. The selection of certain kinds of paths and circuits through graphs is of considerable interest and application today, and there are algorithmic procedures and computer programs to assist with solutions of very complex graphs with many vertices and edge configurations.

Another kind of circuit in a graph is called a *Hamiltonian circuit*. A Hamiltonian circuit begins and ends at the same vertex, and it passes through each *vertex* only once. For example, a Hamiltonian circuit in the graph of Figure 4.14 would be

$$(a, b, c, f, d, g, e, a)$$

The key difference between Euler and Hamiltonian circuits is that an Euler circuit traverses each *edge* once, whereas a Hamiltonian circuit visits each *vertex* once.

Unlike the case for Euler circuits, little is known about necessary and sufficient conditions under which a Hamiltonian circuit may or may not exist.

As an example of the nature of reasoning that often characterizes proofs in graph theory, we examine a result that is related to (and depends on) Theorem 1. In this result, we take "Euler path" to mean a path containing all edges and visiting all vertices.

Theorem   2

A graph has an Euler path from one vertex v to another vertex w ($v \neq w$) if and only if the graph is connected and v and w are the only vertices in the graph having odd degree.

Proof: ($\rightarrow$)   Suppose G is a graph that has a simple path from vertex v to vertex w ($v \neq w$), and this path contains all edges and vertices. Then clearly G must be connected (all vertices are on this path). We need to show v and w are the only vertices of odd degree. Suppose we add an edge joining v and w. Then our graph has an Euler circuit (take the original path from v to w and the added edge from w back to v). By Theorem 1, every vertex in this graph (with the added edge) must have even degree. Now delete the added edge. This subtracts one edge incident with both v and w (making each of them have odd degree), and all remaining vertices are unchanged (having even degree).

($\leftarrow$) Suppose G is a connected graph with exactly two vertices of odd degree—call them v and w. We wish to show that G has a simple path from v to w containing all vertices and edges. Once again, add a new edge connecting v and w. Call this edge e. This new edge contributes 1 to the degree of each of its vertices (v and w); therefore all vertices now have even degree, and by Theorem 1, the graph has an Euler circuit. Choose the circuit so that e is the beginning edge of the Euler circuit. If we delete the (v, w) edge from this circuit, we obtain our Euler path from v to w.

As a final consideration of the elements of graph theory, there is a particularly interesting relationship of invariance between the number of vertices, edges, and plane regions of separation for certain graphs, as well as an unusual application of proof by mathematical induction to verify it. We now introduce the concept of a *planar graph*.

A *planar graph* is a graph that can be drawn in a plane without its edges crossing. That is, if any two edges intersect, they must intersect in a vertex. For clarification, consider the various graphs in Figure 4.17.

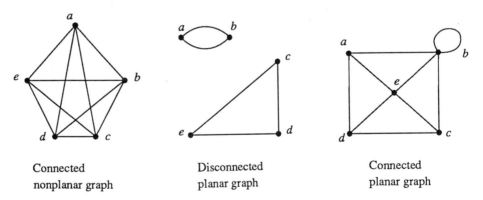

Connected nonplanar graph   Disconnected planar graph   Connected planar graph

Figure 4.17

Suppose we focus our attention now on *connected planar graphs*. Such graphs divide the plane into a number of nonoverlapping regions, where the boundary of each region is some circuit, loop, or pair of parallel edges in the graph. For example, in the connected planar graph of Figure 4.17, the circuit (b, e, c, b) bounds a triangular region, the loop at vertex b bounds one region, and the circuit (a, b, b, c, d, a) forms a boundary of the exterior or outer region. In this graph, there are 5 vertices, 9 edges, and 6 regions of separation (nonoverlapping regions). By exploring various connected planar graphs, we might discover the relationship

$$v - e + r = 2$$

where v, e, and r, respectively, are the number of vertices, edges, and regions (including the unbounded region) in the graph. This relationship is known in graph theory as *Euler's formula for connected planar graphs* (Euler, 1752). Consider the following connected planar graphs with the count of v, e, and r for each (Figure 4.18).

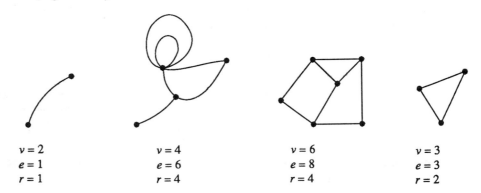

$v = 2$	$v = 4$	$v = 6$	$v = 3$
$e = 1$	$e = 6$	$e = 8$	$e = 3$
$r = 1$	$r = 4$	$r = 4$	$r = 2$

Figure 4.18

In an extreme case, a connected planar graph can consist minimally of a single vertex and no edges. Still, the Euler formula, $v - e + r = 2$, holds. Also, we can extend a connected planar graph in only four different ways or operations —we will call them point, loop, bridge, or tail—as follows:

(1) add a vertex to an existing edge (*point*)

(2) add an edge from an existing vertex back to itself (*loop*)

(3) add an edge connecting two existing vertices (*bridge*)

(4) add an edge and a vertex to an existing vertex (*tail*)

If we are willing to accept the hypothesis that connected planar graphs can be extended by no other operations than these four, then we have the basis for an interesting proof of Euler's formula by stepwise mathematical induction.

Theorem* *[Euler's Formula for Connected Graphs]

For any connected planar graph with v vertices, e edges, and r regions,

$$v - e + r = 2$$

Proof: (by mathematical induction on n = the number of steps (operations) from a single vertex to the existing graph)

(i) If $n = 0$, we have a single vertex, no edges, and 1 region, so $v - e + r = 2$ holds.

(ii) Next, assume the formula $v - e + r = 2$ holds for any connected planar graph created by $n = k$ steps of the form point, loop, bridge, or tail from a single vertex graph.

(iii) Now, consider the effect on $v - e + r$ when we take one more step. We summarize the possibilities in a table where Δv, Δe, Δr, and $\Delta(v - e + r)$ indicate the *change* in each of these respective quantities.

operation	Δv	Δe	Δr	$\Delta(v - e + r)$
point	+1	+1	0	0
loop	0	+1	+1	0
bridge	0	+1	+1	0
tail	+1	+1	0	0

In every possible case, one additional step creates no change in $v - e + r$. We conclude that the Euler formula holds for all connected planar graphs.

Assigned Problems

1. Suppose $S = \{1, 2, 3, 4\}$. Draw digraphs that represent each of the following binary relations on S.

   (a) $D = \{(a, b) | (a, b) \in S \times S \text{ and } a \text{ divides } b\}$

   (b) $U = \{(a, b) | (a, b) \in S \times S \text{ and } a \neq b\}$

   (c) $E = \{(a, b) | (a, b) \in S \times S \text{ and } a + b \text{ is even}\}$

   (d) $O = \{(a, b) | (a, b) \in S \times S \text{ and } a + b \text{ is odd}\}$

   (e) $L = \{(a, b) | (a, b) \in S \times S \text{ and } a < b\}$

2. "Read" each of the digraphs in your response to problem 1, and tell which properties—reflexive, symmetric, transitive, or antisymmetric—each of the relations has.

3. Let R and S be nonempty binary relations on a set A. Prove, using suitable definitions and set/element arguments where necessary, each of the following:

   (a) $R \cup R^{-1}$ is a symmetric relation.

   (b) $R \cap R^{-1}$ is a symmetric relation.

   (c) $(R \cup S)^{-1} = R^{-1} \cup S^{-1}$

   (d) $(R \cap S)^{-1} = R^{-1} \cap S^{-1}$

   (e) $(R \subseteq S) \leftrightarrow (R^{-1} \subseteq S^{-1})$

4. Let $S = \{1, 2, 3\}$. For each set of conditions, construct a digraph that represents a nonempty relation R on S that satisfies the given conditions. Also, express the resulting relation in each case as a set of ordered pairs.

(a) reflexive and symmetric, but not transitive

(b) reflexive and transitive, but not symmetric

(c) symmetric and transitive, but not reflexive

(d) reflexive, but neither symmetric nor transitive

(e) none of the conditions reflexive, symmetric, or transitive

5. Generalize the proof of the second part of Theorem 1. That is, prove that every connected graph having all vertices of even degree must have an Euler circuit. (Use the reasoning of Example 5 to guide your thinking, but don't appeal to a specific graph.)

6. Consider the graph G shown in Figure 4.19.

   (a) Indicate the order and size of the graph, and identify the degree of each vertex.

   (b) Is it possible to trace the graph, beginning and ending at the same point, without lifting your pencil from the paper and without retracing lines? If yes, show how (give a circuit); if not, explain why not.

   (c) Is it possible to trace the graph without lifting your pencil from the paper or retracing any lines if you are not required to begin and end at the same point? If yes, show how (give a path); if not, explain why not.

   (d) Does the graph have a Hamiltonian circuit? Explain why or why not.

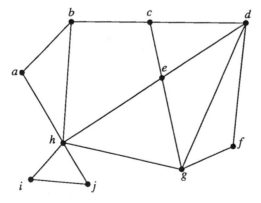

Figure 4.19

7. (a) Prove that the sum of the degrees of vertices in any graph G must be an even number. (*Hint:* Consider the relationship between edges and vertices.)

   (b) Prove that in any graph, there must be an even number of vertices of odd degree.

8. Apply the method outlined in Example 5 to find an Euler circuit in the graph given in Figure 4.20. Show your intermediate circuits and subgraphs.

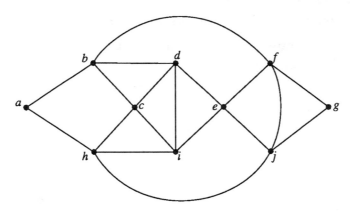

Figure 4.20

9. A graph with no loops or parallel edges in which *every* pair of distinct vertices are connected by an edge is called a *complete graph*. Complete graphs having 3, 4, and 5 vertices are shown in Figure 4.21.
Prove:

(a) The sum of the degrees of the vertices for a complete graph of n vertices is $n(n - 1)$. (Assume $n \geq 1$.)

(b) The number of edges in a complete graph of n vertices is $\frac{1}{2}n(n - 1)$. (Assume $n \geq 1$.)

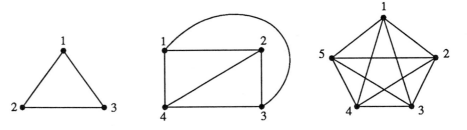

Figure 4.21

10. A *convex polyhedron* is a polyhedron with the property that given any face, the entire figure lies in one of the half-spaces determined by the plane containing that face. Several examples of convex polyhedra (with counts of vertices, edges, and faces) are given in Figure 4.22.

It would appear that a variation of Euler's formula for connected planar graphs also holds for the relationship between vertices, edges, and faces of

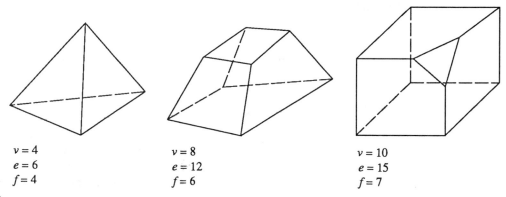

$$v = 4 \qquad\qquad v = 8 \qquad\qquad v = 10$$
$$e = 6 \qquad\qquad e = 12 \qquad\qquad e = 15$$
$$f = 4 \qquad\qquad f = 6 \qquad\qquad f = 7$$

Figure 4.22

convex polyhedra. That is,

$$v - e + f = 2$$

Is this true? If yes, give some justification (at least informally); if no, give a counterexample.

4.6   *Inequality Proofs*

We have seen that the relation $<$ on the set of real numbers R is a linear order. In this section we shall concentrate only on this order relation and others like it ($>$, $\leq$, $\geq$) with respect to the set of real numbers. Any statement involving any of these four order relations is called an *inequality*. Our discussion shall begin with the axioms of inequality, several definitions, and extend to theorems, problems, and proofs involving inequalities on the real numbers. As noted before, this kind of discussion is given for the purpose of practicing our proof techniques, here in the context of an axiomatic system. It is not intended to make rigorous our development of mathematics itself.

Axioms of Inequality

There is a subset R^+ of the set of real numbers R that is characterized by the axioms and definitions below.

Axiom 1 $\quad \forall_{x \in R} \qquad$ exactly one of the following holds:
$$x \in R^+ \quad \text{or} \quad x = 0 \quad \text{or} \quad -x \in R^+$$

Axiom 2 $\quad \forall_{x,\, y \in R} \qquad$ (if $x \in R^+$ and $y \in R^+$, then $x + y \in R^+$)

Axiom 3 $\quad \forall_{x,\, y \in R} \qquad$ (if $x \in R^+$ and $y \in R^+$, then $xy \in R^+$)

Definitions

Definition 1 ($<$) $\forall_{x,\,y \in R}$ $[x < y$ if and only if $(\exists_{z \in R^+}$ such that $x + z = y)]$

Definition 2 ($>$) $\forall_{x,\,y \in R}$ $(x > y$ if and only if $y < x)$

Definition 3 ($\leq$) $\forall_{x,\,y \in R}$ $[x \leq y$ if and only if $(x < y$ or $x = y)]$

Definition 4 ($\geq$) $\forall_{x,\,y \in R}$ $[x \geq y$ if and only if $(x > y$ or $x = y)]$

The axioms above are really defining the subset of "positive" real numbers R^+; the definitions introduce the relational symbolisms ($<$, $>$, $\leq$, $\geq$) in terms of the positive reals and the operation of addition. Other conventional subsets of the real numbers are:

$$R^- \text{ (Negative reals)} = R - [R^+ \cup \{0\}]$$

$$\text{Nonnegative reals} = R^+ \cup \{0\}$$

$$\text{Nonpositive reals} = R^- \cup \{0\}$$

We can visualize these subsets on the number line shown in Figure 4.23.

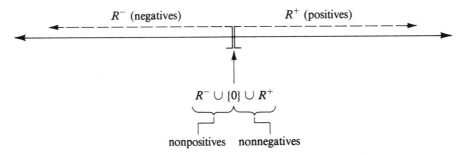

Figure 4.23

From these primitive beginnings, we can deduce many theorems about inequalities, some of which are discussed and proved next. It is assumed that the reader is already familiar with the real number system, its operations of addition and multiplication, and the well-known properties of those operations that characterize the system $(R, +, \cdot)$ as a field (e.g., commutative, associative, identity, and inverse properties).

Theorems on Inequality

$$O_1 \quad 0 \notin R^+$$

Proof: Axiom 1. If $0 \in R^+$, then we would have $0 = 0$ and $0 \in R^+$, which contradicts Axiom 1.

$$O_2 \quad 1 \in R^+$$

Proof: From algebra, $1 \neq 0$. By Axiom 1, either $1 \in R^+$ or $-1 \in R^+$. Suppose $-1 \in R^+$. By Axiom 3, $(-1) \cdot (-1) \in R^+$. Thus $1 \in R^+$. On the other hand, if we assume $1 \in R^+$, we have $1 \in R^+$. By proof by cases, $1 \in R^+$. (*Now* we can conclude by Axiom 1, $-1 \notin R^+$.)

$$O_3 \quad \forall_{x \in R} \ (x > 0 \leftrightarrow x \in R^+)$$

Proof: Let x be any real number.

$$x > 0 \leftrightarrow 0 < x \qquad \text{(definition 2)}$$
$$\leftrightarrow \exists_{z \in R^+} \text{ such that } 0 + z = x \qquad \text{(definition 1)}$$
$$\leftrightarrow \exists_{z \in R^+} \text{ such that } z = x \qquad \text{(algebra)}$$
$$\leftrightarrow x \in R^+ \qquad \text{(logic)}$$
$$(x = z \text{ and } z \in R^+, \therefore x \in R^+)$$

$$O_4 \quad \forall_{x, y \in R} \ (x < y \leftrightarrow y - x > 0)$$

Proof: Let x, y be any real numbers

$$x < y \leftrightarrow \exists_{z \in R^+} \text{ such that } x + z = y \qquad \text{(definition 1)}$$
$$\leftrightarrow \exists_{z > 0} \text{ such that } x + z = y \qquad (O_3)$$
$$\leftrightarrow \exists_{z > 0} \text{ such that } y - x = z \qquad \text{(algebra)}$$
$$\leftrightarrow y - x > 0 \qquad (y - x = z \text{ and } z > 0)$$

$$O_5 \quad \forall_{x \in R} \ (x < 0 \leftrightarrow -x > 0)$$

Proof: Let x be any real number

$$x < 0 \leftrightarrow 0 - x > 0 \qquad (O_4)$$
$$\leftrightarrow -x > 0 \qquad \text{(algebra)}$$

We are now ready to prove the theorem which, in effect, says that the relation $<$ is a total order on R.

O_6 (a) *Trichotomy*: $\forall_{x,\,y\,\in\,R}$ exactly one of the following holds:

$$x < y \qquad x = y \qquad x > y$$

(b) *Transitivity*: $\forall_{x,\,y,\,z\,\in\,R}$ (if $x < y$ and $y < z$, then $x < z$)

Proof (a): (trichotomy)

Let x, y be any real numbers. Consider the number $x - y$. By Axiom 1, we know exactly one of the following possibilities holds:

$$x - y \in R^+ \quad \text{or} \quad x - y = 0 \quad \text{or} \quad -(x - y) \in R^+$$

By Theorem O_3, these translate to

$$x - y > 0 \quad \text{or} \quad x - y = 0 \quad \text{or} \quad -(x - y) > 0$$

By algebra, the last inequality is equivalent to $y - x > 0$. Applying Theorem O_4 and algebra, we have that exactly one of the following holds:

$$y < x \quad \text{or} \quad x = y \quad \text{or} \quad x < y$$

Finally, by definition 2 and rearrangement, we obtain

$$x < y \quad \text{or} \quad x = y \quad \text{or} \quad x > y$$

Proof (b): (transitivity)

Assume x, y, and z are any real numbers, and $x < y$ and $y < z$. We show $x < z$. Since $x < y$, then $y - x > 0$ by O_4. By O_3, $y - x \in R^+$. Similarly, since $y < z$, $z - y > 0$ (O_4) and $z - y \in R^+ (O_3)$. Now, by Axiom 2,

$$(y - x) + (z - y) \in R^+$$
$$\therefore \quad z - x \in R^+$$
$$\therefore \quad z - x > 0 \qquad (O_3)$$
$$\therefore \quad x < z \qquad (O_4)$$

O_7 (Addition/Cancellation)

$$\forall_{x,\,y,\,z\,\in\,R} \quad (x < y \leftrightarrow (x + z < y + z))$$

This theorem says ($\rightarrow$) given any inequality, any real number can be added to both sides without changing the sense of the inequality; also ($\leftarrow$) says the same real number added to each side of an inequality can be "cancelled."

Proof: Let x, y, z be any real numbers,

$$x < y \leftrightarrow y - x > 0 \qquad\qquad (O_4)$$
$$\leftrightarrow (y + z) - (x + z) > 0 \quad \text{(algebra)}$$
$$\leftrightarrow x + z < y + z \qquad\qquad (O_4)$$

An alternate proof of O_7 could be given straightforward from the definition of $<$ and algebra as follows:

Proof: ($\rightarrow$) Let x, y, and z be any real numbers. Assume $x < y$, then $\exists_{h \in R^+}$ such that $x + h = y$. Therefore, $(x + z) + h = y + z$, adding z to both sides. Hence,

$$x + z < y + z$$

($\leftarrow$) Assume $x + z < y + z$. Then $\exists_{k \in R^+}$ such that $(x + z) + k = y + z$. Now, subtracting z from both sides of the equation, we have

$$x + k = y$$

from which $x < y$.

The reader should note from the above proofs that there are at least two perspectives for investigating many inequalities:

1. Use axioms, definitions, and other theorems to help deduce the theorem at hand. (This was the approach demonstrated above for deducing theorems O_1–O_7.)

2. Use the definition of $<$. This converts any " $<$ " statement into a statement of equality. Then one can apply known algebraic properties of equations, and eventually (appealing to inequality definitions) translate back to statements of inequality. (This was the alternate approach for deducing Theorem O_7.)

O_8 (Multiplication/Cancellation)

(a) $\forall_{x, y \in R} \, \forall_{z > 0} \, (x < y \leftrightarrow xz < yz)$

(b) $\forall_{x, y \in R} \, \forall_{z < 0} \, (x < y \leftrightarrow xz > yz)$

This theorem (a) states the properties about multiplying (or cancelling) a positive number relative to both sides of an inequality without changing the sense of the inequality; also (b) deals with multiplying (or cancelling) a negative number with respect to both sides of an inequality and the effect is to change the sense of the inequality.

We prove part (b). Part (a) is similar and left as a problem.

Proof (b): Let $x, y \in R$ and $z < 0$.

$$x < y \leftrightarrow y - x > 0 \qquad (O_4)$$

Now

$$z < 0 \leftrightarrow -z > 0 \qquad (O_5)$$

Also a consequence of O_3 and Axiom 3 is that $(y - x)(-z) > 0$. Thus, by algebra, assuming $z < 0$,

$$x < y \leftrightarrow xz - yz > 0$$
$$\leftrightarrow yz < xz \qquad (O_4)$$
$$\leftrightarrow xz > yz \qquad (\text{definition } 2)$$

O_9 (Positive Product/Negative Product)

(a) $\forall_{x, y \in R} \qquad (xy > 0) \leftrightarrow \begin{bmatrix} (x > 0 \text{ and } y > 0) \\ \text{or} \\ (x < 0 \text{ and } y < 0) \end{bmatrix}$

(b) $\forall_{x, y \in R} \qquad (xy < 0) \leftrightarrow \begin{bmatrix} (x > 0 \text{ and } y < 0) \\ \text{or} \\ (x < 0 \text{ and } y > 0) \end{bmatrix}$

The proof of (a) will be given here; (b) is left as a problem.

Proof (a): The easier direction of proof is ($\leftarrow$) and we shall start there. Assume ($x > 0$ and $y > 0$) or ($x < 0$ and $y < 0$). If $x > 0$ and $y > 0$, then as a consequence of Theorem O_3 and Axiom 3, $xy > 0$. On the other hand, if ($x < 0$ and $y < 0$), then ($-x > 0$ and $-y > 0$) by Theorem O_5. Thus, by Theorem O_3 and Axiom 3, $(-x)(-y) > 0$, from which $xy > 0$ again. We have thus proved two statements having logical forms $P \rightarrow Q$ and $R \rightarrow Q$. By a tautology in logic, we have

$$(P \lor R) \rightarrow Q$$

which is the form we desired.

($\rightarrow$) This direction of proof is somewhat more tedious since it involves a number of different cases. We start by assuming x, y are real numbers and

$$xy > 0$$

Now by trichotomy

$$x < 0 \quad \text{or} \quad x = 0 \quad \text{or} \quad x > 0$$

and

$$y < 0 \quad \text{or} \quad y = 0 \quad \text{or} \quad y > 0$$

This would lead to examination of nine separate combinations for x and y. However, we can simplify the argument considerably by dispensing with all cases involving $x = 0$ or $y = 0$ (or both) immediately. By algebra,

$$(x = 0 \quad \text{or} \quad y = 0) \leftrightarrow xy = 0$$

Thus, we can conclude (in view of our assumption $xy > 0$) that $x \neq 0$ and $y \neq 0$. That leaves us with four cases. Actually, if we argue by contradiction, we can reduce this proof to only two cases—eliminating the possibilities we feel are not true.

Case 1:   Assume $x > 0$ and $y < 0$.
Then $x > 0$ and $-y > 0$ by Theorem O_5.
Hence $x(-y) > 0$ by Theorem O_3 and Axiom 3.
$\therefore$ $-(xy) > 0$ by algebra, and
$\therefore$ $xy < 0$ by Theorem O_5.

Since we are assuming $xy > 0$ and we have deduced that $xy < 0$, we have a contradiction to the trichotomy theorem O_6. Therefore, it is *not* the case that $x > 0$ and $y < 0$.

Case 2:   Assume $x < 0$ and $y > 0$.
This situation is perfectly symmetric to the argument in case 1, interchanging the roles of x and y. It leads to the same contradiction. Therefore, it is *not* the case that $x < 0$ and $y > 0$.

Since there are only two possibilities left, we cite them as the conclusion to our proof:

$$(x > 0 \text{ and } y > 0) \quad \text{or} \quad (x < 0 \text{ and } y < 0)$$

In the foregoing, we have examined carefully a number of axioms, definitions, and theorems involving the order relation ($<$) on the real numbers. It should be noted that most of the theorems (O_4–O_8) hold when the " $<$ " symbol is replaced by $>$, $\leq$, or $\geq$. In *using* statements to construct proofs of inequalities, it is not as important to recall the reference label of the axiom or theorem or definition (e.g., A_3, O_5, Definition 2 . . .), or even the formal statement itself, as it is to *translate the statement into your own words*. If you do that, you will develop a "feeling" for the theorems and you will be able to apply them readily in proof situations. For more complex theorems and definitions in mathematics, it is essential to be quite familiar with the hypotheses involved and the precise formulations. But for general properties like those discussed in this section, it suffices to have a good intuitive grasp of the ideas.

In words, then, much of the foregoing can be summarized as follows:

- $<$ and $>$ have the properties of trichotomy and transitivity (O_6).

- One number is less than another if and only if the second number is greater than the first (Definition 2).

- One number is less than another if you can find a *positive* number which, when added to the first, produces the second (Definition 1).

- Being positive means the same as being greater than 0 (O_3).

- One number is less than another if their difference is positive (O_4).

- Being negative means the same as being less than 0 (O_5 and trichotomy).
- The sum of two positive numbers is always positive (Axiom 2).
- The sum of two negative numbers is always negative (Axiom 2 and O_5).
- The product of two positive numbers is always positive (O_9).
- The product of two negative numbers is always positive (O_9).
- The product of a positive and a negative number is always negative (O_9).
- If a product of two real numbers is positive, then both are positive or both are negative (O_9).
- If a product of two real numbers is negative, then one is positive and the other negative (O_9).
- You can add (or subtract) the same number to both sides of an inequality without changing the sense (direction) of inequality (O_7).
- You can multiply (or divide) both sides of an inequality by a positive number without changing the sense of the inequality (O_8).
- You can multiply (or divide) both sides of an inequality by a negative number, and this changes the sense of the inequality (O_8).

Assigned Problems

1. Prove Theorem O_8 (a).

2. Prove Theorem O_9 (b).

3. Prove the following:
$$\forall_{x,\,y\,\in\,R}\ (\text{if } x \leq y \text{ and } y \leq x, \text{ then } x = y)$$

4. Prove:
$$\forall_{x,\,y,\,u,\,v\,\in\,R}\ (\text{if } x < y \text{ and } u < v, \text{ then } x + u < y + v)$$

5. Prove each of the following:

(a) $\forall_{x\,\in\,R}$ (if $x \neq 0$, then $x^2 > 0$)

(What is a logical equivalent to this statement?)

(b) $\forall_{x,\,y\,\in\,R}$ (if $x \neq y$, then $x^2 + y^2 > 2xy$)

(Restate this in a logically equivalent form. Can you deduce $x^2 + y^2 \geq 2xy$ for all $x, y \in R$?)

(c) $\forall_{x\,\in\,R}\ \left(\text{if } x > 0, \text{ then } x + \dfrac{1}{x} \geq 2\right)$

6. Prove: If a and b are two nonnegative real numbers such that $a \leq b$, then there exists a real number r such that $0 \leq r \leq 1$ and $a = br$.

7. Prove that $\forall_{a, b \in R}$ $\left(\text{if } 0 < a < b, \text{ then } \dfrac{1}{a} > \dfrac{1}{b}\right)$.

   (*Hint:* Apply the definition of $<$, using differences.)

8. Show that for any positive integer N, there exists an integer n such that

$$\sum_{i=1}^{n} \frac{1}{i} > N$$

   (*Hint:* Recall the proof technique cued by "for any positive integer . . . .") (Also, how are $1/(n + 1)$ and $1/2n$ related?)

9. (a) Prove that if a, b, c, d are all positive, and $a < b$ and $c < d$, then $ac < bd$.

   (b) Extend the theorem in part (a) to the following: If $a_i, b_i > 0$ $\forall_{i \in I}$ and $a_1 < b_1, a_2 < b_2, \ldots a_n < b_n$ for any positive integer n, then

$$\prod_{i=1}^{n} a_i < \prod_{i=1}^{n} b_i$$

   That is, $(a_1 \cdot a_2 \cdot a_3 \cdot \cdots \cdot a_n) < (b_1 \cdot b_2 \cdot b_3 \cdot \cdots \cdot b_n)$ for any positive integer n. $\left(\displaystyle\prod_{i=1}^{n} x_i \text{ means the generalized product } x_1 \cdot x_2 \cdot x_3 \cdot \cdots \cdot x_n.\right)$

10. Is the following inequality true for all positive real numbers a, b? Explain why or why not.

$$\left(\frac{2}{\dfrac{1}{a} + \dfrac{1}{b}}\right)^2 \le ab$$

4.7   Divisibility Proofs

We have seen that the divisibility relation ($|$) on the set of positive integers is a partial order. In this section we shall formally define this relation and investigate some of its basic properties. We begin by extending the divisibility relation to all integers.

Divisibility

Suppose a and b are integers with $a \ne 0$. We say a *divides* b, denoted $a|b$, if and only if

$$\exists_{k \in I} \text{ such that } b = ka$$

In this case, *a* is called a *factor* or *divisor* of *b*, and *b* is called a *multiple* of *a*. Note that this definition of the divisibility relation is equivalent to our earlier one (see Section 4.4), where $ak = b$.

Prime Number

A *prime number* is an integer *p* greater than 1 whose only positive divisors are 1 and *p* itself. Any positive integer other than 1 that is not prime is said to be *composite*.

Thus, the set of positive integers I^+ can be regarded as the union of three disjoint sets—the set of primes, the singleton {1}, and the set of composites. Note that 1 is considered neither prime nor composite. Also, an immediate consequence of the definition of prime/composite is the following:

A positive integer *n* is composite ↔ there exist positive integers *a*, *b* such that $n = ab$, where neither *a* nor *b* is 1 or *n*.

The reader's attention is called to an analogy between the order relation ($\leq$) on the real numbers and the divisibility relation ($|$) on the integers. The definition of $\leq$ essentially shifts inequalities into addition (or subtraction) statements, and we apply algebra to them to prove theorems on inequalities. Similarly, the definition of $|$ shifts statements about divisibility into statements about multiplication (or division), and we apply algebra to them to prove theorems about divisibility.

$$D_1 \quad \forall_{a \in I} \; [\text{if } a \neq 0, \text{ then } (a|0 \text{ and } a|a)].$$

Proof: Let *a* be any nonzero integer.
$a|0$ since $0 = a \cdot 0$.
$a|a$ since $a = a \cdot 1$ (this is the *reflexive* property).

$$D_2 \quad \forall_{b \in I} \; (1|b).$$

Proof: Let *b* be any integer.
$1|b$ since $b = 1 \cdot b$.

$$D_3 \quad \forall_{a, b, c \in I} \; (\text{if } a|b, \text{ then } a|bc).$$

Proof: Let *a*, *b*, and *c* be any integers.
Assume $a|b$. Then ∃ an integer *k* such that

$$b = ka$$

Now, multiplying both sides by c, we have

$$bc = (ck)a$$

where ck is an integer (since $c, k \in I$).
Thus, $a|bc$.

D_4 $\forall_{a,\,b \in I}$ [if $(a|b$ and $b|a)$, then $a = \pm b$].

If we restrict our attention to just the positive integers, or just the negative integers, this theorem says the relation | is antisymmetric. Note that | is therefore *not* an order relation on all integers.

Proof: Let a and b be any integers, and assume $a|b$ and $b|a$.

$$a|b \rightarrow \exists_{k_1 \in I} \text{ such that } b = k_1 a$$

$$b|a \rightarrow \exists_{k_2 \in I} \text{ such that } a = k_2 b$$

Therefore, substituting, we obtain

$$b = k_1(k_2 b)$$
$$= (k_1 k_2)b$$

from which $k_1 \cdot k_2 = 1$. The only integers for which the last equation holds are

$$k_1 = k_2 = 1 \text{ where } a = b$$
$$k_1 = k_2 = -1 \text{ where } a = -b$$
$$\therefore \quad a = \pm b$$

D_5 (Transitivity)

$\forall_{a,\,b,\,c \in I}$ [if $(a|b$ and $b|c)$, then $a|c$].

Proof: Let a, b, c be any integers, and assume

$$a|b \text{ and } b|c$$
$$a|b \rightarrow \exists_{k \in I} \text{ such that } b = ka$$
$$b|c \rightarrow \exists_{j \in I} \text{ such that } c = jb$$

Therefore, $c = jb$
$$= j(ka)$$
$$= (jk)a$$

So $a|c$ since $jk \in I$ if $j, k \in I$.

D_6 $\forall_{a,\,b,\,c,\,x,\,y \in I}$ [if $(a|b$ and $a|c)$, then $a|(bx + cy)$].

This theorem says that if one integer divides each of two other integers, then it divides *any* integral linear combination of the two.

Proof: Let a, b, and c be any integers and assume

$$a|b \quad \text{and} \quad a|c$$
$$a|b \rightarrow \exists_{k \in I} \text{ such that } b = ka$$
$$a|c \rightarrow \exists_{j \in I} \text{ such that } c = ja$$

Now let x, y be any integers, and consider the linear combination

$$bx + cy$$
$$bx + cy = (ka)x + (ja)y$$
$$= (kx)a + (jy)a$$
$$= (kx + jy)a$$

where $kx + jy$ is an integer since $k, x, j, y \in I$. Therefore, we have $a|(bx + cy)$.

Theorem D_6 is particularly helpful in solving many divisibility problems. A special case of this theorem would be

If $a|b$ and $a|c$, then $a|(b + c)$ and $a|(b - c)$.

Moreover, this theorem extends (via mathematical induction) to sums of more than two terms.

Greatest Common Divisor

If a, b, and d are integers where $d|a$ and $d|b$, then d is said to be a *common divisor* of a and b. If d is the largest common divisor of a and b, it is called the *greatest common divisor* of a and b, and denoted by

$$\gcd(a, b)$$

(It is assumed, for the existence of greatest common divisor, that a and b are not both 0.)

Since 1 divides every integer, it is clear that $\gcd(a, b)$, for any two integers, must be greater than or equal to 1. If $\gcd(a, b) = 1$, we say a and b are *relatively prime*. Note that $\gcd(8, 15) = 1$ but neither 8 nor 15 are prime numbers. Thus two integers being relatively prime says nothing about whether they are individually prime or not.

The following theorems are stated here without proof. The interested reader is advised to consult any text on number theory for formal proofs. These theorems are used to help solve numerous divisibility problems. Often, they are tacitly assumed in such solutions.

$$D_7 \quad \forall_{a, b, c \in I} \text{ [if } (a|bc \text{ and } \gcd(a, b) = 1), \text{ then } a|c].$$

D_8 $\forall_{a, b \in I}$ [if (p is a prime number and $p|ab$), then ($p|a$ or $p|b$)].

D_9 $\forall_{a, b, c \in I}$ [if ($a|c$ and $b|c$ and gcd(a, b) = 1), then $ab|c$].

Following are some examples intended to illustrate the reasoning behind the solution of certain divisibility problems.

Example 1   Prove that the product of three consecutive integers, the second one being odd, is divisible by 24.

Discussion:

The prime factorization of 24 is $2^3 \cdot 3$. If we can show that 2^3 divides the product and 3 divides the product, then since gcd(8, 3) = 1, theorem D_9 will guarantee that 24 divides the product. Since the second number is odd, the other two must be even. In fact, one of them must be a multiple of 4, since every second even number is a multiple of 4. Thus our product must be divisible by 8. Now since there are three *consecutive* integers involved, then one of them must be divisible by 3. (The latter statement could be argued by observing that if *none* were divisible by 3, then by the Division Algorithm they would all have to be of the form $3k + 1$ or $3k + 2$, and taking cases, a contradiction is easily produced.) Therefore, the product is divisible by 24.

Example 2   Suppose m is an integer greater than 1 such that for some integer n,

$$m|(35n + 26) \quad \text{and} \quad m|(7n + 3)$$

Prove that $m = 11$.

Proof   Since $m|(35n + 26)$ and $m|(7n + 3)$, then by theorem D_6, m divides any linear combination of the two expressions. In particular,

$$m|[(35n + 26) - 5(7n + 3)]$$

$$\therefore \quad m|11$$

Now 11 is a prime number, hence its only divisors are 1 and 11. It was given that $m > 1$. Therefore, we conclude $m = 11$.

Example 3   Let c be any nonzero integer. Show that if a is divisible by b, then ca is divisible by cb for any integers a, b.

Proof   This is almost a straightforward consequence of the definition of divisibility.

$$b|a \rightarrow \exists_{k \in I} \text{ such that } a = bk$$

Therefore, $ca = c(bk)$

$$= (cb)k$$

Hence, by the definition of divisibility,

$$cb|ca$$

Example 4   Suppose a, b, and c are positive integers such that a is a factor of b and a is a factor of $b + c$. Prove that a must also be a factor of c.

Proof   Let $a, b, c \in I^+$. Since $a|b$ and $a|b + c$, then by theorem D_6 (the linear combination theorem), $a|[1(b + c) + (-1)b]$. Therefore, $a|c$.

Assigned Problems

1. Prove the converse of Example 4. That is, let a, b, and c be positive integers such that a is a factor of b. Show if a is a factor of c, then a is a factor of $b + c$.

2. What conclusion can be drawn in each case? Why? Assume a, b, and c are positive integers.

   (a) a is a factor of b and c is a multiple of b

   (b) a is a factor of b and a is a multiple of c

   (c) a is a multiple of c and b is a multiple of c

   (d) a is a factor of b and b is a multiple of c

3. Prove: If n is any integer, then $n^5 - n$ is a multiple of 30. (*Hint:* What are the possible remainders upon division by 5?)

4. Compute the gcd(1274, 3087, 1085).

5. Find counterexamples to each of these false assertions:

   (a) If p is a prime number, then $p + 1$ is composite.

   (b) If p is a prime number, then $p^2 + 1$ is composite.

   (c) Let $a, b, c \in I$. If $a|c$ and $b|c$, then either $a|b$ or $b|a$.

6. Prove:

   (a) If an integer b is divisible by a^3, then b^2 is divisible by a^3.

   (b) If an integer b is divisible by a, then b^3 is divisible by a^3.

7. Show, by mathematical induction on n, that for any positive integer n,

$$6 \cdot (10)^n + 8 \cdot (10)^{n-1} + 4$$

is divisible by 9.

8. Show that any integer $n \geq 2$ can be expressed in the form

$$n = 2a + 3b$$

where a and b are nonnegative integers.

9. Verify that if n is an odd integer, then $n^9 - n$ is divisible by 32.

10. Show why if n is a positive integer greater than 1 and

$$n|(15j + 13) \text{ and } n|(j + 1)$$

for some integer j, then $n = 2$.

5

Functions

5.1   Function Perspectives

There are three major kinds of relations in mathematics—*equivalence relations*, *order relations*, and *functions*. The first two are relations defined *on* a set A; the third are relations defined *from* a set A to a set B. It is this latter class of relations—namely *functions*—to which we now turn our attention. The concept of *function* is so important and so far-reaching throughout all branches of mathematics that it is worthwhile for us to take a moment here, in this section, to reflect on the many different ways of looking at mathematical functions. There are at least six different perspectives we shall consider. In each, we will be talking about the same general idea—that of function—but each time we will be taking a slightly different view.

The first perspective is the most formal and is the standard mathematical definition for the concept of function. We shall use it to introduce notation and ideas related to functions that will be referred to in the remainder of the course.

Perspective 1: Function as a Binary Relation Meeting Certain Conditions

A *function f* from A to B, denoted

$$f\colon A \to B$$

is a binary relation from A to B (that is, a subset of $A \times B$) meeting two conditions:

1. *Existence:* $\forall_{x \in A} \; \exists_{y \in B}$ such that $(x, y) \in f$
   (Every element in A has a correspondent in B.)

2. *Uniqueness:* $\forall_{(x_1, y_1), (x_2, y_2) \in f}$ (if $x_1 = x_2$, then $y_1 = y_2$)
   (Every element in A has only one correspondent in B.)

241

Note that in the formal definition, a function is a *set of ordered pairs*, a relation, having certain specified conditions. When we talk about functions, the set of all first components in the ordered pairs of the function is called *the domain of the function*, and the elements in this set are frequently referred to as *preimages*. On the other hand, the set of all second components in the ordered pairs of the function is called *the range of the function*, and the elements in this set are called *images*. Using this terminology, the existence and uniqueness conditions for the function can be restated:

1. *Existence:* For every preimage, there is an image.

2. *Uniqueness:* For every preimage, there is only one image.
   (If two preimages are equal, then their images are equal.)

In mathematics, there are several notations for functions with respect to ordered pairs, preimages, and images. The following three are the most popular:

$(x, y) \in f$	set-theoretic notation
$f(x) = y$	analytic notation
$(x)f = y$ or simply $xf = y$	algebraic notation

Each notation has advantages and disadvantages, and we shall adopt the *analytic* notation for discussing functions. This is the notation you are probably most familiar with from calculus. That is, instead of saying $(x, y) \in f$, we will say $f(x) = y$, where x is some preimage (first component) and y is its image (second component). Using our adopted notation, the existence and uniqueness conditions for a function from A to B become:

1. *Existence:* $\forall_{x \in A} \exists_{y \in B}$ such that $f(x) = y$

2. *Uniqueness:* $\forall_{x_1, x_2 \in A}$ (if $x_1 = x_2$, then $f(x_1) = f(x_2)$)

The above remarks on notation are summarized in the following statements. Assume f is a function (set of ordered pairs with existence and uniqueness conditions holding) and $x \in \text{dom } f$.

1. $y = f(x) \leftrightarrow (x, y) \in f$

2. $y \in \text{ran } f \leftrightarrow \exists_{x \in \text{dom } f}$ such that $y = f(x)$

3. $\forall_{x \in \text{dom } f} (x, f(x)) \in f$

4. $f = \{(x, f(x)) | x \in \text{dom } f\}$

5. $\text{ran } f = \{f(x) | x \in \text{dom } f\}$

Equality of Functions

Two functions f and g are equal $\leftrightarrow$ (a) dom $f =$ dom g, and

$$\text{(b) } f(x) = g(x) \; \forall_{x \in \text{dom } f, g}$$

This definition for $f = g$ arises directly from the fact that functions are sets of ordered pairs. It says that two functions are equal if and only if their domains are the same set *and* every element in the domain has the same image under *both* functions. These two conditions force equality of all the ordered pairs in the two functions. Note that it is *not* sufficient to simply require that the functions have equal domains and equal ranges because the ranges may be *setwise* equal without being *pointwise* equal (the latter depending on the correspondence).

Functions vs. Mappings

Strictly speaking, functions are sets of ordered pairs having the existence and uniqueness conditions. The notation

$$f: A \to B$$

usually denotes a relation called a *mapping* from A to B.

There is a slight (usually overlooked) difference between *functions* and *mappings*. If we start with two sets A and B, the range of the function f (as a relation from A to B) is a *subset of B*,

$$\text{ran } f \subseteq B$$

It is not necessarily always true that ran $f = B$. For this reason, we call B the *image space* of the function f and differentiate it from the *range* of f. In order to specify a function, all we need to do is specify *one* set (its domain) and some rule of correspondence. For example, if $A = \{\text{real numbers}\}$ and $f(x) = x^2$, then we know precisely the nature of the *function f*. On the other hand, we have not specified the image space. It could be all real numbers, all nonnegative reals, the interval $[-5, +\infty)$, or any subset of the real numbers that includes the range of the function. When we specify the function *f and* the image space B, then we are really talking about a *mapping*.

We shall generally use the terms *function* and *mapping* as synonymous, consistent with standard current mathematical practice. The student should be aware that the above distinction does exist, however.

Example 1   (a) Let dom $f = \{1, 2, 3\}$ and

$$f = \{(1,4), (2,4), (3,5)\}$$

This set of just three ordered pairs is a function since every element in the domain $\{1, 2, 3\}$ has a correspondent in the range $\{4, 5\}$, and no element in the domain has more than one correspondent in the range. Note that the range element 4 has two distinct preimages, however. This does not contradict uniqueness, which says if two *preimages* are equal, then their images must be equal (not conversely).

$$\text{Here } f(1) = 4 \text{ or } (1,4) \in f$$
$$f(2) = 4 \text{ or } (2,4) \in f$$
$$f(3) = 5 \text{ or } (3,5) \in f$$

(b) Let $g = \{(1,2),(2,3),(1,4),(5,2)\}$ where dom $g = \{1,2,5\}$.
Is g a function? No, since the domain element 1 has two distinct images—2 and 4. Therefore, the uniqueness condition is not holding. That is, the statement

$$\forall_{x_1, x_2 \in \text{dom } g} \ (\text{if } x_1 = x_2, \text{ then } g(x_1) = g(x_2))$$

is not true.

(c) Let $h = \{(1,2),(2,3),(3,4)\}$, where dom $h = \{1,2,3,4\}$.
Is h a function? No, since not every domain element is used in the correspondence. That is, 4 has no image. The existence condition fails.

Example 2   The following are examples of relations from one set to another. Test the existence and uniqueness conditions mentally to see if they hold and whether the relation is a function. Correct responses are given at the end of this example.

(a) $S = \{\text{all humans}\}$; $N =$ Natural numbers

$$\forall_{x \in S} \ [(x, n) \in f \text{ means "person } x \text{ has age } n\text{," where } n \in N]$$

(b) $A = \{1,2,3,4\}$

$$f = A \times A$$

(c) $I = \{\text{integers}\}$

$$\forall_{n \in I} \ [(n, r) \in f \text{ means "} r \text{ is the remainder } n \text{ leaves upon}$$
$$\text{division by 3," where } r \in I]$$

(d) $A = \{1,2,3,\ldots 100\}$; $H = \{\text{human beings}\}$

$$\forall_{a \in A} \ [(a, h) \in f \text{ means "} h \text{ is the person whose height, rounded to the nearest}$$
$$\text{inch, is } a\text{"}]$$

(e) $S = [0,1]$

$$f = \{(x^2, x) | x \in [0,1)\}$$

(f) $C = \{\text{all cities in Wisconsin}\}$; $I^+ = \{\text{positive integers}\}$

$$\forall_{c \in C} \ [(c, n) \in f \text{ means "} n \text{ is the distance (in integral kilometers)}$$
$$\text{from city } c \text{ to Chicago"}]$$

(Examples a, c, and f are functions; b, d, and e are nonfunctions.)

Perspective 2: Graph of a Function

Another perspective that helps to clarify the character of functions is the *graph*. This is a picture that displays the domain and range of the function together with some visual interpretation of how they are functionally related. You

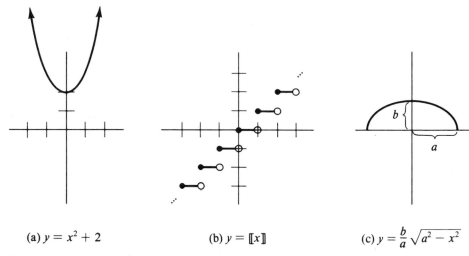

(a) $y = x^2 + 2$ (b) $y = [\![x]\!]$ (c) $y = \dfrac{b}{a}\sqrt{a^2 - x^2}$

Figure 5.1

are probably familiar with the graphs of many $R \to R$ functions (real-valued functions of a single real variable) from calculus (see Figure 5.1).

We can generalize these pictures to any functions—not necessarily numerical—if we simply display the domain along one axis and the range along a perpendicular axis; then the collection of all points (domain element, range correspondent) visually describes the function (see Figure 5.2).

The chief advantages of graphs of functions are the display of trends and characteristics not readily observable from sets of ordered pairs, equations, or other descriptions. One can see the entire function "at a glance," making decisions about monotonicity, concavity, asymptotes, extrema, points of inflection, roots, domain, range, and a host of other pertinent data.

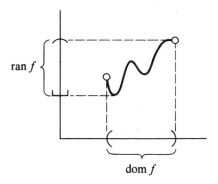

Figure 5.2

Perspective 3: Relationship between Variables

There are at least two distinct ways of thinking about *variables*. These views are psychologically different ways of referring to the same mathematical object.

1. Static View: a variable is a symbol that acts as a placeholder or "blank spot" and stands for replacement by objects from some understood set. For example,

$$\square + 3 = 8$$
$$y = x^2 + 5$$
$$\sin x + \cos y = \frac{\pi}{2}$$

In each instance, we might think of the equation as being a form. The content (e.g., variables) could change, but the form remains. That is, the equation

$$\triangle = \square^2 + 5$$

is really no different mathematically than

$$y = x^2 + 5$$

but it *looks* different. When we replace the variable(s) by elements from the set(s) under consideration, we obtain a statement that can be judged true or false.

In the static sense, a variable plays a role much like a pronoun in language grammar. A pronoun represents a noun, just as a variable represents a member of some prescribed set.

2. Dynamic View: a variable is an entity that varies or ranges over a set, much like a jukebox selector swings back and forth over the set of potential record indicators and then selects one.

Here, we think of the variable as though it were in motion, searching for its values. If two variables are involved in a functional relationship, one is allowed to vary in a controlled fashion over some set, and we look for the effects on the other variable correspondent in this variation. The first is called the *independent* variable, and the second the *dependent* variable. The values of the second variable *depend on* the values of the first. In an experimental situation, the experimenter manipulates the independent variable, and the situation being studied causes values of the dependent variable to change for changes in the independent variable.

Thus, when we consider the functional equation $y = f(x)$, we may concentrate on the variables x and y and the nature of their dependency. If we think of x as a symbol standing for replacement from members of some set A, and the function generally as a set of ordered-pair outcomes that correspond to a collection of true statements from a series of true-false tests, then we are using the static view of function. On the other hand, if we think of x and y as changing quantities, the changes in y depending on the changes in x as x varies throughout the set A, then we perceive the function in a dynamic sense.

Modern abstract mathematics has leaned toward the static view because it does not introduce additional complexities of time and change, and it is more adaptable to rigorous proof. The concept of function itself, however, arose historically from dynamic considerations. In the present day, mathematicians apply both views but favor the static perspective, and scientists tend to favor the dynamic view. Both have merits and you should be familiar with each. The dynamic perspective is more intuitive, but it can be of significant assistance when you try to discover relationships; the static view is somewhat more sterile and formal, but it can help greatly when verification or proof is warranted.

The following examples provide dynamic and static perspectives of the time-distance function for an object falling freely in a vacuum.

Example 3   *Dynamic View*

A physicist conducts a series of experiments in which various objects are dropped for fixed amounts of time (in seconds). The distances fallen (in feet) are recorded. A relationship between time t (independent) and distance s (dependent) is sought. It is discovered that s appears to vary directly with the square of t, with some multiplied constant of proportionality (which turns out to be one-half the acceleration due to gravity). The table below records these observations for integral times, as well as the apparent dependency between variables.

t	s
0	0
1	16
2	64
3	144
4	256
$\vdots$	$\vdots$
t	$16t^2$

As t ranges through a set of nonnegative numbers (possibly integers if the stopwatch accommodates these best), s varies with t in the systematic fashion

$$s = 16t^2$$

Example 4   **Static View**

We begin with the functional equation

$$s(t) = 16t^2$$

This says "s is a function of t, specifically $16t^2$." If t is replaced by various nonnegative integers $t = 0, 1, 2, 3, \ldots, s(t)$ takes values, respectively, $0, 16, 64, 144, \ldots$. So the function can be regarded as the set of ordered pairs that make the equation above true; namely,

$$\{(0,0), (1,16), (2,64), (3,144), \ldots (t, 16t^2), \ldots\}$$

Perspective 4: Rule of Correspondence

To completely specify a function f (or particularly, a mapping) we need two sets—dom f and image space f—and some *rule of correspondence*. For example, if dom f = image space $f = R$ (reals), and the rule is

> *Take a real number, multiply it by* 5, *then subtract* 1.

we are really describing the function whose equation is

$$f(x) = 5x - 1$$

and where x is the original real number.

Very frequently in mathematics (especially calculus), we do not start from an ordered pair approach, or with a graph, or by investigating some dependency between variables, but rather we simply state an equation and imply a function. For example,

$$y = \sqrt{x - 3}$$

where it is "understood" what set of values x can take to make y meaningful; in this case, $x \in [3, +\infty)$. Is this equation a function? Strictly speaking, no. However, it *describes* a function if we agree on the domain and image space. Therefore, it is not unusual to hear mathematicians speak of "the function $y = 2x^3 + 5$," or "the greatest integer function." In this sense or perspective, they are thinking of the concept of function embodied by a pair of sets and a rule connecting the two sets.

The functions described by the equations above have all been *explicitly defined*. That is, we can tell the nature of the function by looking at the equation. Moreover, if we are given the domain and image space of the function, we could construct the function formally as a set of ordered pairs.

There are instances in mathematics where functions are defined *implicitly* by an equation. In a sense, they are embedded in equations that describe general

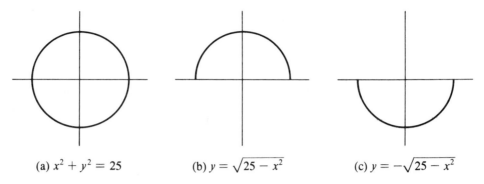

(a) $x^2 + y^2 = 25$ (b) $y = \sqrt{25 - x^2}$ (c) $y = -\sqrt{25 - x^2}$

Figure 5.3

relations but hide the nature of the function or functions involved. For example,

$$x^2 + y^2 = 25$$

is an equation of a relation (a nonfunction). You may recognize it as the equation of a circle (graphical perspective) with center at the origin and radius 5 (see Figure 5.3a). However, this equation defines at least two functions implicitly. Solving for y, we obtain

$$y = \sqrt{25 - x^2} \qquad x \in [-5, 5]$$

or

$$y = -\sqrt{25 - x^2} \qquad x \in [-5, 5]$$

which correspond to the upper semicircle and lower semicircle, respectively (see Figure 5.3b and c).

Perspective 5: Transformation between Sets

Another way to think about functions is to see them as *transformations* from one set to another. We specify what the domain of the function is, then, as shown in Figure 5.4, we draw a variation of a Venn diagram in which the domain is one region, the image space is another, and arrows connect domain elements with their transformed images. In a sense, we trace the path of a representative element as it starts in one set and "changes into" an element in the other set.

If $y = f(x)$, we think of f as an operator. Given $x \in \text{dom } f$, apply f to x to obtain the "transformed object" $f(x)$; or "x is transformed into y via the function f."

There are many alternate ways of picturing functions as transformations. If the sets involved are finite, we might use a vertical (or horizontal) arrangement of both sets with connecting arrows. In Figure 5.5, the function being described is $f = \{(1, b), (2, a), (3, c), (4, c)\}$.

The transformation perspective is a valuable device for studying multiple actions of several functions, or *composition* of functions, as we shall see later.

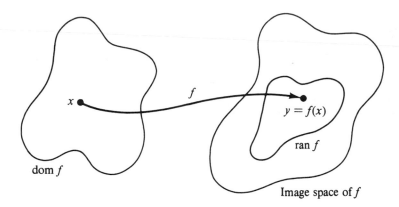

Figure 5.4

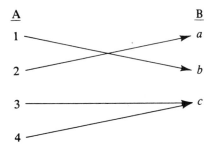

Figure 5.5

Perspective 6: Input-Output Machine

This last perspective, thinking of a function as a machine (graphically depicted in Figure 5.6), is a variation of the transformation idea discussed above. We start with a representative object x from the domain, drop it into the function machine f, and out comes the transformed object—its correspondent in the range.

The function here is regarded mechanically as an input-output device that accepts only domain objects as inputs, and produces outcomes that are corresponding range objects. In this perspective, we lose the sense of function as a set of ordered pairs, but we gain insight into the "operator" nature of function—a mechanical processor that changes information from one state to another.

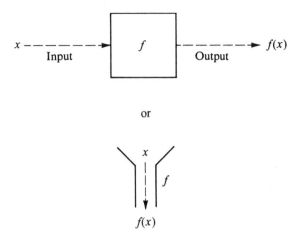

Figure 5.6

The input-output machine perspective is especially useful for studying the nature of *inverse* functions, as we shall see later. For example, if we take the function

$$f(x) = \sqrt[3]{5x^2 - 7}$$

with domain $f = R$ (reals), we can "break it apart" into a series of smaller operations (composition of functions), and create a machine diagram to depict its workings (see Figure 5.7). Now, to get a feeling for the inverse of a function (as a function of x), we start with x and run each part of the machine "backwards," proceeding from the last part to the first (see Figure 5.8). Thus, we have

$$\text{if } f(x) = \sqrt[3]{5x^2 - 7} \qquad (f \text{ expressed as a function of } x)$$

then

$$f^{-1}(x) = \sqrt{\frac{x^3 + 7}{5}} \qquad (f^{-1} \text{ expressed as a function of } x)$$

To summarize this section, we consider one final example that illustrates all six perspectives for a given function.

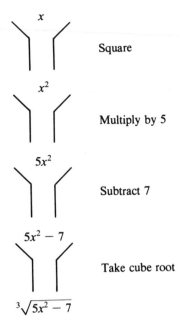

x

Square

x^2

Multiply by 5

$5x^2$

Subtract 7

$5x^2 - 7$

Take cube root

$\sqrt[3]{5x^2 - 7}$

Figure 5.7

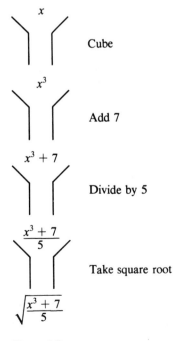

x

Cube

x^3

Add 7

$x^3 + 7$

Divide by 5

$\dfrac{x^3 + 7}{5}$

Take square root

$\sqrt{\dfrac{x^3 + 7}{5}}$

Figure 5.8

Example 5   Consider the function f, which associates each of $1, 2, 3, 4$ with its cube; respectively, $1, 8, 27, 64$.

(a) *Special kind of binary relation, or set of ordered pairs*

$$f = \{(1,1), (2,8), (3,27), (4,64)\}$$

or

$$f = \{(x, y) | y = x^3 \text{ where } x = 1, 2, 3, 4\}$$

(b) *Graph*

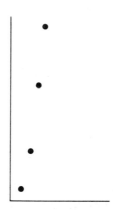

(c) *Relationship between variables*

(Static) $y = x^3$
Replace x, in turn, by 1, 2, 3, and 4; $y = 1, 8, 27, 64$, respectively; $(1,1), (2,8), (3,27), (4,64)$ make the equation true.

(Dynamic)   As x varies across 1, 2, 3, and 4, what happens to x^3?
x^3 goes through values 1, 8, 27, and 64, respectively.

(d) *Rule of correspondence*

Start with the set $\{1, 2, 3, 4\}$; take each element and cube it. We are describing the function

$$y = x^3 \text{ explicitly.}$$

(e) *Transformation between sets*

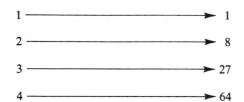

(f) *Input-output machine*

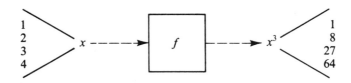

Assigned
Problems

1. Which of the following rules define functions from $R \to R$? ($R = \{$real numbers$\}$.) If the domain is not all of R, describe it and the range.

(a) $f(x) = \dfrac{x^2 - 4}{x - 2}$ *Range All Zero's*
    except 4

(b) $f(x) = |x - 1|$

(c) $f(x) = \sqrt{x^2 - 1}$

(d) $f(x) = \begin{cases} 0 \text{ if } x \text{ is rational} \\ 1 \text{ if } x \text{ is irrational} \end{cases}$

(e) $f(x) = \begin{cases} 0 \text{ if } x \text{ is irrational} \\ 1 \text{ if } x^2 \text{ is rational} \end{cases}$

2. Describe the understood domain and range of each of the following common functions from calculus.

(a) $f(x) = \sin x$ (c) $h(x) = x^3 + 5$

(b) $g(x) = \cos^{-1} x$ (d) $r(x) = \dfrac{1}{(x - 3)(x - 2)}$

3. Sketch a graph for each of the functions given in Problem 2.

4. Consider each of the statements below:

True because functions

(a) If $x = y$, $e^x = e^y$

(b) If $x = y$, $\log x = \log y$

(c) If $x = y$, $\sin x = \sin y$

(d) If $x = y$, $x^3 - 3x^2 + 2x = y^3 - 3y^2 + 2y$

Why are they all true?

5. Which of the following sets of ordered pairs are functions? Why or why not?

   (a) $\{(1, 1), (2, 1), (1, 2), (2, 3)\}$

   (b) $\{(1, 2), (2, 2), (3, 2), (4, 2), (5, 2)\}$

   (c) $\{(x, y) \mid x, y \in R \text{ and } y^2 = x\}$

   (d) $\{(x, y) \mid x \text{ is a triangle in the plane and } y \text{ is its perimeter}\}$

   (e) $\{(x, y) \mid y \text{ is a triangle in the plane and } x \text{ is its perimeter}\}$

6. Take the greatest integer function defined over $[-2, 2]$ and describe it in terms of each of the six perspectives outlined in this section.

7. Let $A = \{a, b, c\}$ and $B = \{1, 2\}$. There are exactly eight different mappings $f: A \to B$. Draw transformation diagrams for each of these.

8. Let A and B be finite sets with $n(A) = k$ and $n(B) = m$. Prove that there are precisely m^k different mappings $f: A \to B$. (We are assuming dom $f = A$.)

9. Consider the function described by

$$y = \sqrt[5]{x^3 - 2x^2 + 5} \qquad \text{where } x \in R$$

Construct an input-output machine that analyzes the operation of this function into simpler components.

10. Let A and B be finite sets with $n(A) = k$ and $n(B) = m$. Prove that there are precisely $(m + 1)^k$ different functions g with $g \subseteq (A \times B)$. (We are *not* assuming dom $g = A$, but rather dom $g \subseteq A$.)

5.2   *Types of Mappings*

Let f be a mapping from set A to set B. We shall denote this by

$$f: A \to B$$

and assume $A = $ dom f and $B = $ image space of f. Therefore, range f is a subset of B. We know that any function is a relation (subset of $A \times B$) that is characterized by two important properties:

1. *Existence:* $\forall_{x \in A} \ \exists_{y \in B}$ such that $f(x) = y$

2. *Uniqueness:* $\forall_{x_1, x_2 \in A}$ (if $x_1 = x_2$, then $f(x_1) = f(x_2)$)

We also know there are many different perspectives for thinking about and representing functions and mappings. We shall draw freely from these perspectives as we see need for emphasizing various points.

In this section, we shall investigate mappings from one set A to another set B that meet additional conditions beyond existence and uniqueness. We know that the range of a function is a subset of the image space. What if a function "uses up" *all* of the image space in the correspondence? That is, what if

range f = image space of f

Should we distinguish that kind of function from others? Also, what if the converse of the uniqueness condition holds? That is, suppose whenever two images are equal ($f(x_1) = f(x_2)$), then it always is true that their preimages must be equal ($x_1 = x_2$). Should such functions be set apart and classified separately from others? These questions lead to the following classifications or "types" of mappings.

Onto Mapping (Surjection)

Let $f: A \rightarrow B$. The mapping f is said to be *onto*, denoted

$$f: A \xrightarrow{\text{onto}} B$$

if and only if, in addition to existence and uniqueness conditions, the following additional condition is met:

3. $\forall_{y \in B}\ \exists_{x \in A}$ such that $f(x) = y$

This condition says that *every element in the image space has a preimage*, or, equivalently,

range f = image space of f

or, equivalently, *the correspondence exhausts (uses up) all of the image space*. In a transformation diagram an *onto* mapping is pictured as shown in Figure 5.9.

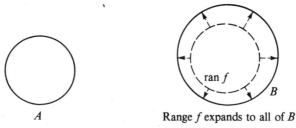

A

Range f expands to all of B

Figure 5.9

Such mappings are also called *surjective mappings*, or *surjections* in the mathematical literature (particularly algebra). In this course, we shall simply refer to any mapping that meets the additional condition (3) above as an *onto mapping*.

Example 1   Consider the mapping $f: I \rightarrow I$ defined by

$$f(n) = n + 2$$

Is f an onto mapping? If not, show why not; if so, prove it.

Solution   The question we should ask ourselves is: "Given any integer, is there an integer (domain element) that is associated with it under this function?"

Suppose we suspect the mapping is onto, and we try to prove it. Let k be *any* integer; can we produce an integer n such that

$$f(n) = k$$

that is,

$$n + 2 = k$$

Clearly, we choose $n = k - 2$. Since k is an integer, then $k - 2$ is also an integer. Thus,

$$f(k - 2) = (k - 2) + 2 = k$$

and the entire set of integers is used up as the range for this mapping. Therefore, f *is* onto.

Example 2   Modify the function f of Example 1 to

$$f(n) = 3n + 2$$

where $f = I \rightarrow I$, and answer the same question.

If we start by suspecting f is onto, we ask ourselves: "Given any $k \in I$, is there an $n \in I$ such that $f(n) = k$?" That is, $3n + 2 = k$.

If there were such an n, it would have to satisfy the condition

$$n = \frac{k - 2}{3}$$

So, our question really boils down to asking, "If k is an integer, is $(k - 2)/3$ always an integer?"

A bit of reflection shows the answer is no. For example, take $k = 3$. Then

$$\frac{k - 2}{3} = \frac{1}{3}$$

not an integer. In other words, the integer 3 has no preimage under the function defined by

$$f(n) = 3n + 2$$

Therefore, f is *not* onto. It can be shown that f is a function, however, by checking the existence and uniqueness conditions.

1. *Existence:* Let $n \in I$. Is $3n + 2 \in I$? Yes.

2. *Uniqueness:* Let $n_1 = n_2$. Is $3n_1 + 2 = 3n_2 + 2$? Yes, again.

If we take several special cases of (domain element, range element) pairs to "get a feeling" for this function, we see why it is *not* onto.

$$(0,2),(1,5),(2,8),(3,11),\dots$$
$$(-3,-7),(-2,-4),(-1,-1),\dots$$

The range components are always three units apart (this is graphically evident since the equation has the form that corresponds to a set of lattice points along a line). Therefore, not all integers are used in the range.

One-to-One Mapping (Injection)

Let $f: A \to B$. The mapping f is said to be *one-to-one*, denoted

$$f: A \xrightarrow{\text{1-1}} B$$

if and only if, in addition to existence and uniqueness conditions, the following additional condition is met:

4. $\forall_{x_1, x_2 \in A}$ (if $f(x_1) = f(x_2)$, then $x_1 = x_2$)

This condition says that *whenever two images are equal, their preimages are equal,* or equivalently,

Every element in the range has a unique preimage in the domain.

Equivalently, if any two elements in the domain are distinct, their correspondents in the range must also be distinct. That is,

$$\forall_{x_1, x_2 \in \text{dom} f} \left(\text{if } x_1 \neq x_2, \text{ then } f(x_1) \neq f(x_2) \right)$$

If the two sets A and B are finite, a one-to-one mapping f is indicated by every arrow pointing to a member of the range coming from only one element in the domain. For example, see Figure 5.10. Note that the domain of f is all of

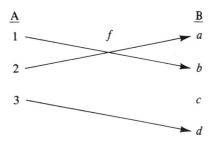

Figure 5.10

$A = \{1, 2, 3\}$; the range of f is $\{a, b, d\}$. Since c has no preimage, the function is not onto. However, it *is* one-to-one since we see that any two distinct elements in the domain have distinct images in the range; we never have two different preimages matched with the same image.

One-to-one functions are also called *injective mappings*, or *injections* in the literature. Moreover, it is interesting to note that the one-to-one condition corresponds graphically to a "horizontal line test." If for every range element, a horizontal line intersects the graph of the function in exactly one point, then the function is one-to-one; otherwise not.

Example 3   Prove that the function $f: I \rightarrow I$ defined by

$$f(n) = 3n + 2$$

is a one-to-one function.

Proof   We shall assume that f is indeed a function (the existence and uniqueness conditions are met—see Example 2). We must show:

$$\forall_{n_1, n_2 \in I} \left(\text{if } f(n_1) = f(n_2), \text{ then } n_1 = n_2 \right)$$

So, assume $f(n_1) = f(n_2)$, where n_1 and n_2 are any arbitrary integers. That is, $3n_1 + 2 = 3n_2 + 2$. Then

$$3n_1 = 3n_2 \quad \text{(subtract 2)}$$

and

$$n_1 = n_2 \quad \text{(divide by 3)}$$

Thus, we have proved that f is one-to-one. (Technically, conditional proof allows us our desired conclusion.)

Example 4   Is the function $f: R \rightarrow R$ defined by

$$f(x) = 3x^2 - 1$$

a one-to-one function? Explain why or why not.

We shall attack this question in two ways—from the definition, and by sketching a graph of the function and applying the "horizontal line test."

(a) (by definition)

If f is one-to-one, the following statement must be true:

$$\forall_{x_1, x_2 \in R} \left(\text{if } f(x_1) = f(x_2), \text{ then } x_1 = x_2 \right)$$

So, assume $f(x_1) = f(x_2)$, where x_1 and x_2 are any arbitrary real numbers. That is,

$3x_1{}^2 - 1 = 3x_2{}^2 - 1$. Then

$$3x_1{}^2 = 3x_2{}^2$$

and

$$x_1{}^2 = x_2{}^2$$

Can we conclude from this last step that $x_1 = x_2$? No, since we could have $x_1 = 2$ and $x_2 = -2$. Thus, we have a counterexample to the assertion that f is one-to-one. Note the negation to the one-to-one condition.

$$\exists_{x_1,\, x_2 \in R} \text{ such that } f(x_1) = f(x_2) \text{ and } x_1 \neq x_2$$

Therefore, f is *not* one-to-one.

(b)   (by horizontal line test)

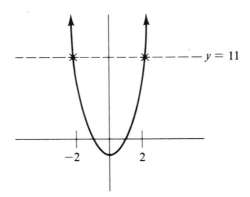

Sketch a graph of $f(x) = 3x^2 - 1$.

According to the horizontal line test, there are lots (all except -1) of range elements for which two distinct preimages exist. In particular, the diagram shows that

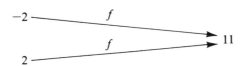

Therefore, f is not one-to-one.

One-to-One and Onto (One-to-One Correspondence, or Bijection)

> Let $f: A \to B$. If f is a mapping that is *both* one-to-one and onto, we denote f by
>
> $$f: A \xrightarrow[\text{onto}]{1\text{-}1} B$$
>
> and we say f is a *one-to-one correspondence* between A and B.

A one-to-one correspondence between two sets is the strongest possible mapping connecting the two sets (without considering any operational structure that might be involved). To prove a relation f from one set A to another set B is a one-to-one correspondence, we need to prove *four* things. First, f must be a function; that is, show existence and uniqueness. Next, f must have the two additional properties of being onto and one-to-one. The four properties are stated below. The reader should note carefully the similarities and differences between these four properties. It should be apparent that a one-to-one correspondence f: $A \to B$ is really a function "both ways," from A to B *and* from B to A. This is the basis of inverse functions, to be discussed later.

Let $f: A \xrightarrow[\text{onto}]{1\text{-}1} B$. Then

1. *Existence:* $\forall_{x \in A} \, \exists_{y \in B}$ such that $f(x) = y$.
   (Every preimage has an image.)

2. *Uniqueness:* $\forall_{x_1, x_2 \in A}$ (if $x_1 = x_2$, then $f(x_1) = f(x_2)$).
   (If any two preimages are equal, then their images must be equal.)

3. *Onto:* $\forall_{y \in B} \, \exists_{x \in A}$ such that $f(x) = y$.
   (Every image has a preimage.)

4. *One-to-one:* $\forall_{x_1, x_2 \in A}$ (if $f(x_1) = f(x_2)$, then $x_1 = x_2$).
   (If any two images are equal, then their preimages must be equal.)

Example 5   Prove that the mapping $f: R^+ \to R^+$ defined by

$$f(x) = \frac{3}{2x}$$

is a one-to-one correspondence.

Proof   We are given that f is a mapping (existence and uniqueness properties hold). Therefore, we need to show that f is onto and one-to-one.

(a) (*onto*)
We must show: $\forall_{y \in R^+} \exists_{x \in R^+}$ such that $f(x) = y$. We begin by selecting any element y in the positive real numbers and letting x represent the desired preimage. Then we try to establish what x would have to be in terms of y. Let $y \in R^+$, and set $f(x) = y$. That is,

$$\frac{3}{2x} = y$$

Now, is there such an x in R^+? Solve for x:
First,

$$\frac{2x}{3} = \frac{1}{y}$$

Then

$$x = \frac{3}{2y}$$

Since $y \in R^+$,

$$\frac{3}{2y} \in R^+$$

Hence

$$\frac{3}{2y}$$

is the x we want.

(b) (*One-to-one*)
We must show: $\forall_{x_1, x_2 \in R^+}$ (if $f(x_1) = f(x_2)$, then $x_1 = x_2$). Assume $f(x_1) = f(x_2)$, where x_1 and x_2 are any positive real numbers. That is,

$$\frac{3}{2x_1} = \frac{3}{2x_2}$$

Therefore,

$$2x_1 = 2x_2$$

and

$$x_1 = x_2$$

We have shown, by conditional proof, that equality of the function images implies equality of the preimages. Hence, f is one-to-one. Thus, f maps $R^+ \to R^+$ both onto and one-to-one. Hence, f is a one-to-one correspondence. Note, if we had drawn the graph of f below, that it passes both the vertical and horizontal line tests.

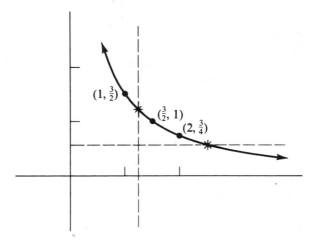

There are two other kinds of one-to-one correspondences that play such important roles in mathematics that they deserve special mention here. These are *permutations* and certain *sequences*.

Permutations

A *permutation* of a set A is a one-to-one, onto mapping of A onto itself. That is, a permutation is a function

$$f: A \xrightarrow[\text{onto}]{1\text{-}1} A$$

Permutations of finite sets are particularly interesting, since they are simply a "shuffling around" of the members in the set.

For example, if $A = \{1, 2, 3\}$, then the two functions illustrated in Figure 5.11 are permutations of A.

There are six different permutations of the set A above, including the "identity" permutation, which maps every element to itself. (Why?)

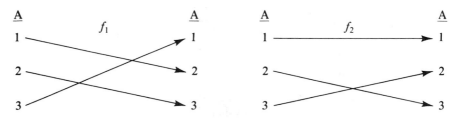

Figure 5.11

One of the classical notations for permutations of finite sets is to use a set of parentheses with the domain elements appearing along the top row and their respective correspondents appearing directly below each in the second row. The two permutations above could be thus represented by the symbolisms

$$f_1 = \begin{pmatrix} 1 & 2 & 3 \\ 2 & 3 & 1 \end{pmatrix} \quad \text{and} \quad f_2 = \begin{pmatrix} 1 & 2 & 3 \\ 1 & 3 & 2 \end{pmatrix}$$

Sequences

Some (not all) infinite *sequences* of real numbers are one-to-one, onto mappings of the set of natural (or sometimes whole) numbers onto some subset of the real numbers. In the schematic below, the first row is the set of natural numbers in horizontal array, and the second row is the classical harmonic sequence. Note that the correspondence

$$n \leftrightarrow \frac{1}{n}$$

is indeed a one-to-one correspondence.

n:	1	2	3	4	5	...	n	...
H_n:	1	$\frac{1}{2}$	$\frac{1}{3}$	$\frac{1}{4}$	$\frac{1}{5}$	...	$\frac{1}{n}$	...

The standard notation for infinite sequence is

$$\{a_n\}_{n=1}^{\infty}$$

which represents $a_1, a_2, a_3, \ldots a_n, \ldots$.

Each "term" in the above sequence is associated with exactly one natural number, and vice-versa. There are, of course, sequences like $1, -1, 1, -1, 1, -1, \ldots$. Therefore, not every sequence is a one-to-one correspondence, but all sequences are functions.

Assigned Problems

1. For a *finite* set S, a mapping $f\colon S \to S$ that is one-to-one must also be onto, and a mapping $g\colon S \to S$ that is onto must also be one-to-one. This is not the case for *infinite* sets.

   (a) Define a function $f\colon N \to N$ that is one-to-one but not onto. ($N = \{1, 2, 3, 4, \ldots\}$).

   (b) Define a function $g\colon N \to N$ that is onto but not one-to-one.

2. Consider each of the following mappings $f\colon R \to R$. Decide whether each is one-to-one, onto, neither, or both. Domains and image spaces are all real numbers in each case.

(a) $f(x) = x$ (identity function)

(b) $f(x) = 5$ (constant function)

(c) $f(x) = 7x + 5$ (a linear function)

(d) $f(x) = \cos x$

(e) $f(x) = \dfrac{3}{1 + x^2}$

(f) $f(x) = \tan^{-1} x$

3. Other than the functions given in Problems 1 and 2, give examples of mappings $f: R \to R$ that are:

(a) one-to-one but not onto

(b) onto but not one-to-one

(c) neither one-to-one nor onto

(d) both one-to-one and onto

4. Prove that the exponential function exp: $R \to R^+$ defined by

$$\forall_{x \in R} \; \exp(x) = e^x$$

is a one-to-one correspondence.

5. If $n(A) = k$, $(k \in I^+)$, prove that the number of permutations of A is $k!$

6. Consider the $R \to R$ functions below and answer the pertinent questions.

(a) $f(x) = \sqrt{x - 3}$

f is *not* a mapping $R \xrightarrow{\text{onto}} R$. Why?

(b) $g(x) = |x| + 2$

g is *not* a one-to-one function. Why?

(c) $h(x) = x^{2/3}$

h is *not* a one-to-one correspondence from $R^+ \to R$. Why?

7. Consider the function f that maps every integer to its remainder upon division by 7. That is, by the division algorithm, if n is any integer, there exist unique integers q and r such that

$$n = 7q + r \qquad \text{where } 0 \le r < 7$$

Then we are considering $f(n) = r$ where r is related to n as above. Is this mapping a one-to-one correspondence from I to I? Show why or why not.

8. Consider a collection of sets $\{M_i\}$ indexed by the natural numbers N. Can indexing of sets in this way be regarded as a one-to-one correspondence? Discuss why or why not.

9. Let f: {points in a plane} $\rightarrow R$ (real numbers) be defined by

$$\forall_p \, [f(p) = \text{the distance from } p \text{ to } q]$$

where p is any point in the plane and q is some fixed point in the same plane. (If $p = (x, y)$ and $q = (0, 0)$, $f(p) = \sqrt{x^2 + y^2}$.)

(a) Prove f is a function.

(b) Is f onto? Explain.

(c) Is f one-to-one? Explain.

10. Suppose X is a set and f is a function with domain X.

(a) Define a relation R on X by:

$$\forall_{x, y \in X} \, (xRy \leftrightarrow f(x) = f(y))$$

Prove R is an equivalence relation on X.

(b) Let E be an equivalence relation on a set X. Consider the induced partition $X/E = \{\bar{x} | x \in X\}$ where

$$\bar{x} = \{y | y \in X \text{ and } xEy\}$$

Define a function p: $X \rightarrow X/E$ by

$$\forall_{x \in X} \, (p(x) = \bar{x})$$

Prove that p is a function and p is onto. This function is referred to in the mathematical literature as the *canonical surjection*.

5.3   Composition of Functions

Throughout nature and human technology, many actions occur as sequences of events, one contingent upon another. The sun rises in the morning, its rays strike a solar collector, the water inside is heated, circulated throughout the house, and cool water replaces the solar heated water for the cycle to repeat itself. Most machinery is composed of an organization of interrelated, multiply dependent parts operating simultaneously to perform some function. In a symphony, each section and musician play dependent roles to produce a synthesis of harmony. Human endeavor is replete with situations involving complex networks and chains of dependent events.

In mathematics, one reflection of dependency is the function. If a variable y depends on u, under certain conditions we have a functional relationship. This relationship reflects dependency between y and u because for any value selected for u, there is one and only one value that y takes. The variable y is called the *dependent* variable, and u is called the *independent* variable. In this manner, we can use a set of functions to represent a multiple contingency, or chain of events, each dependent on another. Suppose u is an independent variable with respect to y, but u *depends* on some other variable x in a functional relationship. Then y depends on x as well as u. So we have

$$
\begin{array}{ll}
y \text{ depends on } u & [y = f(u)] \\
u \text{ depends on } x & [u = g(x)] \\
\textit{Therefore}, \ y \text{ depends on } x & [y = f(g(x))]
\end{array}
$$

This kind of multiple function dependency is very important in mathematics, because we are frequently interested in breaking down complex functions into simpler "parts," and synthesizing simpler functions to represent multiple dependencies. This latter activity is called *composition of functions*, and the result of putting two or more functions together to act one after another is called a *composite function*. We shall prove that it is indeed a function if the component "parts" are functions and their domains and ranges "fit" together properly, but before doing so let us take a closer look at the operation of composing functions from our various function perspectives. We begin with the formal definition.

Composition of Functions

Let $f\colon B \to C$ and $g\colon A \to B$, where dom $f \subseteq B$, dom $g = A$, and ran $g \subseteq$ dom f. Then

$$\forall_{x \in \text{dom } g} \ (f \circ g)(x) = f(g(x))$$

That is to say, as sets of ordered pairs,

$$f \circ g = \{(x, z) | \exists_{y \in B} \text{ such that } (x, y) \in g \text{ and } (y, z) \in f\}$$

The function $f \circ g$ is called the *composite of f with g*.

In terms of dependency, this latter definition says that the composite function of f with g is the set of all ordered pairs (x, z), or "x-z contingencies" such that there is an "intermediate variable y in B" where x and y are dependent via g, and y and z are dependent via f.

Example 1   Suppose $A = \{1, 2, 3, 4\}$
$\qquad\qquad B = \{a, b, c\}$
$\qquad\qquad C = \{\triangle, \square\}$

Also $\qquad f = \{(a, \triangle), (b, \square), (c, \square)\}$
$\qquad\qquad g = \{(1, a), (2, a), (3, b), (4, b)\}$
Then, $f \circ g = \{(1, \triangle), (2, \triangle), (3, \square), (4, \square)\}$

Several other function perspectives applied to composite functions are given in the following examples. Note in each case that the notation we have adopted, namely

$f \circ g$

is actually in the reverse order with respect to the order in which the functions act. That is, g "acts" first, taking an element x to an image $g(x)$. Next, f "acts" on $g(x)$, taking it to an image $f(g(x))$. Therefore, while we read left to right, the action of the functions is right to left.

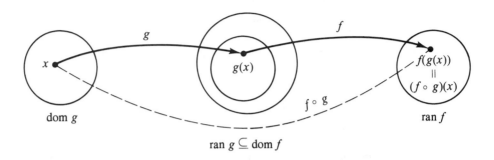

ran $g \subseteq$ *dom* f

Example 2   Transformation Diagram of Function Composition

In the composition $(f \circ g)(x)$, x is transformed under g to $g(x)$, which in turn is transformed under f to $f(g(x))$. It is essential that ran g be a subset of dom f, otherwise f could not "act" on $g(x)$ for some $x \in$ dom g.

Using $R \to R$ functions from calculus, suppose

$$f(x) = \ln x \qquad \text{(natural logarithm function, base } e\text{)}$$

$$g(x) = x^2 + 2$$

Note that dom $f = (0, +\infty)$ ran $f = (-\infty, +\infty)$
         ran $g = [2, +\infty)$ dom $g = (-\infty, +\infty)$

The diagram below illustrates the transformation $f \circ g$.

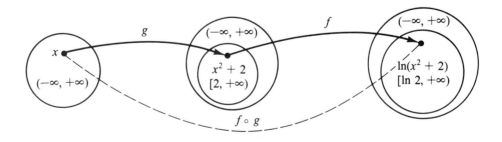

Thus, $\forall_{x \in (-\infty, +\infty)} \ (f \circ g)(x) = \ln(x^2 + 2)$

Example 3   Graphical Perspective of Composition

Suppose f and g are $R \rightarrow R$ functions, not algebraically specified, but having graphs that look like those below. Suppose we note where endpoints go under each function.

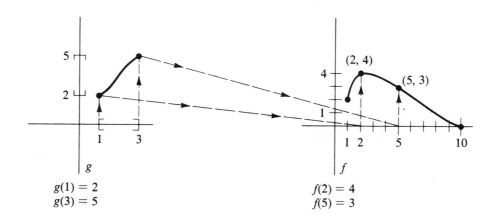

$g(1) = 2$
$g(3) = 5$

$f(2) = 4$
$f(5) = 3$

Then

$$(f \circ g)(1) = f(g(1)) = f(2) = 4$$
$$(f \circ g)(3) = f(g(3)) = f(5) = 3$$

Schematically,

$$1 \xrightarrow{g} 2 \xrightarrow{f} 4$$

and

$$3 \xrightarrow{g} 5 \xrightarrow{f} 3$$

Example 4   Rule of Correspondence View of Composition

Let $f(x) = x^3 + 2x$
  $g(x) = 5x - 3$

Then f and g can be regarded as rules:

For f: 1. Take any real number x.
      2. Cube it. (x^3)
      3. Multiply it by 2. $(2x)$
      4. Add the results of steps 2 and 3. $(x^3 + 2x)$

For g: 1. Take any real number x.
      2. Multiply it by 5. $(5x)$
      3. Subtract 3. $(5x - 3)$

Now $f \circ g$ means g acts first, followed by f.

So $f \circ g$:

1. Take any real number x.
2. Multiply it by 5. (obtain $5x$)
3. Subtract 3. $(5x - 3)$
4. Cube the result of step 3. $(5x - 3)^3$
5. Multiply result of step 3 by 2. $[2(5x - 3)]$
6. Add results of steps 4 and 5. $(5x - 3)^3 + 2(5x - 3)$

Thus, we have $(f \circ g)(x) = (5x - 3)^3 + 2(5x - 3)$

Example 5   Composition of Functions as a Machine

We can view a function as an input-output machine. If we "link-up" two or more functions in a composition, we can model the composition as another input-output machine.

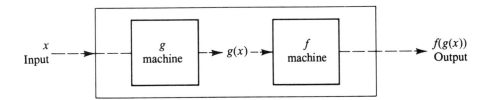

In this model, x is input into g, output as $g(x)$; $g(x)$, in turn, is input into f, output as $f(g(x))$.

An alternate input-output model might be

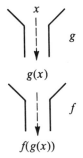

The foregoing examples illustrate many different ways of visualizing and thinking about the operation of composition of functions. All of these models, as well as the formal definition of composition, extend to situations involving more than two functions.

If $f_1, f_2, f_3, \ldots f_n$ are functions for which

$$\text{ran } f_i \subseteq \text{dom } f_{i+1} \qquad \forall_{i=1,2,\ldots n-1}$$

then

$$(f_1 \circ f_2 \circ \cdots \circ f_n)(x) = f_1(f_2(\cdots f_{n-1}(f_n(x)) \cdots))$$

Note that using this notation, the functions are subscripted 1 through n reading left to right, but their order of action is f_n first, f_{n-1} second, $\ldots f_1$ last.

An interesting application of function composition is *permutation multiplication*. Recall that a *permutation* is a one-to-one correspondence mapping a set A onto itself. Also, if A is finite, we have a special notation for permutations on A, in which the elements of A are listed in one row and their images directly below respective elements, all contained in a set of parentheses. For example, let $A = \{1, 2, 3, 4\}$ and $f, g: A \to A$ be permutations of A, and suppose

$$f = \begin{pmatrix} 1 & 2 & 3 & 4 \\ 4 & 2 & 1 & 3 \end{pmatrix} \qquad g = \begin{pmatrix} 1 & 2 & 3 & 4 \\ 3 & 1 & 4 & 2 \end{pmatrix}$$

This says
$$\begin{array}{ll} f(1) = 4 & g(1) = 3 \\ f(2) = 2 & g(2) = 1 \\ f(3) = 1 & g(3) = 4 \\ f(4) = 3 & g(4) = 2 \end{array}$$

We define *permutation multiplication* as ordinary function composition except the permutation that appears first acts first, and so on, the order of action being the same as the order of appearance (left to right). That is,

$$\begin{pmatrix} 1 & 2 & 3 & 4 \\ 4 & 2 & 1 & 3 \end{pmatrix} \cdot \begin{pmatrix} 1 & 2 & 3 & 4 \\ 3 & 1 & 4 & 2 \end{pmatrix} \qquad \text{means} \quad g \circ f$$

In other words, the permutation f appears first on the left, so it acts first, followed by the permutation g. Under their composed action, we see that "1 goes to 4, which goes to 2," "2 goes to 2, which goes to 1," and so on. Therefore,

$$\begin{pmatrix} 1 & 2 & 3 & 4 \\ 4 & 2 & 1 & 3 \end{pmatrix} \cdot \begin{pmatrix} 1 & 2 & 3 & 4 \\ 3 & 1 & 4 & 2 \end{pmatrix} = \begin{pmatrix} 1 & 2 & 3 & 4 \\ 2 & 1 & 3 & 4 \end{pmatrix}$$

If we reverse the order of the permutations, we obtain $f \circ g$, or

$$\begin{pmatrix} 1 & 2 & 3 & 4 \\ 3 & 1 & 4 & 2 \end{pmatrix} \cdot \begin{pmatrix} 1 & 2 & 3 & 4 \\ 4 & 2 & 1 & 3 \end{pmatrix} = \begin{pmatrix} 1 & 2 & 3 & 4 \\ 1 & 4 & 3 & 2 \end{pmatrix}$$

It should be clear from this observation that *function composition in general* (*permutation multiplication in particular*) is not *a commutative operation*. That is, it is generally true that for most functions f, g

$$f \circ g \neq g \circ f$$

There are some important properties of function composition worth remembering. We start with the most basic, that the composition of two functions is a function itself, providing the range of the first to act is a subset of the domain of the second to act.

Theorem   (Composition of Functions Is a Function)

Let f and g be functions such that

$$f: B \to C$$
$$g: A \to B$$

and (range $g \subseteq$ domain f). Then

(1) $f \circ g$ is a function

(2) $\mathrm{dom}(f \circ g) = \{x \mid x \in \mathrm{dom}\ g \text{ and } g(x) \in \mathrm{dom}\ f\}$

(3) $\mathrm{ran}(f \circ g) = \{f(y) \mid y \in \mathrm{range}\ g\}$

We shall prove (1) here and leave (2) and (3) to the reader.

Proof of (1):

We must show that $f \circ g$ meets the existence and uniqueness conditions for a function. We assume that f and g are individually functions.

(a) *Existence:* We must show $\forall_{x \in A}\ \exists_{z \in C}$ such that

$$(f \circ g)(x) = z$$

Let x be any element of A. Since g is a function, there exists an element $y \in B$ such that $g(x) = y$. Now by hypothesis, since ran g is a subset of dom f, $g(x)$ is an element in dom f. Therefore, since f is a function, there exists an element z in C such that $f(y) = z$. Thus, given any element x in A, we have shown the existence of element z in C such that

$$(f \circ g)(x) = z$$

(b) *Uniqueness:* We must show $\forall_{x_1, x_2 \in A}$

$$\left(\text{if } x_1 = x_2, \text{ then } (f \circ g)(x_1) = (f \circ g)(x_2)\right)$$

Assume $x_1 = x_2$, where x_1 and x_2 are any arbitrary elements in A. Then,

$$g(x_1) = g(x_2)$$

since g is a function and

$$f(g(x_1)) = f(g(x_2))$$

since f is a function. Therefore,

$$(f \circ g)(x_1) = (f \circ g)(x_2) \qquad \text{(definition of composition)}$$

Thus, we have proved *if* f and g are functions and ran $g \subseteq$ dom f, *then* $f \circ g$ is a function.

We noted earlier that function composition is generally *not commutative*. It is, however, an *associative* operation. This is the content of the next theorem.

Theorem   *(Function Composition Is Associative)*

Let f, g, and h be any three functions for which the compositions $(f \circ g) \circ h$ and $f \circ (g \circ h)$ make sense. Then

$$(f \circ g) \circ h = f \circ (g \circ h)$$

Proof: We are required to show that two *functions* are equal. By the definition of function equality, this means we must show two things:

(1) $\mathrm{dom}[(f \circ g) \circ h] = \mathrm{dom}[f \circ (g \circ h)]$

(2) $\forall_{x \in \mathrm{dom}(f \circ g \circ h)}\ ([(f \circ g) \circ h](x) = [f \circ (g \circ h)](x))$

We note that $x \in \mathrm{dom}[(f \circ g) \circ h]$ if and only if $x \in \mathrm{dom}\ h$, $h(x) \in \mathrm{dom}\ g$, and $g(h(x)) \in \mathrm{dom}\ f$. Exactly the same condition holds for the $\mathrm{dom}[f \circ (g \circ h)]$, so we conclude (1) is true.

Next, let $x \in \mathrm{dom}[(f \circ g) \circ h]$, where x is any arbitrary element therein.

$$
\begin{aligned}
[(f \circ g) \circ h](x) &= (f \circ g)(h(x)) &&\text{(definition composition—}f \circ g \text{ and } h) \\
&= f[g(h(x))] &&\text{(definition composition—}f \text{ and } g) \\
&= f[(g \circ h)(x)] &&\text{(definition composition—}g \text{ and } h) \\
&= [f \circ (g \circ h)](x) &&\text{(definition composition—}f \text{ and } g \circ h)
\end{aligned}
$$

Therefore, function composition *is* associative.

Because of this theorem, we can write compositions of three or more functions without parentheses, since it makes no difference which way the functions in a composition are grouped as long as the order remains invariant.

In the next theorem we investigate compositions of functions having special properties one-to-one or onto.

Theorem   *(Composition of One-to-One and Onto Functions)*

(1) Any meaningful composition of two one-to-one functions is a one-to-one function.

(2) Any meaningful composition of two onto functions is an onto function.

(3) Any meaningful composition of two one-to-one correspondences is a one-to-one correspondence.

Proof of (1):

Suppose f and g are one-to-one functions, where

$$f: B \xrightarrow{\ 1\text{-}1\ } C$$

$$g: A \xrightarrow{\ 1\text{-}1\ } B$$

We claim $(f \circ g)\colon A \to C$ is also one-to-one. To show this, we must prove (using the definition of one-to-one)

$$\forall_{x_1, x_2 \in \text{dom } g} \ \left(\text{if } (f \circ g)(x_1) = (f \circ g)(x_2), \text{ then } x_1 = x_2\right)$$

Assume

$$(f \circ g)(x_1) = (f \circ g)(x_2)$$

for any arbitrary x_1, x_2 in dom g. Then,

$$f(g(x_1)) = f(g(x_2)) \qquad \text{(definition of composition—} f \text{ and } g\text{)}$$

and so,

$$g(x_1) = g(x_2) \qquad \text{(} f \text{ is one-to-one)}$$

and finally,

$$x_1 = x_2 \qquad \text{(} g \text{ is one-to-one)}$$

Therefore, $f \circ g$ is one-to-one, if both f and g are one-to-one.

Proof of (2):

Suppose f and g are onto functions, where

$$f\colon B \xrightarrow{\text{onto}} C$$

$$g\colon A \xrightarrow{\text{onto}} B$$

We claim $(f \circ g)\colon A \to C$ is also an onto function. To show this, we must prove (using the definition of onto), for any element z in C, there exists a preimage x in A such that

$$(f \circ g)(x) = z$$

Let z be any element of C. Then,

$$\exists_{y \in B} \text{ such that } f(y) = z \quad \text{(} f \text{ is onto)}$$

and

$$\exists_{x \in A} \text{ such that } g(x) = y \quad \text{(} g \text{ is onto)}$$

Therefore,

$$\exists_{x \in A} \text{ such that } f(g(x)) = z$$

or

$$(f \circ g)(x) = z \quad \text{(definition of composition—} f \text{ and } g\text{)}$$

Therefore, $f \circ g$ is onto if both f and g are onto.

Proof of (3):

This is a direct consequence of parts (1) and (2).

The previous theorem tells us what we can conclude about a composition if we know the two functions involved are both one-to-one or both onto. Suppose we turn the question around. What if we know the composition $f \circ g$ is a one-to-one function, or the composition $f \circ g$ is onto, or both? Can we conclude anything about the individual functions involved? The next and final theorem of this section says we can.

Theorem   (One-to-One and Onto Compositions of Functions)

Let $f: B \to C$ and $g: A \to B$, and consider $f \circ g: A \to C$.

(1) If $f \circ g$ is one-to-one, then g is one-to-one.

(2) If $f \circ g$ is onto, then f is onto.

(3) If $f \circ g$ is one-to-one and onto, then f is onto and g is one-to-one.

Proof of (1):

Assume $f \circ g$ is one-to-one. This says

$$\forall_{x_1, x_2 \in A} \left(\text{if } (f \circ g)(x_1) = (f \circ g)(x_2), \text{ then } x_1 = x_2 \right)$$

That is, if $f(g(x_1)) = f(g(x_2))$, then $x_1 = x_2$. Now we wish to prove that g is one-to-one. That is,

$$\text{if } g(x_1) = g(x_2), \text{ then } x_1 = x_2$$

So suppose we start with the assumption that

$$g(x_1) = g(x_2) \qquad (\text{where } x_1 \text{ and } x_2 \text{ are any elements in } A)$$

Then

$$f(g(x_1)) = f(g(x_2)) \qquad (\text{uniqueness for } f)$$

Therefore,

$$(f \circ g)(x_1) = (f \circ g)(x_2) \qquad (\text{definition of composition—} f \text{ and } g)$$

and we have

$$x_1 = x_2 \qquad (\text{since } f \circ g \text{ is one-to-one})$$

Hence, if $f \circ g$ is one-to-one, then g must be one-to-one (by conditional proof).

Proof of (2):

Assume $f \circ g$ is onto. This says:

$$\forall_{z \in C} \exists_{x \in A} \text{ such that } (f \circ g)(x) = z$$

Now we wish to prove that f must be onto; that is,

$$\forall_{z \in C} \exists_{y \in B} \text{ such that } f(y) = z$$

To do this, let z be any element of C. By the hypothesis that $f \circ g$ is onto, there is an element x in A such that

$$(f \circ g)(x) = z$$

Thus,

$$f(g(x)) = z \qquad (\text{definition of composition—} f \text{ and } g)$$

But $g(x)$ is an element in B. Therefore, $g(x)$ is the y we want. Therefore, we have shown that for any element z in C, there exists an element y in B (namely, $y = g(x)$) such that $(f \circ g)(x) = f(g(x)) = f(y) = z$. Therefore, f is onto. We conclude that if $f \circ g$ is onto, then f is onto (conditional proof).

Proof of (3):

This is an immediate consequence of parts (1) and (2).

Assigned
Problems

1. Prove, by indirect strategy, that if $f: B \rightarrow A$ and $g: A \rightarrow B$, and $g \circ f$ is the identity function on B [i.e., $(g \circ f)(x) = x, \forall_{x \in B}$], then f is one-to-one and g is onto.

2. Consider the finite set $A = \{1, 2, 3\}$. Let S_3 denote the set of all permutations on A. There are six such permutations, and they are defined below.

$$f_1 = \begin{pmatrix} 1 & 2 & 3 \\ 1 & 2 & 3 \end{pmatrix} \quad \text{(identity permutation)}$$

$$f_2 = \begin{pmatrix} 1 & 2 & 3 \\ 2 & 1 & 3 \end{pmatrix}$$

$$f_3 = \begin{pmatrix} 1 & 2 & 3 \\ 3 & 2 & 1 \end{pmatrix}$$

$$f_4 = \begin{pmatrix} 1 & 2 & 3 \\ 1 & 3 & 2 \end{pmatrix}$$

$$f_5 = \begin{pmatrix} 1 & 2 & 3 \\ 2 & 3 & 1 \end{pmatrix}$$

$$f_6 = \begin{pmatrix} 1 & 2 & 3 \\ 3 & 1 & 2 \end{pmatrix}$$

Construct a "multiplication table" for S_3 as shown below where each entry is the result of a composition of two permutation functions (product of two permutations).

$\cdot$	f_1	f_2	f_3	f_4	f_5	f_6
f_1						
f_2						
f_3						
f_4						
f_5						
f_6						

3. Suppose you have a cardboard equilateral triangle with vertices labeled 1, 2, and 3, respectively, and axes of symmetry (fixed) l_1, l_2, l_3, as shown below.

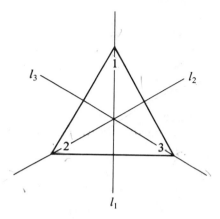

Now define a set of six transformations of this triangle, as follows.

R_0 rotation of 0°   (identity transformation)

R_1 rotation of 120° counterclockwise

R_2 rotation of 240° counterclockwise

F_1 flip over axis l_1

F_2 flip over axis l_2

F_3 flip over axis l_3

The composition of two such transformations is defined by finding a single transformation that has the same effect on the vertices as two acting together, one after the other. For example,

$F_1 \circ R_2$ means "flip over axis l_1, then rotate 240° counterclockwise." Under F_1, vertex 1 remains in place, and vertices 2 and 3 interchange; under R_2, (original vertices) 1 goes to 3, 2 to 1, and 3 to 2.

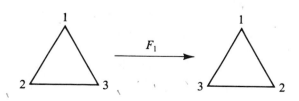

Next, applying R_2 *to the result* of F_1, we obtain

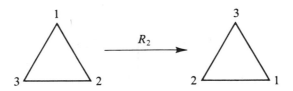

The net effect of applying F_1 first, followed by R_2 could have been accomplished by the single transformation F_2. Therefore,

$$F_1 \circ R_2 = F_2$$

(a) Compute at least five more compositions.

(b) Is there any relationship with permutations? Explain.

4. Let $A = \{1, 2, 3\}$. Suppose $f \colon A \to A$ is given by $f(1) = 3$, $f(2) = 1$, and $f(3) = 2$. There are 27 different mappings (*not* all permutations, of course) $g \colon A \to A$. Find all such mappings $g \colon A \to A$ such that

$$f \circ g = g \circ f$$

5. Give an example of a pair of functions $f \colon B \to C$ and $g \colon A \to B$ such that $f \circ g \colon A \to C$ is one-to-one, but f is not one-to-one.

6. Give an example of a pair of functions $f \colon B \to C$ and $g \colon A \to B$ such that $f \circ g \colon A \to C$ is onto, but g is not onto.

7. Give an example of a pair of functions $f \colon A \to A$ and $g \colon A \to A$ such that $f \circ g$ is a one-to-one correspondence, but neither f nor g is a one-to-one correspondence.

8. Consider the two functions

$$f(x) = \frac{x^2}{x^2 - 1} \qquad (\text{where } x \in R - \{-1, 1\})$$

$$g(x) = \begin{cases} 1 & \text{if } x \neq 2 \\ 0 & \text{if } x = 2 \end{cases}$$

It is easy to see that neither f nor g is one-to-one, that $\mathrm{dom}(f \circ g) = \{2\}$, that $\mathrm{ran}(f \circ g) = \{0\}$, and therefore $f \circ g$ is one-to-one. (Check these statements out.) Does this contradict the last theorem (1) in this section?

9. Prove that the composition of two linear functions is always a linear function. (A linear function is a function $f \colon R \to R$ defined by $f(x) = mx + b$ where $x \in R$ and m, b are real constants.)

10. The chain rule in calculus is a rule for differentiating compositions of functions. It states that if f and g are differentiable $R \to R$ functions (domains and ranges are subsets of the real numbers) and $f \circ g$ is a meaningful composition, then

$$\forall_{x \in \text{dom}(f \circ g)} \left[(f \circ g)'(x) = f'(g(x))(g'(x)) \right]$$

(a) Let $h(x) = \sin\left[\ln\left(\dfrac{1}{x} \right) \right]$.

Express h as a composition of individual functions and apply the chain rule to differentiate h.

(b) If $f(x) = \tan x$

$$g(x) = x^3 + 2x$$

$$h(x) = \sqrt{x}$$

Define the compositions $f \circ g \circ h$, $f \circ h \circ g$, and $g \circ f \circ h$ so that they are meaningful, and differentiate each.

5.4   Inverse of a Function

Every *function* is a *relation*, but not every *relation* is a *function*. As a relation, a function is a set of ordered pairs; but more than that, it must satisfy the existence and uniqueness conditions—every first component must be associated with one and only one second component. Suppose f is a function; then f is a relation, and its inverse f^{-1} is a relation. To form the f^{-1} relation, we simply reverse the order of components in each pair. That is,

If $f = \{(x, y) | x \in A$ and $y = f(x)\}$,
then $f^{-1} = \{(y, x) | (x, y) \in f\}$.

Thinking graphically with respect to $R \to R$ functions, it is this interchange of x and y components that produces symmetry in the graphs of f and f^{-1} with respect to the line $y = x$.

The central question we pose next is, "Under what conditions will the inverse (relation) of a function be a function itself?" Suppose

$$f: A \to B$$

where we assume dom $f = A$. Then f is a relation from A to B and f^{-1} is a relation from B to A. Now, if the existence condition holds for f^{-1}, this would mean *every* element in B should have an image (with respect to f^{-1}) in A. The only way this can happen is if every element in B has a preimage (with respect to f) in A. *This says f must be onto.* Another way of saying this is to note that all of B must be used up in the f relation so that when the ordered pairs are reversed to form the f^{-1} relation, every first component (element of B) has a correspondent (element of A). Additionally, if f^{-1} is to be a function, then f^{-1} must meet the uniqueness condition. What does this imply about f? Uniqueness for f^{-1} requires that every first component (element in B) have only one correspondent (element in A). The only way that can happen is if no two elements in A have the same correspondent (under f) in B. *This says f must be one-to-one.* Thus, if f is a function, then f^{-1} is a function if and only if f is one-to-one and onto. We formalize this discussion in the following theorem.

Theorem   (Existence of Inverse Function)

Let f be a function mapping $A \rightarrow B$ with dom $f = A$. The inverse of f, denoted f^{-1}, is a function if and only if f is both one-to-one and onto.

Proof: ($\rightarrow$) Assume f^{-1} is a function. By the definition of f^{-1} and using functional notation, we can relate f and f^{-1} by the equation

$$(f^{-1}(y) = x) \leftrightarrow (f(x) = y) \forall_{x \in A, y \in B}$$

We shall use this relationship in our subsequent argument. We wish to prove that f is one-to-one and onto.

(a) Claim: f is onto

We need to show $\forall_{y \in B} \exists_{x \in A}$ such that $f(x) = y$. Let y be any element in B. Then since f^{-1} is a function, by the existence condition on f^{-1}, there exists an image x in A. That is,

$$f^{-1}(y) = x$$

But this x is the element we want, since

$$f(x) = y$$

Therefore, f is onto.

(b) Claim: f is one-to-one

Here we must show

$$\forall_{x_1, x_2 \in A} \left(\text{if } f(x_1) = f(x_2), \text{ then } x_1 = x_2 \right)$$

So, we start by assuming $f(x_1) = f(x_2)$. Suppose we introduce notation

$$f(x_1) = y_1$$
$$f(x_2) = y_2$$

Now y_1 and y_2 are elements in B, and since f^{-1} is a function, by the uniqueness condition on f^{-1}, we have: if $y_1 = y_2$, then $f^{-1}(y_1) = f^{-1}(y_2)$. But this says $x_1 = x_2$, since

$$f^{-1}(y_1) = x_1$$
$$f^{-1}(y_2) = x_2$$

Therefore, f is one-to-one.

Thus, if f^{-1} is a function, *then f is both one-to-one and onto.*

($\leftarrow$)   Now assume f is one-to-one and onto. We prove f^{-1} is a function. To do this, we must prove that the existence and uniqueness conditions hold for f^{-1}.

(a) Existence:

We must show $\forall_{y \in B} \exists_{x \in A}$ such that $(y, x) \in f^{-1}$. (We use this set-theoretic notation since f^{-1} is a relation and we're trying to *prove* it is a function.) Now f is onto, which means

$$\forall_{y \in B} \exists_{x \in A} \text{ such that } (x, y) \in f$$

But by the definition of f^{-1}

$$(x, y) \in f \leftrightarrow (y, x) \in f^{-1}$$

Thus, we have

$$\forall_{y \in B} \exists_{x \in A} \text{ such that } (y, x) \in f^{-1}$$

Therefore, f^{-1} meets the existence condition.

(b) Uniqueness:

We must show $\forall_{y_1, y_2 \in B}$ if $(y_1, x_1), (y_2, x_2) \in f^{-1}$ and $y_1 = y_2$, then $x_1 = x_2$. So, suppose $y_1 = y_2$ and consider the ordered pairs

$$(y_1, x_1), (y_2, x_2) \in f^{-1}$$

This means the ordered pairs

$$(x_1, y_1), (x_2, y_2) \in f$$

by the definition of f. Now f is one-to-one, which means

$$\text{if } y_1 = y_2, \text{ then } x_1 = x_2$$

and we are assuming $y_1 = y_2$. Therefore, $x_1 = x_2$; hence f^{-1} meets the uniqueness condition.

Therefore, we have f^{-1} is a function. In part ($\leftarrow$) we have shown

if f is both one-to-one and onto, then f^{-1} is a function

Therefore, the theorem is proved. A necessary and sufficient condition for f^{-1} to be a *function* is that f be both one-to-one and onto. In the cases where f and f^{-1} are both functions, it is easy to see that *both* f and f^{-1} are one-to-one and onto functions. Also, f^{-1} fails to exist (as a function) if f is not one-to-one or f is not onto.

Example 1   Let $f: R \rightarrow R$ and $\forall_{x \in R}$ $f(x) = x^k$, where k is an odd positive integer. Show that f^{-1} exists (is a function).

To do this, we show f is both one-to-one and onto. First, f is *onto*, since taking any real number y in R, the real number $\sqrt[k]{y}$ is its preimage. (Note that the kth root exists for *any* real number since k is *odd*. If k were even and y negative, this would not be the case.) So,

$$f\left(\sqrt[k]{y}\right) = \left[\sqrt[k]{y}\right]^k = y$$

Also, f is *one-to-one*. To demonstrate this, suppose x_1 and x_2 are elements of R and $f(x_1) = f(x_2)$. This says,

$$x_1^k = x_2^k$$

Now take the kth root of both sides (it exists as a real number and is unique since k is *odd*), and we have

$$x_1 = x_2$$

Thus, f is one-to-one. Therefore, f is both onto and one-to-one, so by the theorem on existence of inverse functions, f^{-1} is a function. Indeed, f^{-1} is the kth root function. We have shown here that all power functions having odd positive integral exponents are *invertible* (i.e., their inverses are functions). We might also note the general graph characteristics for this class of functions.

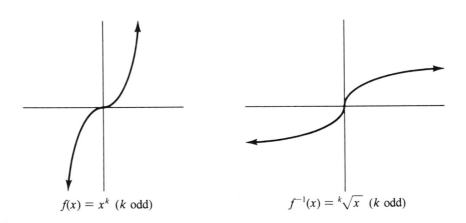

$$f(x) = x^k \ (k \text{ odd}) \qquad\qquad f^{-1}(x) = \sqrt[k]{x} \ (k \text{ odd})$$

Example 2   Does the trigonometric function $f(x) = \sin x$ have an inverse (that is a function)?

Discussion:

The domain of the sine function is all reals, its range is $[-1, 1]$. Since the sine function is periodic (it repeats its graph over intervals of length 2π), it is easy to see it is *not* one-to-one. For example,

$$\sin \frac{\pi}{2} = \sin \frac{5\pi}{2}, \quad \text{but} \quad \frac{\pi}{2} \neq \frac{5\pi}{2}$$

A glance at the graph verifies the failure of the one-to-one property via the horizontal line test.

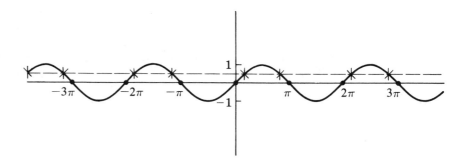

Therefore, $\sin^{-1}$ is *not* a function. However, in mathematics we talk about the "inverse sine function" or "Arcsin" function. This is really not the inverse of the sine function, but rather the inverse of the *sine restricted to* $[-\pi/2, \pi/2]$ function. In other words, we create a new function from the sine function (call it the Sin function) by restricting the domain to some interval (usually $[-\pi/2, \pi/2]$) where the restricted function Sin is one-to-one and onto. Thus,

$$\text{Sin: } \left[-\frac{\pi}{2}, \frac{\pi}{2} \right] \to [-1, 1]$$

and

$$\text{Sin}^{-1}: [-1, 1] \to \left[-\frac{\pi}{2}, \frac{\pi}{2} \right]$$

where

$$\left(\text{Sin}^{-1} y = x \right) \leftrightarrow \left[(y = \sin x) \text{ and } x \in \left[-\frac{\pi}{2}, \frac{\pi}{2} \right] \right]$$

Thus, Sin^{-1} exists as a function (also called Arcsin) and the graphs of Sin and Sin^{-1} are shown below.

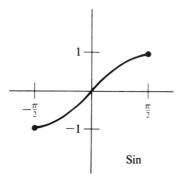

Sin

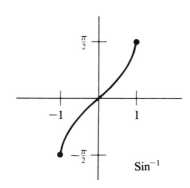

Sin^{-1}

An important property of an inverse function f^{-1} involves its relationship with the original function f under composition. If we follow the path of an element x in dom f as it maps under f to some element y in ran f, then y returns under f^{-1} to x, we see that the resultant effect of composing a function with its inverse (function) is the *identity* function on the original domain.

Theorem   (Composition of a Function with its Inverse Function)

Let $f: A \to B$ and $f^{-1}: B \to A$. Then

$f \circ f^{-1}$ is the identity function on B

and

$f^{-1} \circ f$ is the identity function on A

Proof: Suppose we let i_A and i_B represent the identity functions on A and B respectively. Recall that this means

$$\forall_{x \in A} \left(i_A(x) = x \right)$$

and

$$\forall_{y \in B} \left(i_B(y) = y \right)$$

Now assume that f and f^{-1} are functions as described in the hypothesis. Consider $f \circ f^{-1}$ first. We show $f \circ f^{-1} = i_B$.

The domains are the same set, namely B. Let y be any element of B where $f^{-1}(y) = x$. That is, x is the image of y under f^{-1}. Hence, y is the image of x under f, because f and f^{-1} are inverses of each other.

$$\begin{aligned}
(f \circ f^{-1})y &= f(f^{-1}(y)) &&(\text{definition of composition} - f \text{ and } f^{-1}) \\
&= f(x) &&(\text{existence } f^{-1}) \\
&= y &&(f, f^{-1} \text{ inverses})
\end{aligned}$$

Therefore,

$$f \circ f^{-1} = i_B$$

Next consider $f^{-1} \circ f$. We show $f^{-1} \circ f = i_A$. Both functions have domain A. Let x be any element of A. Then

$$\begin{aligned}
(f^{-1} \circ f)(x) &= f^{-1}(f(x)) &&(\text{definition of composition} - f^{-1} \text{ and } f) \\
&= f^{-1}(y) &&(\text{existence } f) \\
&= x &&(f, f^{-1} \text{ inverses})
\end{aligned}$$

Therefore,

$$f^{-1} \circ f = i_A$$

The two theorems proved in this section deal with the conditions for existence of inverse functions and the composition of a function with its inverse.

They do not tell us (assuming a function *has* an inverse) how to define the inverse function. This is not always easy to do. One way to do this is usually illustrated in calculus for some functions. If a function can be analyzed into a composition of simpler functions, each of which is invertible, then we can find its inverse by the following method.

1. Determine each function in the composition in order.

2. Reverse the order of the composition.

3. Replace each function by its inverse.

Example 3   Let $f: R \rightarrow R$ be defined by

$$f(x) = \frac{2x^3 - 5}{4}$$

Find f^{-1} and express it as a function of x.

Discussion:

If a function f is defined as a function of x (independent variable is x), then its inverse f^{-1} is really not a function of x, but rather a function of the dependent variable for f (usually called y). However, for convenience in graphing and other reasons, we traditionally prefer to express x as the independent variable for any function we deal with. Therefore, the formation of f^{-1} usually is accompanied by a step where y and x are interchanged (and the equation is solved for the other variable).

Method 1:

Start with: $f(x) = \dfrac{2x^3 - 5}{4}$

Replace $f(x)$ with: $y = \dfrac{2x^3 - 5}{4}$

Solve for x: $x = \sqrt[3]{\dfrac{4y + 5}{2}}$

Interchange variables: $y = \sqrt[3]{\dfrac{4x + 5}{2}}$

Thus, we say: $f^{-1}(x) = \sqrt[3]{\dfrac{4x + 5}{2}}$

Method 2:

Break f down into a composition of simpler functions and apply the "reverse order/inverse function procedure." We illustrate this by an input-output machine.

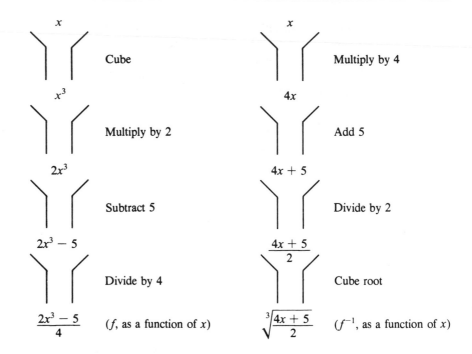

Note in Method 2 that each function in the composition, namely,

$$g(x) = x^3, \qquad h(x) = 2x, \qquad k(x) = x - 5, \qquad r(x) = \frac{x}{4}$$

has an inverse. These inverse functions are:

$$r^{-1}(x) = 4x, \qquad k^{-1}(x) = x + 5, \qquad h^{-1}(x) = \frac{x}{2}, \qquad g^{-1}(x) = \sqrt[3]{x}$$

Method 2 is a consequence of the following theorem and its generalization to any number of invertible functions.

Theorem   (Inverse of a Composition)
_____

Let f and g be functions whose inverses f^{-1} and g^{-1} are also functions. Then

$$(f \circ g)^{-1} = g^{-1} \circ f^{-1}$$

Proof: We are required to verify equality of two functions. First, we show they have the same domain. To do this, we observe the following two sets and the relationship of a function with its inverse.

$$\text{dom}(f \circ g)^{-1} = \{z \mid \exists_{y \in \text{dom} f} \text{ such that } f(y) = z$$

$$\text{and } \exists_{x \in \text{dom} g} \text{ such that } g(x) = y\}$$

$$\text{dom}(g^{-1} \circ f^{-1}) = \{z \mid \exists_{y \in \text{ran} f^{-1}} \text{ such that } f^{-1}(z) = y$$

$$\text{and } \exists_{x \in \text{ran} g^{-1}} \text{ such that } g^{-1}(y) = x\}$$

Since the domain of a function is the same as the range of its inverse, we see these two sets are equal. Next, for simplicity in reference, suppose

$$f: B \to C \qquad g: A \to B$$
$$f^{-1}: C \to B \qquad g^{-1}: B \to A$$

Then we observe

$$(f \circ g)^{-1} \circ (f \circ g) = i_A$$

Also,

$$\begin{aligned}
(g^{-1} \circ f^{-1}) \circ (f \circ g) &= g^{-1} \circ [f^{-1} \circ (f \circ g)] &&\text{(composition is associative)} \\
&= g^{-1} \circ [(f^{-1} \circ f) \circ g] &&\text{(composition is associative)} \\
&= g^{-1} \circ [i_B \circ g] &&(f^{-1} \circ f = i_B) \\
&= g^{-1} \circ g &&(i_B \circ g = g) \\
&= i_A &&(g^{-1} \circ g = i_A)
\end{aligned}$$

Thus, we have

$$(f \circ g)^{-1} \circ (f \circ g) = (g^{-1} \circ f^{-1}) \circ (f \circ g)$$

from which it follows that

$$(f \circ g)^{-1} = g^{-1} \circ f^{-1}$$

since the inverse of a function, if it exists, is unique (proof of this last statement is left to the reader).

Assigned Problems

1. (a) Prove that the function $f: R \to R^+$ defined by $\forall_{x \in R} \; f(x) = e^{-x}$ has an inverse (function).

(b) Prove that the function $g: R \to R^+$ defined by

$$\forall_{x \in R} \; g(x) = e^{(-x^2)}$$

is not invertible.

2. Use either method described in this section to define an inverse function (as a function of x) for

$$h(x) = 2(x-1)^5 + 7$$

3. Verify that the inverse of a function f, if it exists, is unique.

4. Extend, by mathematical induction on n, the last theorem of this section to n functions. That is, if $f_1, f_2, \ldots, f_n$ are n invertible functions such that compositions make sense, then show

$$(f_1 \circ f_2 \circ \cdots \circ f_n)^{-1} = f_n^{-1} \circ f_{n-1}^{-1} \circ \cdots \circ f_1^{-1}$$

(*Note:* Assume domains and ranges are compatible, and use the result of the theorem for two functions.)

5. Give an example to illustrate each of the following:

(a) $g: A \xrightarrow[\text{onto}]{1\text{-}1} B$, $\quad f: B \xrightarrow[\text{onto}]{1\text{-}1} A$ but $f \circ g \neq i_A$ and $g \circ f \neq i_B$

(b) $g: A \to B$, $\quad f: B \to A$, $f \circ g = i_A$ but $g \circ f \neq i_B$

6. Prove, for any invertible function f, $(f^{-1})^{-1} = f$.

7. It was tacitly assumed in the proof of a theorem in this section that if $f: A \to B$, then

$$i_B \circ f = f \quad \text{and} \quad f \circ i_A = f$$

Prove these assertions.

8. Let $f: A \to B$ and $g: B \to A$ be functions. Prove that if $f \circ g = i_B$ and $g \circ f = i_A$, then f is one-to-one and onto, and $g = f^{-1}$.

(*Note:* This is the converse of the theorem on *compositions of a function with its inverse*.)

9. Consider the $R \to R$ polynomial function f defined by

$$\forall_{x \in R} \ f(x) = x^3 + 3x^2 + 5$$

Is f invertible? (Use calculus to help answer this question. A function from one continuous set to another is one-to-one and onto if and only if its graph is increasing across the first set (with a range of the second set) or decreasing. How can you tell, using calculus, if a function is monotonic?)

10. Discuss carefully why, for $R \to R$ functions f whose inverses are functions, the graphs of both functions are symmetric with respect to the line $y = x$ in the plane.

5.5 Binary Operations

Imagine for a moment the phenomenon of addition. What exactly is it that we do when we add numbers? We take a *pair* of numbers, and using some device (two sets, counting, memory, calculator), we assign a number called the *sum* to that pair. For every pair of numbers we take, we would like to always "have a sum," and for any pair of numbers we take, we expect the sum to be unique. It would certainly be strange arithmetic if the pair $(3,4)$ had a sum of 5 on Monday, 7 on Tuesday, no sum on Wednesday, 23 on Thursday, etc. In other words, the process of normal addition has the characteristics of a functional relationship between pairs of numbers and other numbers called sums. This is precisely the essence of *binary operations*.

Binary Operation

Let S be a set. A *binary operation on S* is a mapping: $S \times S \to S$. Usually, we designate a symbol such as "$*$" as the function representative, and adopt the notation

$$\forall_{(a,b) \in S \times S} \quad *(a, b) = a * b$$

Hence, instead of using formal function notation with the function symbol placed in front of the domain element, for binary operations we place the function symbol ($*$) *between* the components of the ordered pair (domain element from $S \times S$). If the operation were addition, we would have "$a + b$" rather than "$+(a, b)$." This is conventional notation for binary operations. The reader should be aware, however, that a binary operation is a *function* that associates *ordered pairs* (from $S \times S$) with individual elements of S. Since it is a function, in a formal sense it is itself a set of ordered pairs, but now the first component is an *ordered pair* in $S \times S$ and the second is an element of S. As such,

$$* = \{((a, b), c) | (a, b) \in S \times S \quad \text{and} \quad c \in S, \text{where}$$

$$(a, b) \text{ bears some functional relationship with } c \}$$

Note also the interpretations of the existence and uniqueness conditions for functions when the function in question is a binary operation mapping: $S \times S \to S$.

1. *Existence:* For every ordered pair in $S \times S$ (every pair of elements taken from S), there is an element of S that is its image.

   $$\forall_{(a,b) \in S \times S} \exists_{c \in S} \quad \text{such that } a * b = c$$

   (This says that the "result of operating" on a pair of elements in S is always an element of S.)

2. *Uniqueness:* Every ordered pair in $S \times S$ (every pair of elements taken from S) has a unique image under $*$.

$$\forall_{(a, b), (c, d) \in S \times S} \text{ (if } (a, b) = (c, d), \text{ then } a * b = c * d)$$

(This is equivalent to saying: if $a = c$ and $b = d$, then $a * b = c * d$.)

Binary operations are not the only kind of operations on sets. There are also *unary operations*, which associate individual elements from a set S with elements of that set—in short, functions: $S \to S$. For example, "taking square roots" is a unary operation from nonnegative reals to nonnegative reals. So are "adding 5," "taking the absolute value," and "cubing." Additionally, just as in the case of relations (there are unary, binary, ternary relations, and so on), there are also ternary, quaternary, etc. *operations* on sets. We can generalize the definition of operation on a set to that of *n*-ary operation.

> An *n-ary operation* on a set S is a mapping from $(S \times S \times \cdots \times S) \to S$, where the domain is the Cartesian product of S taken n times.

Since *binary operations* play such an important role in mathematics, and particularly in the study of algebraic systems, we shall restrict our attention to this kind of operation. Most of these operations with which we are familiar are binary operations. For example, it is difficult to think of adding three *numbers* simultaneously (e.g., $2 + 3 + 4$). While we may think we are adding three numbers, invariably what we do mentally is add two of the numbers, obtain a sum, then add the remaining number and that sum. In other words, there are really two stages of pairwise addition, which could be diagrammed as in Figure 5.12.

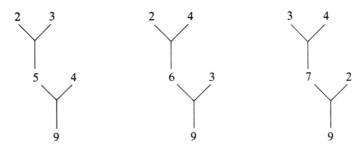

Figure 5.12

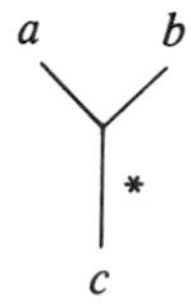

Figure 5.13

We shall occasionally use such diagrams as an input-output machine representation for binary operations. Note that there will always be two inputs (an ordered pair) and one output (see Figure 5.13).

Example 1   Investigate the four "standard" operations of arithmetic—addition, subtraction, multiplication, and division—with respect to various number sets.

(a)   *Addition:* Let N = natural numbers. Define $*: N \times N \rightarrow N$ by

$$\forall_{(a,b) \in N \times N} \; (a * b = a + b)$$

This is a binary operation on N since every pair of natural numbers has one and only one sum. Addition is also a binary operation on W (whole numbers), I (integers), Q (rationals), R (reals), and C (complex numbers).

(b)   *Subtraction:* Is standard subtraction, defined by $a * b = a - b$, a binary operation on N?
   No, since the ordered pair $(2, 3)$, for example, has no image in N. That is,

$$2 - 3 \notin N$$

Therefore, subtraction is not a mapping from $N \times N \rightarrow N$ since the existence condition fails to hold. Subtraction is also *not* a binary operation on W, but it *is* a binary operation on I, Q, R, and C. For example, the difference of any two integers is always an integer, and that integer is unique.

(c)   *Multiplication:* Let N = natural numbers. Define $*: N \times N \rightarrow N$ by

$$\forall_{(a,b) \in N \times N} \; (a * b = a \cdot b)$$

Since the product of any two natural numbers is always a natural number, and it is unique, then multiplication is a binary operation on N. It is also a binary operation on W, I, Q, R, and C.

(d)   *Division:* Consider the process of standard division of natural numbers. If a and b are natural numbers, is their quotient $(a \div b)$ also a natural number?
   Clearly, if $a = 2$ and $b = 3$, $(a \div b)$ is *not* a natural number. Therefore since not *every* pair of natural numbers has an image (*in the natural numbers*) under division, then division is *not* a binary operation on N. If we consider W, we must exclude 0, since division by 0 leads to inconsistencies (normally, we say it is "undefined"). Even so, we have the same problem as with N. Similarly, the quotient of two integers is not necessarily an integer. However, if we modify the set of rational numbers Q to exclude 0, then

division *is* a binary operation on $Q - \{0\}$. That is,

$$\div : ([Q - \{0\}] \times [Q - \{0\}]) \to [Q - \{0\}]$$

Similarly, division is a binary operation on $R - \{0\}$, and $C - \{0 + 0i\}$.

Thus, we see from Example 1 that whether or not a potential operation is a binary operation on a set S depends not only on the nature of the function but also on the *set*. The same operation (e.g., subtraction) is a binary operation on some sets and fails to be a binary operation on other sets. When testing a potential operation to see if it is a *binary operation on a set S*, we should ask ourselves two questions:

Q1   Given *any* pair of elements from the set, is the result of performing the indicated operation always a member of that set?

Q2   Is the result unique? Is it possible to obtain two or more different results for the same pair of elements?

The next examples provide practice with the thinking outlined above, since the potential operations are somewhat unfamiliar.

Example 2   Consider the set N of natural numbers, and the potential operation $*$ defined on N by

$$\forall_{(a,b) \in N \times N} \ a * b = a^b$$

Is this a binary operation on N? If so, prove it; it not, produce a counterexample.

Discussion:

We test existence and uniqueness as embodied in the two questions ($Q1$ and $Q2$) above. Take any pair of natural numbers a, b, and ask

Is a^b a natural number?

To answer this question, we need to think about what a^b means. We might test several special cases to get a feeling for the way the operation acts.

$$2^3 = 8, \quad 3^2 = 9, \quad 5^4 = 625$$

By definition of exponential forms,

$$a^b = \underbrace{a \cdot a \cdot a \cdot \cdots \cdot a}_{b \text{ times}}$$

Since a is a natural number, and multiplication is a binary operation on N, then the result of multiplying a by itself b times shall always yield a natural number. Therefore, we have proved existence.

What about uniqueness? Can we ever expect several different outcomes for the same base a and exponent b (if a and b are fixed natural numbers)? Intuitively, we guess no. How shall we *prove* this? Looking back at the interpretation of uniqueness for binary operations, we see, for any $(a, b), (c, d) \in N \times N$,

If $(a, b) = (c, d)$, then $a * b = c * d$.

In this particular instance, what we need to ask is:

If $a = c$ and $b = d$, is $a^b = c^d$?

The answer is clearly yes since the bases and exponents are the same (and multiplication has the uniqueness property). Therefore the operation $*$ is indeed a binary operation on N.

Example 3 Suppose P is the set of all points in a plane. We define an operation $*$ on P by

$$\forall_{p_1, p_2 \in P} \quad (p_1 * p_2 = \text{the distance from } p_1 \text{ to } p_2)$$

$$\left(Note: \text{This is the same as saying } \forall_{(p_1, p_2) \in P \times P}. \right)$$

Test $*$ to see if it is a binary operation on P.

Discussion:

We start with the question of *existence*. Given any two points in the plane, is the outcome of performing operation $*$ on them always a *point in the plane*? Hardly so, since the definition of $*$ involves assigning *distances* (not points) to pairs of points. Therefore, this "potential" operation is *not* a binary operation on P.

Example 4 Let $R = $ real numbers. Define an operation $\triangle$ on R by

$$\forall_{x, y \in R} \quad x \triangle y = \max\{x, y\}$$

This operation assigns to each pair of unequal real numbers the larger number of the pair, and if both components are the same, it assigns one of the components. For example,

$$2 \triangle 3 = 3$$
$$\pi \triangle e = \pi$$

and

$$1.5 \triangle 1.5 = 1.5$$

Is $\triangle$ a binary operation on R?

Discussion:

(a) *Existence:* Clearly, for any pair of real numbers x, y, if $x \neq y$, then $x < y$ or $y < x$. Thus,

if $x < y$, then $x \triangle y = y$

and

if $y < x$, then $x \triangle y = x$

If $x = y$, then $x \triangle y = x \triangle x = x$. In any event, the outcome is always a real number.

(b) *Uniqueness:* Suppose $x = w$ and $y = z$; is it true that $\max\{x, y\} = \max\{w, z\}$? Yes. Replace w and z with x and y and obviously we're asking about the maximum of two sets of equal numbers. Therefore, $\triangle$, as defined on R, is a binary operation on R.

Often when we talk about binary operations on sets, we hear the term *closure*, or "*this set is closed with respect to that operation.*" What this means is equivalent to the *existence* condition for binary operations—namely, the result of performing an operation (binary) on any two elements of a set is *always* a member of that set (or stays within the set). As we have observed, one can change the set for a given binary operation and no longer have a binary operation on the new set. In particular, it is of some interest in mathematics to investigate *subsets* of sets, the latter having binary operations defined on them. We then ask, "If S is a set and T is a subset of S, and $*$ is a binary operation on S, is it necessarily also a binary operation on T?" The answer is no, as we have already seen in the case of standard subtraction, which *is* a binary operation on the integers I, but *not* on the subset of positive integers I^+. This leads us to the definition of *closed set*.

Closed Set (with respect to a binary operation)

> Let S be a set and $*$ be a binary operation S. A subset T of S is said to be *closed* with respect to $*$ if and only if
>
> $$\forall_{t_1, t_2 \in T} \ (t_1 * t_2) \in T$$
>
> (That is, for any pair of elements in T, the result of performing the operation on them is always still a member of T.)

Example 5 (a) Let I be the set of integers, O be the subset of *odd* integers. Is O closed with respect to the binary operation of addition on I? No, because the sum of any two odd integers is not always an odd integer (in fact, it *never* is; it is always even). Thus, by performing the operation $(+)$ on odd numbers, we do not necessarily "stay within the set of odd numbers." Formally, we can prove "O is not closed" by providing one counterexample:

$3 \in O$ and $5 \in O$, but $3 + 5 = 8$ and $8 \notin O$

(b) Let $P = \{$all $R \to R$ polynomial functions$\}$. Consider the subset L in P of all *linear* functions. Is L closed with respect to composition of functions?

This is equivalent to asking: "If we take any two linear functions and compose them, do we always obtain a linear function?" Suppose we find out. Let

$f(x) = m_1 x + b_1$ (linear function)

and

$$g(x) = m_2 x + b_2 \qquad \text{(linear function)}$$

Now consider $f \circ g$.

$$\forall_{x \in R} \ (f \circ g)(x) = f(g(x))$$
$$= f(m_2 x + b_2)$$
$$= m_1(m_2 x + b_2) + b_1$$
$$= (m_1 m_2) x + (m_1 b_2 + b_1)$$

Therefore, $f \circ g$ is a *linear* function, and L is closed with respect to function composition.

Whenever a set is finite and not too large (two to ten elements), a binary operation can conveniently be defined or displayed by constructing an *operation table*. This is a schematic much like the familiar "addition tables" from elementary school. The members of the set are listed in a column along the left and top borders of the table and each entry in the table is obtained by operating on an ordered pair having the order (left column, top row). That is, the ij position in the table is the result $a_i * a_j$ where a_i is the ith element from the top in the left column and a_j is the jth element from the left in the top row.

$*$	a_1	a_2	$\cdots$	a_n
a_1	$(a_1 * a_1)$	$(a_1 * a_2)$	$\cdots$	$(a_1 * a_n)$
a_2	$(a_2 * a_1)$	$(a_2 * a_2)$	$\cdots$	$(a_2 * a_n)$
$\vdots$	$\vdots$	$\vdots$		$\vdots$
a_n	$(a_n * a_1)$	$(a_n * a_2)$	$\cdots$	$(a_n * a_n)$

Example 6   Let $W =$ whole numbers $\{0, 1, 2, 3, \ldots\}$. Recall the relation $\equiv \pmod 5$ defined on W by

$$\forall_{a, b \in W} \ a \equiv b \pmod 5 \text{ if and only if}$$
(a and b leave the same remainder upon division by 5).

This was proved to be an equivalence relation on W (see Section 4.2). Moreover, the induced partition of W consisted of exactly five classes, one corresponding to each possible remainder upon division by 5.

$$\bar 0 = \{0, 5, 10, 15, 20, \ldots\}$$
$$\bar 1 = \{1, 6, 11, 16, 21, \ldots\}$$
$$\bar 2 = \{2, 7, 12, 17, 22, \ldots\}$$
$$\bar 3 = \{3, 8, 13, 18, 23, \ldots\}$$
$$\bar 4 = \{4, 9, 14, 19, 24, \ldots\}$$

Now consider the partition, or set of equivalence classes. We shall call this W_5.

$$W_5 = \{\bar{0}, \bar{1}, \bar{2}, \bar{3}, \bar{4}\}$$

We define binary operation $\oplus$ on W_5 as follows:

$$\forall_{x, y \in W_5} \ \bar{x} \oplus \bar{y} = \overline{x + y}$$

This is an *addition of classes*. What it means is, take any two classes in W_5 and "add" them ($\oplus$) by adding (normally) their representatives and finding the class that sum represents. The outcome is the class that $\oplus$ assigns to the pair of classes. For example,

$$\bar{2} \oplus \bar{3} = \bar{5} = \bar{0}$$

(Given classes $\bar{2}$ and $\bar{3}$, add $2 + 3$, get 5; look for the class containing 5; it's $\bar{0}$. Then $\bar{2} \oplus \bar{3} = \bar{0}$.)

To display this operation on W_5, we construct the operation table below.

$\oplus$	$\bar{0}$	$\bar{1}$	$\bar{2}$	$\bar{3}$	$\bar{4}$
$\bar{0}$	$\bar{0}$	$\bar{1}$	$\bar{2}$	$\bar{3}$	$\bar{4}$
$\bar{1}$	$\bar{1}$	$\bar{2}$	$\bar{3}$	$\bar{4}$	$\bar{0}$
$\bar{2}$	$\bar{2}$	$\bar{3}$	$\bar{4}$	$\bar{0}$	$\bar{1}$
$\bar{3}$	$\bar{3}$	$\bar{4}$	$\bar{0}$	$\bar{1}$	$\bar{2}$
$\bar{4}$	$\bar{4}$	$\bar{0}$	$\bar{1}$	$\bar{2}$	$\bar{3}$

Note that a quick glance at the table tells us that $\oplus$ is a binary operation on W_5 since every entry in the table is an element in W_5 (existence), and no part of the table contains two or more distinct entries (uniqueness). This example creates a mathematical system called the *mod 5 arithmetic* or the *residue-class arithmetic modulo 5*.

Assigned Problems

1. Suppose S is a finite set and $n(S) = k$ for some positive integer k. How many different binary operations could be defined on S? Explain.

2. Test each of the following potential operations on the given sets to see whether they are binary operations. If so, prove it; if not, demonstrate with a counterexample.

   (a) $R = \{\text{reals}\}$. Define $*$ on R by

   $$\forall_{a, b \in R} \ a * b = \text{the average of } a \text{ and } b, \text{ namely, } \left(\frac{a + b}{2}\right)$$

   (b) $I = \{\text{integers}\}$. Define $\triangle$ on I by

   $$\forall_{m, n \in I} \ m \triangle n = 3m + n \ (\text{``}+\text{''} \text{ is normal addition})$$

   (c) $W = \{\text{whole numbers}\}$. Define $*$ on W by

   $$\forall_{a, b \in W} \ a * b = 0$$

(d) $R = \{\text{reals}\}$. Define $\circ$ on R by

$$\forall_{x, y \in R} \; x \circ y = x + y - xy$$

(e) X is any set. $S = \{f \mid f: X \to X\}$. Define $*$ on S by

$$\forall_{f, g \in S} \; f * g = f \circ g$$

3. Let S be any set and $\mathscr{P}(S)$ its power set. Discuss whether or not the following are binary operations on $\mathscr{P}(S)$.

(a) (joining) $\forall_{A, B \in \mathscr{P}(S)} \; A * B = A \cup B$

(b) (intersecting) $\forall_{A, B \in \mathscr{P}(S)} \; A * B = A \cap B$

(c) (relative complementation) $\forall_{A, B \in \mathscr{P}(S)} \; A * B = A - B$

(d) (taking symmetric difference) $\forall_{A, B \in \mathscr{P}(S)} \; A * B = (A \cup B) - (A \cap B)$

4. Which of the following operations are binary operations on the set P of all points in a plane? Discuss why or why not.

(a) $p_1 * p_2 =$ the line through p_1 and p_2

(b) $p_1 * p_2 =$ any point equidistant from p_1 and p_2

(c) $p_1 * p_2 = p_1$

(d) $p_1 * p_2 = \begin{cases} \text{midpoint of segment } \overline{p_1 p_2} & (\text{if } p_1 \neq p_2) \\ p_1 & (\text{if } p_1 = p_2) \end{cases}$

(e) $p_1 * p_2 = \begin{cases} p_3 & \text{where } p_1, p_2, p_3 \text{ are collinear and } p_2 \\ & \text{bisects segment } \overline{p_1 p_3} \; (\text{if } p_1 \neq p_2) \\ p_1 & (\text{if } p_1 = p_2) \end{cases}$

5. Consider the binary operation of standard addition $(+)$ on the set of whole numbers W $(+: W \times W \to W)$. Prove or disprove each of the following.

(a) Addition (as a function) is one-to-one.

(b) Addition (as a function) is onto.

(c) There is a one-to-one, onto function $f: W \times W \to W$.

6. Reread Example 6 (the residue-class arithmetic modulo 5), and construct operation tables for

(a) Multiplication of classes (in mod 5) defined by

$$\bar{x} \odot \bar{y} = \overline{x \cdot y}$$

(b) Addition of classes (in mod 7) defined by

$$\bar{x} \oplus \bar{y} = \overline{x + y}$$

(c) Addition and multiplication of classes in mod 4

7. Let $F = \{$all $R \to R$ functions for which composition both ways is meaningful$\}$. Which of the following subsets of F are closed with respect to function composition? Explain why or why not.

   (a) The set of all constant functions

   (b) The set of all continuous functions in F

   (c) The set of all differentiable functions in F

   (d) $\{f \mid f \in F$ and $f(1) = 0\}$

   (e) $\{f \mid f \in F$ and $f(0) = 0\}$

   (f) $\{f \mid f \in F$ and $f(1) = 1\}$

   (g) $\{f \mid f \in F, \ f$ is continuous, and $\int_0^1 f = 0\}$

8. The midpoint operation (see Problem 4d) is a binary operation on the set of points in a plane. Which of the following subsets of the plane are closed with respect to that operation?

   (a) A line   (e) A disk (circle $\cup$ interior)

   (b) A line segment   (f) The interior of a convex polygon

   (c) A ray   (g) The interior of any simple closed curve

   (d) A circle

9. Consider the set A of four complex numbers $\{1, -1, i, -i\}$, where $i^2 = -1$.

   (a) Is A closed with respect to multiplication of complex numbers?

   (b) Construct an operation table for A with respect to multiplication.

   (c) Show how the entire system—elements of A under multiplication—could be represented as powers of one of the elements of A.

10. In calculus, addition, subtraction, multiplication, and division of $R \to R$ functions are defined as shown below. Are these binary operations on the set F of all $R \to R$ functions? Explain.

   (a) Addition of functions
   $$\forall_{f, g \in F} \ [(f + g)(x) = f(x) + g(x)]$$

   (b) Subtraction of functions
   $$\forall_{f, g \in F} \ [(f - g)(x) = f(x) - g(x)]$$

   (c) Multiplication of functions
   $$\forall_{f, g \in F} \ [(f \cdot g)(x) = f(x) \cdot g(x)]$$

(d) Division of functions

$$\forall_{f,g \in F}\left[\left(\frac{f}{g}\right)(x) = \frac{f(x)}{g(x)}\right] \qquad (\forall_{x \in R} \text{ such that } g(x) \neq 0)$$

5.6 *Morphism Functions*

We have seen that a *function* is a certain kind of *relation* between two sets, or a *mapping* from one *set A* to another *set B*. In this final section on functions, we shall examine functions that map from one *system* to another *system*. Since a mathematical (or algebraic) system $(S, *)$ consists of a set S and at least one binary operation $*$ defined on the set S, then any such "intersystem" function might be expected to be concerned not only with the set but also with the *operation* on the set. This is indeed the case with the class of functions we shall call *morphism functions*.

The word *morphism* means *structure*; the prefix *homo* means *similar*. Thus, a *homomorphism* is a function that maps one algebraic system $(S, *)$ to another $(T, \circ)$ in such manner as to *preserve the similarity of the system*. It is this preservation of similarity among certain structural characteristics of a system that sets *morphism* functions apart from other kinds of functions. To give a more formal mathematical definition, we have the following:

Homomorphism

Suppose $(S, *)$ and $(T, \circ)$ are algebraic systems, and f: $S \to T$ is a mapping from S to T. The function f is called a *homomorphism* from S to T if and only if f has the following (morphism) property:

Morphism: $\forall_{s_1, s_2 \in S} [f(s_1 * s_2) = f(s_1) \circ f(s_2)]$

In words, what the morphism, or structure-preserving characteristic of a homomorphism, means is that if we take any two elements s_1 and s_2 in the system S, operate on them via the binary operation $*$ to produce the element $s_1 * s_2$ (in S), then map that element under f to produce an element in T (namely $f(s_1 * s_2)$), we obtain the same result as if we were to take s_1 and s_2 individually, map *each one* under f (*without* first operating in S using $*$), obtain $f(s_1)$ and $f(s_2)$ as elements of T, and then, operating in T on these two latter elements using T's operation $\circ$, producing $f(s_1) \circ f(s_2)$.

We can view the process schematically by using the diagram in Figure 5.14.

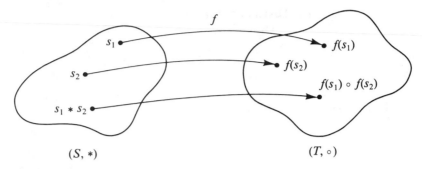

Figure 5.14

If f is a homomorphism, there are two ways of proceeding through the diagram.

①

1. Choose elements s_1, s_2
2. Operate in S: $s_1 * s_2$
3. Map that to T: $\underbrace{f(s_1 * s_2)}$

②

1. Choose elements s_1, s_2
2. Map individually to T: $f(s_1), f(s_2)$
3. Operate in T: $\underbrace{f(s_1) \circ f(s_2)}$

$\longrightarrow$ same element in $T \longleftarrow$

Example 1   Consider the two systems:

1. $(R, +)$: Real numbers under operation of standard addition

2. $(R^+, \cdot)$: Positive real numbers under operation of standard multiplication

Define the function $f\colon R \to R^+$ by

$$\forall_{x \in R} \ f(x) = e^x$$

In other words, f is the well-known exponential function having base e.

Is f a *homomorphism*? We know that f is a function and thus we shall not deal here with the problem of establishing existence and uniqueness properties. (We shall treat well-known functions this way, and verify the *function* properties only in the case of unfamiliar situations.) Therefore, to prove f is a homomorphism, we need only show that the *morphism* property holds; that is,

$$\forall_{x, y \in R} \ f(x + y) = f(x) \cdot f(y)$$

To establish this, we look at the nature of f itself.

$$f(x + y) = e^{x+y}$$
$$f(x) = e^x$$
$$f(y) = e^y$$

Thus, we are really asking ourselves, does

$$e^{x+y} = e^x \cdot e^y$$

for all real numbers x and y? We know this is a true statement for any real numbers x and y by the standard laws of exponents. Therefore, the exponential function f is indeed a homomorphism mapping $(R, +)$ to $(R^+, \cdot)$.

Note that whenever we discuss *morphism* functions, we include also the structure of the systems involved. This is important, because in Example 1 above, had we switched the operations, we would not change the *function* (it would still be the e^x function), but it would no longer be a homomorphism. That is, is the following statement true? If

$$f: (R, \cdot) \rightarrow (R^+, +)$$

is defined by $\forall_{x \in R} \ f(x) = e^x$, then do we have

$$\forall_{x, y \in R} \ [f(x \cdot y) = f(x) + f(y)] \quad ?$$

This would say

$$\forall_{x, y \in R} \ e^{xy} = e^x + e^y$$

which is *not* a true statement. For a counterexample, take $x = 1$, $y = 2$, and, obviously

$$e^{1 \cdot 2} \neq e^1 + e^2$$

Example 2   Suppose P is the set of all nonzero polynomials of one variable with real coefficients, and $(P, \cdot)$ is this system with respect to multiplication of polynomials. As a second system, take $(W, +)$, the set of whole numbers under standard addition. Now consider the function $d: P \rightarrow W$ defined by

$$\forall_{p \in P} \ d(p) = \text{the degree of polynomial } p$$

Is $d: (P, \cdot) \rightarrow (W, +)$ a homomorphism?

Discussion:

First, since this is not a "well-known" function, it makes sense to at least ponder whether the standard function properties—existence and uniqueness—hold for d.

(a) *Existence:* Does every polynomial $p \in P$ have an image under d? This is the same as asking, "Does every polynomial in P have a degree?" Clearly, the answer is yes, since nonzero constant polynomials have degree 0 and the degree of a polynomial with nonzero coefficients is the largest exponent appearing on the variable in a polynomial expression. For example, the degree of the polynomial

$$\sum_{i=0}^{n} a_i x^i \quad (a_n \neq 0)$$

is n.

(b) *Uniqueness:* Is the degree of a polynomial unique? Can a fixed polynomial have more than one degree? These questions are equivalent to asking

$$\forall_{p, q \in P} \ (\text{if } p = q, \text{ then } d(p) = d(q))$$

By definition of equality among polynomials (two polynomials are equal if and only if their degrees are the same, and all corresponding coefficients are equal), we see that equality of degree is required.

Thus, we see that the "degree relation" is really a *function*. Next, we show it has the morphism property. That is, we ask ourselves, is it true that

$$\forall_{p,\,q \in P} \; d(p \cdot q) = d(p) + d(q)$$

This means: Take any two polynomials p and q, multiply them together to produce a new polynomial $p \cdot q$, and map that under d. That is, assign a degree to the polynomial $p \cdot q$. Now look at the degrees of p and q individually—$d(p)$ and $d(q)$. If we *add* the degrees —$d(p) + d(q)$—is this the same as the degree of the product polynomial? The answer is *yes*; in other words, *the degree of a product of two nonzero polynomials of one variable is the same as the sum of the individual degrees*. Therefore, d is a homomorphism.

Having considered two examples so far, let us look back and once again state what it means to prove that a function f mapping one system to another is a homomorphism. We must show it is a *function* that *preserves structure* (*morphism*). That is, we take any two elements of the first system S, operate ($*$) on them in that system S, and note the image of the result under f; then we take their individual images under f, operate ($\circ$) on them in the other system T, and the result should be the same (*if* the function is a homomorphism).

As you might expect, there are other kinds of *morphism* functions, and they vary by the properties of one-to-one, onto, and whether or not they map to the same or different systems. These characteristics are reflected in the prefixes used.

Other Morphism Functions

- An *onto* homomorphism is called an *epimorphism*.

- A one-to-one homomorphism is called a *monomorphism*.

- A one-to-one and onto homomorphism is called an *isomorphism*.

- An isomorphism of a set onto itself is called an *automorphism*.

The term *homomorphism* is a noun describing a function from one system $(S, *)$ to another $(T, \circ)$ having the morphism property. If two systems are related via some homomorphism, we use the adjective *homomorphic* to describe the relationship. If a homomorphism is *not* one-to-one, then there are cases in which two or more elements in the system $(S, *)$ map to the same element in $(T, \circ)$. In a larger sense, sometimes we think mathematically of homomorphisms as describing class-to-element relationships, in which whole *classes* of one system "behave like" the *elements* of another system. The next example illustrates this.

Example 3 Let $(W, +)$ be the system of whole numbers under addition, and $(W_5, \oplus)$ be the mod-5 arithmetic under its own special "closed-circuit" addition. Let f be the function defined by

$$\forall_{n \in W} \ f(n) = (\text{the remainder } n \text{ leaves upon division by 5})$$

or, equivalently,

$$f(n) = \bar{r} \text{ where } n = 5q + r \text{ and } 0 \leq r < 5$$

The function f is a homomorphism; in fact, it is an epimorphism. We show it has the morphism property and it is onto.

Morphism: Let $m, n \in W$. Is $f(m + n) = f(m) \oplus f(n)$?

This means: add any two whole numbers and then look at the remainder of that sum upon division by 5. Is that the same as the sum of the individual remainders modulo 5? The answer is yes, since, if

$$m = 5q_1 + r_1 \qquad (0 \leq r_1 < 5)$$

and

$$n = 5q_2 + r_2 \qquad (0 \leq r_2 < 5)$$

then $m + n = 5(q_1 + q_2) + (r_1 + r_2)$ and the remainder that $m + n$ leaves upon division by 5 is $(r_1 + r_2)$ or $(r_1 + r_2) - 5$, since any summand of 5 included in $(r_1 + r_2)$ could be reassociated (by the distributive property) with the term $5(q_1 + q_2)$, leaving the unique remainder as a number between 0 and 4, inclusive.

Onto: Take any element in $W_5 = \{\bar{0}, \bar{1}, \bar{2}, \bar{3}, \bar{4}\}$. Is there a preimage for it under f? Again, we see each element (class) has any representative of the class as its preimage. This is summarized in the diagram below.

Thus, f is an epimorphism mapping $(W, +)$ onto $(W_5, \oplus)$. Note the class $\rightarrow$ element relationship. In each case, an equivalence class (or residue class) of whole numbers is associated with just one whole number. We say each class "behaves" (under addition, in this case) like a whole number, so that the addition of whole numbers modulo 5 is identified as the "addition" of entire classes of whole numbers. This example hints at a

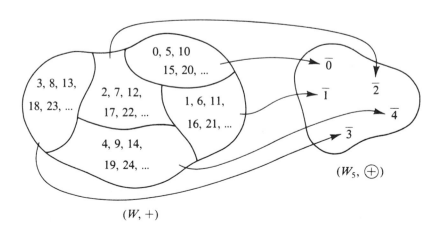

connection between equivalence relations and functions which we shall not explore further in this course. Nevertheless, there is a close relationship, and it is used as a tool in the proofs of many theorems in algebra.

Just as a homomorphism provides us with a view of a "class-to-element" morphism function, an *isomorphism* (one-to-one and onto homomorphism) is an "element-to-element" morphism function. Two systems $(S, *)$ and $(T, \circ)$ that are connected by some isomorphism are said to be *isomorphic*. This latter condition has a special notation reserved for it:

$$(S, *) \cong (T, \circ) \text{ means } "(S, *) \text{ is isomorphic to } (T, \circ)."$$

Isomorphism is the strongest possible relationship between two systems. The word *isomorphism* derives from *iso*, meaning *same*, and *morphism*, meaning *structure*. Any two systems that are isomorphic are "algebraically indistinguishable," or have precisely the *same structure*. That is to say, the elements of one system S behave (with respect to operation $*$) exactly the same as the elements of T (with respect to operation $\circ$). The only difference between two isomorphic systems is a superficial one—the symbols, names for objects, and operational symbols may appear different—but once the elements are matched with their correspondents in one-to-one, onto fashion, then results in either system can be used to predict results in the other system. The idea of isomorphism is presented in the next example.

Example 4   Consider the two systems $(C, \cdot)$ where $C = \{1, -1, i, -i\}$ under complex number multiplication, and $(W_4, \oplus)$, the mod-4 arithmetic under addition. For simplicity, the two operation tables are given below. We also remove the bars from W_4 element-symbols for simple notation.

$\cdot$	1	-1	i	$-i$
1	1	-1	i	$-i$
-1	-1	1	$-i$	i
i	i	$-i$	-1	1
$-i$	$-i$	i	1	-1

$\oplus$	0	1	2	3
0	0	1	2	3
1	1	2	3	0
2	2	3	0	1
3	3	0	1	2

Are these systems isomorphic? Do the elements of $(C, \cdot)$, with respect to the operation of complex number multiplication, behave like the elements of $(W_4, \oplus)$, with respect to the operation of addition modulo 4? To establish an isomorphism, we shall need to define it element-by-element, and test to see if the one-to-one, onto function we produce has the morphism property. Looking down the main diagonal of each table, we can see a pattern. Two elements, 1 and -1, in $(C, \cdot)$ are "self-invertible"; that is, they each multiply by themselves to produce the identity element. There are two elements, 0 and 2, in $(W_4, \oplus)$ that play a similar role. It makes sense to match the identity elements, and given the observation above, to match -1 with 2. This leaves two elements, i and $-i$, of $(C, \cdot)$ to be matched with two elements, 1 and 3, of $(W_4, \oplus)$. Thus there are two different ways to match up elements of these systems with an eye toward "structure-preservation." We define two one-to-one, onto functions f and g.

$$f \qquad\qquad g$$
$$1 \leftrightarrow 0 \qquad\quad 1 \leftrightarrow 0$$
$$-1 \leftrightarrow 2 \qquad -1 \leftrightarrow 2$$
$$i \leftrightarrow 1 \qquad\quad i \leftrightarrow 3$$
$$-i \leftrightarrow 3 \qquad -i \leftrightarrow 1$$

Obviously, f and g are one-to-one and onto functions. Suppose we test f to see if the morphism property holds.

There are 16 different combinations of element-pairs, one for each entry in the table. These are given below.

$$f(1 \cdot 1) = f(1) \oplus f(1) \qquad 0 = 0 \oplus 0$$
$$f(1 \cdot -1) = f(1) \oplus f(-1) \qquad 2 = 0 \oplus 2$$
$$f(1 \cdot i) = f(1) \oplus f(i) \qquad 1 = 0 \oplus 1$$
$$f(1 \cdot -i) = f(1) \oplus f(-i) \qquad 3 = 0 \oplus 3$$
$$f(-1 \cdot 1) = f(-1) \oplus f(1) \qquad 2 = 2 \oplus 0$$
$$f(-1 \cdot -1) = f(-1) \oplus f(-1) \qquad 0 = 2 \oplus 2$$
$$f(-1 \cdot i) = f(-1) \oplus f(i) \qquad 3 = 2 \oplus 1$$
$$f(-1 \cdot -i) = f(-1) \oplus f(-i) \qquad 1 = 2 \oplus 3$$
$$f(i \cdot 1) = f(i) \oplus f(1) \qquad 1 = 1 \oplus 0$$
$$f(i \cdot -1) = f(i) \oplus f(-1) \qquad 3 = 1 \oplus 2$$
$$f(i \cdot i) = f(i) \oplus f(i) \qquad 2 = 1 \oplus 1$$
$$f(i \cdot -i) = f(i) \oplus f(-i) \qquad 0 = 1 \oplus 3$$
$$f(-i \cdot 1) = f(-i) \oplus f(1) \qquad 3 = 3 \oplus 0$$
$$f(-i \cdot -1) = f(-i) \oplus f(-1) \qquad 1 = 3 \oplus 2$$
$$f(-i \cdot i) = f(-i) \oplus f(i) \qquad 0 = 3 \oplus 1$$
$$f(-i \cdot -i) = f(-i) \oplus f(-i) \qquad 2 = 3 \oplus 3$$

These 16 computations verify that f indeed is an isomorphism between $(C, \cdot)$ and $(W_4, \oplus)$. Therefore, although the systems *appear* to be different, they are really indistinguishable with respect to structure. It turns out that g is another isomorphism between the two systems (verification left to the reader). There are at least three observations we can make from this example:

1. Two systems may *look* quite different, yet be isomorphic.

2. To show two systems are isomorphic, it is necessary to *define* a function relating the two systems and then show the function is an isomorphism. In this case, we

used a "blast-out" approach—defining the function element by element and image by image, then testing all possible operation combinations in the morphism pattern. Had we had a nice, concise algebraic definition for our function, we could have appealed to that for our verification of isomorphism properties.

3. It is possible, in some cases, to produce more than one isomorphism between two systems. This does not change the fact that the systems are isomorphic; it simply demonstrates that they are structurally identical in more than one way.

In the next example, we take a function that is defined algebraically and prove it is an isomorphism.

Example 5   Consider the system $(R - \{0\}, \cdot)$, the nonzero real numbers under multiplication. Let r be some fixed nonzero real number. Define a function f mapping this system to itself by

$$\forall_{x \in R - \{0\}} \ f(x) = rxr^{-1}$$

Prove that f is an isomorphism from $(R - \{0\}, \cdot)$ to itself (actually, an *automorphism*) without using the fact that multiplication of real numbers is a commutative operation. (The latter makes this exercise trivial.)

Proof   If we start from a point of minimal information about f, there are five things to show: f is a function (existence and uniqueness), f is one-to-one, f is onto, and f has the morphism property. We treat each briefly in that order.

Existence: Given any $x \in R - \{0\}$, the number rxr^{-1} exists, and is an element in $R - \{0\}$. That is, since r is nonzero, r^{-1} exists and the product rxr^{-1} is a nonzero real number. Thus, every preimage has an image.

Uniqueness: Suppose $x, y \in R - \{0\}$ and $x = y$. Is it true that $f(x) = f(y)$? The answer is yes, since we know we can multiply both sides of the original equation on the left by r to produce

$$rx = ry$$

and then on the right by r^{-1} to obtain

$$rxr^{-1} = ryr^{-1}$$

Therefore, $f(x) = f(y)$.

f is one-to-one: Suppose $f(x) = f(y)$ for any elements x, y in the system. Does it necessarily follow that $x = y$? If $f(x) = f(y)$, then

$$rxr^{-1} = ryr^{-1}$$

Multiply on the left by r^{-1} and on the right by r (use the associative property for multiplication). Thus, $x = y$.

f is onto: Let y be any element in $R - \{0\}$. Can we find a preimage x for y? That is, is there an element $x \in R - \{0\}$ such that $f(x) = y$? This would mean

$$rxr^{-1} = y$$

Multiplying on the left by r^{-1} and on the right by r, we obtain

$$x = r^{-1}yr$$

which is the desired preimage. [$r^{-1}yr$ is a nonzero real number and $f(r^{-1}yr) = y$.] Thus, every image has a preimage.

f has the morphism property: Here, we need to show

$$\forall_{x,\,y \in R - \{0\}} \; [f(x \cdot y) = f(x) \cdot f(y)]$$

$$
\begin{aligned}
f(x \cdot y) &= r(xy)r^{-1} & &\text{(definition of } f) \\
&= (rx)(yr^{-1}) & &\text{(associative property of multiplication in } R) \\
&= (rx)(r^{-1}r)(yr^{-1}) & &\text{(insert } r^{-1}r, \text{ which is 1)} \\
&= (rxr^{-1})(ryr^{-1}) & &\text{(associative property of multiplication in } R) \\
&= f(x) \cdot f(y) & &\text{(definition of } f)
\end{aligned}
$$

Therefore, we have proved that f as defined above is an automorphism from $(R - \{0\}, \cdot)$ onto itself.

We conclude this section with an important theorem about isomorphisms.

Theorem   (The Inverse of an Isomorphism Is an Isomorphism)

If f is an isomorphism mapping $(S, *)$ to $(T, \circ)$, then f^{-1} is an isomorphism mapping $(T, \circ)$ to $(S, *)$.

Proof: First, since f is an isomorphism, then f is a one-to-one, onto function mapping $S \to T$. By previous considerations (see Section 5.4), we know f^{-1} exists (is a function) and f^{-1} is both one-to-one and onto. Thus, it remains only to show that f^{-1} has the morphism property. To do this, we adopt some notation. Let t_1 and t_2 be any elements of T, and

$$f^{-1}(t_1) = s_1.$$

and

$$f^{-1}(t_2) = s_2$$

where s_1 and s_2 are elements in S. Then, since f^{-1} is the inverse function of f, we know

$$f(s_1) = t_1$$

and

$$f(s_2) = t_2$$

Now, we wish to show

$$f^{-1}(t_1 \circ t_2) = f^{-1}(t_1) * f^{-1}(t_2)$$

We know, since f is an isomorphism,

$$f(s_1 * s_2) = f(s_1) \circ f(s_2)$$

Thus,

$$f^{-1}(t_1 \circ t_2) = f^{-1}[f(s_1) \circ f(s_2)]$$
$$= f^{-1}[f(s_1 * s_2)]$$
$$= (f^{-1} \circ f)(s_1 * s_2)$$
$$= s_1 * s_2$$
$$= f^{-1}(t_1) * f^{-1}(t_2)$$

Therefore, we have shown that f^{-1} has the morphism property and, consequently, f^{-1} is an isomorphism.

Assigned
Problems

1. In the following list, identify and verify which of the functions are homomorphisms or isomorphisms. For those that are not morphism functions, explain why not.

(a) $(W, +) \xrightarrow{m_7} (W, +)$ m_7 means multiply by 7; $W = \{0, 1, 2, \dots\}$

(b) $(W, +) \xrightarrow{\exp_2} (W, \cdot)$ $\exp_2$ means exponentiate base two

(c) $(W, \cdot) \xrightarrow{p_4} (W, \cdot)$ p_4 means raise to fourth power

(d) $(W, +) \xrightarrow{a_5} (W, +)$ a_5 means add 5

(e) $(W, \cdot) \xrightarrow{m_7} (W, \cdot)$ m_7 means multiply by 7

(f) $(R, \cdot) \xrightarrow{\text{abs}} (R^+ \cup \{0\}, \cdot)$ abs means take absolute value of; R = reals; $R^+ \cup \{0\}$ = nonnegative reals

(g) $(R^+ \cup \{0\}, +) \xrightarrow{\text{sqrt}} (R^+ \cup \{0\}, +)$ sqrt means square root

(h) $(R^+ \cup \{0\}, \cdot) \xrightarrow{\text{sqrt}} (R^+ \cup \{0\}, \cdot)$ sqrt means square root

(i) $(I, +) \xrightarrow{\text{neg}} (I, +)$ neg means take negative of; I = integers

(j) $(R - \{0\}, \cdot) \xrightarrow{1/x} (R - \{0\}, \cdot)$ $1/x$ means reciprocal of $R - \{0\}$ = reals except 0

(k) $(D, +) \xrightarrow{f'} (F, +)$ f' means differentiate D = differentiable functions F = all functions

(l) $(D, \cdot) \xrightarrow{f'} (F, \cdot)$ f' means differentiate

2. Devise a binary operation $*$ on $7W$ (the set of all whole number multiples of 7) so that m_7 (multiplication by 7) will be a homomorphism: $(W, \cdot) \to (7W, *)$.

3. The function $f: R \to R$ defined by $f(x) = x + 1$ is one-to-one and onto. Devise an operation $\oplus$ on R so that f will be an isomorphism: $(R, +) \to (R, \oplus)$.

4. Verify that the function g of Example 4 is an isomorphism between $(C, \cdot)$ and $(W_4, +)$.

5. Suppose $g: (S, *) \to (T, \circ)$ is an isomorphism and $f: (T, \circ) \to (M, \triangle)$ is an isomorphism. Prove that $f \circ g$ is an isomorphism mapping $(S, *) \to (M, \triangle)$.

6. Suppose $\mathscr{A}$ is a collection of systems. Prove that the relation $\cong$ (is isomorphic to) is an equivalence relation on $\mathscr{A}$.

7. Let $Q\sqrt{2} = \{a + b\sqrt{2} \mid a, b \in Q,$ the rational numbers$\}$ and $Q\sqrt{3} = \{a + b\sqrt{3} \mid a, b \in Q\}$.

   Define a function $f: (Q\sqrt{2}, +) \to (Q\sqrt{3}, +)$, which is an isomorphism, and verify that it is.

8. Let $Q\sqrt{2}$ and $Q\sqrt{3}$ be the sets in Problem 7. Consider the function $g: Q\sqrt{2} \to Q\sqrt{3}$ defined by

$$g(a + b\sqrt{2}) = a + b\sqrt{3}$$

   Prove that g is *not* a homomorphism when the operation involved is multiplication.

9. Let U be a universe, A a fixed set in U, and define a function f on $\mathscr{P}(U)$ by

$$\forall_{C \in \mathscr{P}(U)} \ f(C) = A \cap C$$

   Show that f is a homomorphism mapping $(\mathscr{P}(U), \cup)$ to itself. Note that the operation here is set *union*.

10. The calculations below are from a high school algebra text. Point out where homomorphism is used in each case and identify the nature of the homomorphism.

    (a) $a - b = a + (-b)$
    $\qquad = a + [-(a + b - a)]$
    $\qquad = a + ((-a) + [-(b - a)])$
    $\qquad = [a + (-a)] + [-(b - a)]$
    $\qquad = 0 + [-(b - a)]$
    $\qquad = -(b - a)$

(b) $\dfrac{a}{b} = a \cdot \dfrac{1}{b}$ $(b \neq 0)$

$\phantom{(b) \dfrac{a}{b}} = a \cdot \left[\dfrac{1}{a \cdot \dfrac{b}{a}} \right]$

$\phantom{(b) \dfrac{a}{b}} = a \cdot \left[\dfrac{1}{a} \cdot \dfrac{1}{\dfrac{b}{a}} \right]$

$\phantom{(b) \dfrac{a}{b}} = \left[a \cdot \dfrac{1}{a} \right] \cdot \left[\dfrac{1}{\dfrac{b}{a}} \right]$

$\phantom{(b) \dfrac{a}{b}} = 1 \cdot \left[\dfrac{1}{\dfrac{b}{a}} \right]$

$\phantom{(b) \dfrac{a}{b}} = \dfrac{1}{\dfrac{b}{a}}$

6

Mathematical Systems

6.1 Properties of Operations

An *algebraic* system consists of a nonempty set S and one or more binary operations on S. In this section we shall concern ourselves only with *algebraic systems involving one binary operation* on the set under consideration. Thus, our algebraic systems can be regarded, and shall be denoted, as ordered pairs $(S, *)$ where the first component is some nonempty set S, and the second component is a binary operation $*$ on S. Recall that a binary operation on S is a mapping $*: S \times S \to S$, such that given any pair of elements from S (or ordered pair in $S \times S$) there is exactly one element in S that is the "result" (image) under the operation.

Algebraic systems can be classified by structure that resides in certain properties of the operation(s) and combinations of these properties. We are all familiar with the *associative*, *commutative*, *identity element*, and *inverse of an element* properties in the context of familiar number systems. We shall now discuss these properties in a more general or abstract context. One of the basic processes of abstract algebra is to "abstract" structural characteristics from various specific algebraic systems, use these characteristics to define general systems, and then classify each system by its general structure. There is a distinct advantage in doing things this way. Theorems can be proved with reference to the general system, and the results apply simultaneously to all specific instances of systems sharing those general characteristics. So, rather than proving lots of theorems for each special system we might encounter, thereby proving the same theorem over and over again each time the context changes, we prove the theorems for the general abstract system and note that they apply to all systems having that defining set of structural characteristics.

Associative Property

A binary operation $*$ is said to be an *associative* binary operation on S if and only if

$$\forall_{a,b,c \in S} \; [(a * b) * c = a * (b * c)]$$

The algebraic system $(S, *)$ is said to be *associative* if and only if $*$ is an associative operation.

The associative property deals with *regrouping* (indicated by parentheses); note that the order of the elements is unchanged on both sides of the equality, only the grouping is different—a and b are grouped together on the left, and b and c are grouped together on the right. Also note that by standard algebraic notational convention, the letters a, b, and c need not refer to different elements —any two or all three letters could be replaced by the same element of S. The associative property is schematized in Figure 6.1, followed by two examples.

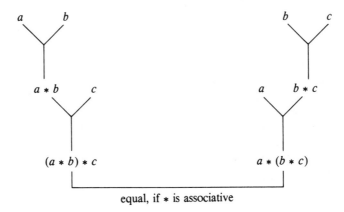

Figure 6.1

Example 1   Let $S = R$ (reals) and define $*$ on S by

$$\forall_{x,y \in S} \; [x * y = x + y - xy]$$

Prove or disprove that $*$ is an associative binary operation on S.

Discussion:

We look first at the nature of $*$. It involves three known operations on real numbers—addition, multiplication, and subtraction. That $*$ is a binary operation is easy to see, since $+$, $\cdot$, and $-$ are all binary operations on R. Thus, the result of performing $*$ is always a unique real number. Next, we consider the associativity question. We know both addition and multiplication are associative, but we also know subtraction is not an associative operation on R. Consider

$$(5 - 3) - 2 \quad \text{versus} \quad 5 - (3 - 2)$$

Thus, it is not easy to predict whether $*$ is associative. We might test several examples in search of a counterexample:

$$(5*3)*2 = 9 \qquad\qquad 5*(3*2) = 9$$
$$(2*4)*10 = 28 \qquad\qquad 2*(4*10) = 28$$
$$(-3*1)*(-7) = 1 \qquad\qquad -3*[1*(-7)] = 1$$
$$(1*0)*1 = 1 \qquad\qquad 1*(0*1) = 1$$

These examples certainly do not verify associativity, but they build our faith that $*$ may indeed be an associative operation. They suggest we might try a proof.

Proof   We shall try to prove

$$\forall_{a,b,c \in S} \ [(a*b)*c = a*(b*c)]$$

Using the definition of $*$, we have

$$(a*b)*c = [(a+b) - ab]*c$$
$$= [(a+b) - ab + c] - [(a+b) - ab] \cdot c$$
$$= a + b - ab + c - ac - bc + abc$$

Rearranging, we have

$$(a*b)*c = a + b + c - ab - ac - bc + abc$$

Next, we go to the other side of the desired equation and apply the definition of $*$.

$$a*(b*c) = a*[(b+c) - bc]$$
$$= [a + (b+c) - bc] - a[(b+c) - bc]$$
$$= a + b + c - bc - ab - ac + abc$$

This last result matches the outcome from the left side; therefore, we have proved that $*$ is an associative operation on S.

Example 2   Consider the system consisting of the set $S = \{a, b, c\}$ and a binary operation $*$ on S defined by the table given below.

$*$	a	b	c
a	a	b	c
b	b	a	b
c	c	a	a

Is $*$ an associative binary operation on S?

Discussion:

Answering this question could conceivably involve testing as many as $3 \cdot 3 \cdot 3 = 27$ different combinations in the "associative pattern"

$$\square * \triangle * \bigcirc$$

In general, if a set has n elements, there are n^3 different equations to test for associativity if nothing else is known about the operation except that its outcomes are displayed in a table. Here, we look strategically around the table and note that each element "starred" with itself produces a. Also, a plays a special role of identity element. Thus, for a potential counterexample to associativity, we might try some combination involving b's and c's with one repeated.

$$(b * b) * c = a * c = c$$

$$b * (b * c) = b * b = a$$

Therefore, since we are assuming that the three elements a, b, and c are distinct, we have shown by counterexample that $*$ is *not* an associative binary operation on S.

Commutative Property

A binary operation $*$ is said to be a *commutative* binary operation on S if and only if

$$\forall_{a, b \in S} \, \left[a * b = b * a \right]$$

The algebraic system $(S, *)$ is said to be *commutative* if and only if $*$ is a commutative binary operation on S.

The commutative property deals with *reordering*. If elements can be operated on in either order with the same result in every case, then the operation is commutative. A scheme for the commutative property is shown in Figure 6.2; examples follow.

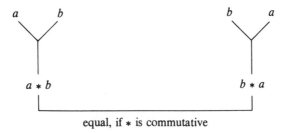

equal, if $*$ is commutative

Figure 6.2

Example 3   Is the system $(S, *)$ of Example 1 commutative?

Discussion:

Suppose we begin with the basic design of a proof and see where it leads us. The binary operation involved is defined

$$\forall_{x, y \in S} \ [x * y = x + y - xy]$$

Let us test $y * x$.

$$y * x = y + x - yx$$

Since addition and multiplication are both commutative operations on the real numbers, we see that the latter result is the same as the outcome of $x * y$ in the definition. Thus, the operation $*$ is commutative, and the system $(S, *)$ of Example 1 is a commutative system.

Example 4   Let $A = \{1, 2, 3\}$ and S_3 be the system of permutations (one-to-one, onto functions from $A \to A$) on A. Recall that the six elements of S_3 can be denoted as follows:

$$\begin{pmatrix} 1 & 2 & 3 \\ 1 & 2 & 3 \end{pmatrix}, \quad \begin{pmatrix} 1 & 2 & 3 \\ 2 & 1 & 3 \end{pmatrix}, \quad \begin{pmatrix} 1 & 2 & 3 \\ 3 & 2 & 1 \end{pmatrix},$$

$$\begin{pmatrix} 1 & 2 & 3 \\ 1 & 3 & 2 \end{pmatrix}, \quad \begin{pmatrix} 1 & 2 & 3 \\ 2 & 3 & 1 \end{pmatrix}, \quad \begin{pmatrix} 1 & 2 & 3 \\ 3 & 1 & 2 \end{pmatrix}$$

The operation involved is function composition or *permutation multiplication*. Since function composition is associative (this is an example of a very important general theorem we proved in Section 5.3 that is now applying to one of many special cases), we know that $(S_3, \circ)$ is an associative system. Is $(S_3, \circ)$ a commutative system? We know that function composition is *not* generally commutative, but we *cannot* deduce from that result that this particular situation is noncommutative. It is certainly possible that some restricted instances involving function composition may be commutative. We start by testing some "products."

$$\begin{pmatrix} 1 & 2 & 3 \\ 2 & 1 & 3 \end{pmatrix} \cdot \begin{pmatrix} 1 & 2 & 3 \\ 3 & 2 & 1 \end{pmatrix} = \begin{pmatrix} 1 & 2 & 3 \\ 2 & 3 & 1 \end{pmatrix}$$

$$\begin{pmatrix} 1 & 2 & 3 \\ 3 & 2 & 1 \end{pmatrix} \cdot \begin{pmatrix} 1 & 2 & 3 \\ 2 & 1 & 3 \end{pmatrix} = \begin{pmatrix} 1 & 2 & 3 \\ 3 & 1 & 2 \end{pmatrix}$$

We immediately see that the permutations on the right are not the same, thus permutation multiplication in S_3 is *not* a commutative operation.

Example 5   There is a very handy visual device for telling whether or not an operation is commutative when it is defined by an operation table. The "main diagonal" of the table is the diagonal array of elements extending from the upper left corner to the lower right corner. The operation is *commutative* if and only if the table is *symmetric* with respect to this main diagonal. That is, if one were to "fold" the table using the main diagonal as a "crease," coincident elements would be identical. Note that what happens along the main diagonal itself is immaterial, since these results arise from elements operating on themselves ($x * x$ types).

$S = \{a, b, c, e\}$ $S = \{a, b, c, d, e\}$

* defined by   * defined by

*	e	a	b	c
e	e	a	b	c
a	a	e	c	b
b	b	c	e	a
c	c	b	a	e

*	a	b	c	d	e
a	b	ⓒ	d	ⓔ	a
b	ⓔ	d	a	ⓒ	b
c	d	a	e	b	c
d	ⓒ	ⓔ	b	a	d
e	a	b	c	d	e

* is commutative.   * is noncommutative.
$$a * b \neq b * a$$
$$d * a \neq a * d$$
$$d * b \neq b * d$$

Existence of Identity Element

Let $(S, *)$ be a system with binary operation ($*$). An element e in S is called an *identity element* (or *neutral* element, or *unit*) for this system if and only if

$$\forall_{x \in S} \ x * e = e * x = x$$

It should be clearly understood that not all systems have such a special element as an identity element, but for those that do, it is an element of the system, it plays the special role of commuting with every element of the system regardless of whether the operation is commutative, and it preserves the "identity" of any element with which it operates.

That is, the outcome is always the original element. As a historical note, the symbol e is usually reserved for an identity element (if such exists in a given system) because of its derivation from the German word *einheit*, meaning *unity*, or *one*. The number 1 plays this role for real numbers under multiplication. Thus, making proofs about identity elements involves these three considerations above. Unlike proofs about commutativity or associativity, proofs about identity elements are *existence* proofs. One searches for a special element that "works" this particular way for the whole system. In some cases there exists such an element; in other cases, not.

Example 6   Consider the system given in Example 1, where the binary operation * is defined on S (reals) by

$$\forall_{x, y \in S} \ [x * y = x + y - xy]$$

Is there an identity element in S relative to *?

Discussion:

Suppose there is such an element—call it *e*. Then we must have

$$\forall_{x \in S} \; x * e = (x + e) - xe \quad \text{(definition *)}$$

$$= x \quad \text{(identity)}$$

Now, can we think of any real number that will play this role for us? It appears that 0 should work.

$$\forall_{x \in S} \; x * 0 = (x + 0) - (x \cdot 0)$$

$$= x$$

However, it must work on *both* sides, and it does.

$$\forall_{x \in S} \; 0 * x = (0 + x) - (0 \cdot x)$$

$$= x$$

Therefore, 0 is the element we want; it is the identity element for *S* relative to operation *.

It is well-known that 0 and 1 play the classical identity elements for standard addition and multiplication of real numbers. However, the identity element (if such exists) can arise in unexpected forms, as the next example illustrates.

Example 7   Let $R = \{\text{real numbers}\}$ and define a binary operation $\triangle$ on R by

$$\forall_{x, y \in R} \; [x \triangle y = xy - x - y + 2]$$

Does *R* have an identity element relative to $\triangle$?

Discussion:

Let us assume there is an identity element. Call it *e*, and see what would have to happen under that assumption.

$$\forall_{x \in R} \; x \triangle e = xe - x - e + 2 \quad \text{(definition of operation $\triangle$)}$$

$$= x(e - 1) - e + 2 \quad \text{(algebra)}$$

and we want

$$x(e - 1) - e + 2 = x$$

so

$$x(e - 2) = e - 2 \quad \text{(algebra)}$$

Now, if $e - 2 \neq 0$, then we can divide both sides by $e - 2$ and obtain $x = 1$. But we started by taking *any* real number x. Thus, we set

$$e - 2 = 0$$

and

$$e = 2$$

This suggests that if there is an identity element in R relative to $\triangle$, it should be 2. We could have taken a different route, and solved the equation for e, obtaining $e(x - 1) = 2(x - 1)$ from which $e = 2$ (if $x \neq 1$, we must consider the case $x = 1$ separately). We test further to establish that 2 is indeed a two-sided identity element.

Let x be any real number

$$\begin{aligned} x \triangle 2 &= x \cdot 2 - x - 2 + 2 \\ &= 2x - x \\ &= x \end{aligned}$$

Also, $\begin{aligned} 2 \triangle x &= 2x - 2 - x + 2 \\ &= 2x - x \\ &= x \end{aligned}$

Finally, in the special case where $x = 1$,

$$\begin{aligned} 1 \triangle 2 &= 1 \cdot 2 - 1 - 2 + 2 \\ &= 1 \end{aligned}$$

and

$$2 \triangle 1 = 1$$

Thus, the real number 2 is the identity element for $\triangle$ in the system of real numbers.

Example 8   Modify the operation $\triangle$ of Example 7 just slightly:

$$\forall_{x,\, y \in R} \; [x \triangle' y = xy - x - y + 3]$$

Does R have an identity element with respect to the modified $\triangle'$? We follow a similar line of reasoning. Suppose one exists and call it e. Then

$$x \triangle' e = xe - x - e + 3$$

and we want

$$xe - x - e + 3 = x$$

Thus,

$$e(x - 1) = 2x - 3 \qquad \text{(by algebra)}$$

Now assume x is any real number except 1; then

$$e = \frac{2x - 3}{x - 1}$$

This says e depends on the value of x and changes whenever x changes. Such is not the case with an identity element. We shall prove (in the problems) that an identity element in a system, if it exists, is unique. Thus, we conclude that there is no such element in the real numbers that plays the role of identity element for $\triangle'$.

Example 9   If an identity element exists in a system $(S, *)$ where $*$ is defined only by an operation table, it is easy to spot. Since it preserves the identity of each element and commutes with each, then all we need to do is look for a row and a column that are identical with the upper labeling row and left labeling column for the table. Consider the two examples below, involving systems $(S_1, *)$ and $(S_2, *)$

*	a	b	c	d
a	c	d	a	b
b	d	c	b	a
c	a	b	c	d
d	b	a	d	c

c is the identity element for $(S_1, *)$.

*	a	b	c	d
a	a	c	d	b
b	b	c	a	d
c	d	c	b	a
d	c	a	d	b

$(S_2, *)$ has no identity element.

You may have noticed that the system whose operation is defined by the left table is really the same system in disguise (more precisely, isomorphic) as the system defined in the left table of Example 5.

Existence of Inverse Elements

Let $(S, *)$ be a system having an identity element e. An element x in S has an *inverse* relative to $*$ if and only if there exists an element $y \in S$ such that

$$x * y = y * x = e$$

The element y is called the "inverse of x relative to $*$" or the " $*$-inverse of x." It shall be denoted by $\bar{x}$. We shall be interested in the characteristic of systems where *every* element has an inverse with respect to the operation $(*)$. Note especially that it makes no sense to search around for *inverses* of elements in a system where it has not yet been established that an *identity element* exists. Also, it is very important to distinguish the nature of the identity element from the inverse of an element. The identity element for a system $(S, *)$, if it exists, is a *single element e* in the system that "works for all elements." That is, $x * e = e * x = x$ *for all* x in S. For example, 1 preserves the identity of every real number under multiplication. On the other hand, *inverses depend on the individual element we select*. If we select 2 and the operation is multiplication, then its inverse is $\frac{1}{2}$, since

$$2 \cdot \tfrac{1}{2} = \tfrac{1}{2} \cdot 2 = 1 \qquad \text{(1 being the multiplicative identity element)}$$

If we select $\frac{3}{5}$, its multiplicative inverse is $\frac{5}{3}$, and so on. As in the case with proofs about the identity element, proofs about inverses of elements are *existence proofs*. Given an element, we are required to come up with a candidate for its inverse, show that the candidate is a member of the system involved, show that the candidate commutes with the given element (even if the system is noncommutative), and that the result of operating with the original element and its inverse is the identity element of the system.

Example 10   Let $I = \{integers\}$, and define operation $*$ on I by

$$\forall_{m, n \in I} \left[m * n = m + n - 3 \right]$$

(a) Does $(I, *)$ have an identity element?

(b) If the answer to (a) is yes, does every element in I have a $*$-inverse?

Discussion:

The integer 3 looks like a good candidate to test for identity element, and it works:

$$\forall_{m \in I} \; m * 3 = m + 3 - 3 = m$$

$$\forall_{m \in I} \; 3 * m = 3 + m - 3 = m$$

Thus, the answer to question (a) is yes, and the identity element for this system (having the operation $*$ defined as above) is the integer 3.

Next, we look to see if every element in I has a $*$-inverse. Let m represent any integer, and suppose m has a $*$-inverse, call it n. We find out what condition(s) n must meet.

$$m * n = m + n - 3$$

and thus we want

$$m + n - 3 = 3$$

Therefore,

$$n = 6 - m$$

It is clear that if m is any integer, then so is $6 - m$. Moreover, we test $6 - m$ to see if it qualifies to be the $*$-inverse of m.

$$m * (6 - m) = m + (6 - m) - 3 = 3$$

$$(6 - m) * m = (6 - m) + m - 3 = 3$$

Therefore, every integer has a $*$-inverse, and in each case it is the integer obtained by subtracting the original integer from 6.

Example 11   A standard convention in mathematics is to use the notation $(-x)$ for the inverse of element x when the operation involved is addition or "addition-like," and to use the notation (x^{-1}) for the inverse of element x when the operation involved is multiplication or "multiplication-like." We shall use this convention below.

Also, whenever the operations are defined by tables, it is easy to spot the inverse of each element (if such exists) by looking for the identity element in each row and column of the table.

$(W_5, \oplus)$ mod-5 arithmetic $(C, \cdot)$ complex minisystem

$\oplus$	0	1	2	3	4
0	0	1	2	3	4
1	1	2	3	4	0
2	2	3	4	0	1
3	3	4	0	1	2
4	4	0	1	2	3

$\cdot$	1	-1	i	$-i$
1	1	-1	i	$-i$
-1	-1	1	$-i$	i
i	i	$-i$	-1	1
$-i$	$-i$	i	1	-1

$\oplus$-inverses are:

$-0 = 0$

$-1 = 4$

$-2 = 3$

$-3 = 2$

$-4 = 1$

$\cdot$-inverses are:

$1^{-1} = 1$

$(-1)^{-1} = -1$

$i^{-1} = -i$

$(-i)^{-1} = i$

Thus, we have discussed four basic properties of operations in this section, namely

1. Associative property

2. Commutative property

3. Existence of identity element

4. Existence of inverses of elements

Some systems have all, none, or some of these properties. The more properties a system possesses, the "richer" its structure, and the more interesting and applicable the theorems become. In the next section we shall investigate many systems and ask ourselves questions about the richness of their structure in each case.

Assigned Problems

1. Prove that if an algebraic system $(S, *)$ has an identity element, then this identity element is *unique*.

2. Prove that if an algebraic system $(S, *)$ has an associative binary operation and identity element e, then whenever an element has an inverse, the inverse is unique.

3. Let $S = \{x \in R | 0 \le x \le 1\}$. Define a binary operation $*$ on S by

$$\forall_{x, y \in S} \left[x * y = \frac{x + y + |x - y|}{2} \right]$$

Prove that:

(a) $*$ is associative.

(b) There exists an identity element for $(S, *)$. Identify it.

(c) Is $*$ commutative? Explain.

4. Recall the "midpoint" operation on the set of all points in a plane (see Section 5.5, Problem 4d) defined by

$$\forall_{p_1, p_2 \in P} \ p_1 * p_2 = \begin{cases} \text{midpoint of segment } \overline{p_1 p_2} & \text{if } p_1 \neq p_2 \\ p_1 & \text{if } p_1 = p_2 \end{cases}$$

Explain why or why not the midpoint operation is associative. Is the midpoint operation commutative? Is there an identity element?

5. Suppose $(S, *)$ and $(T, \circ)$ are algebraic systems, and f is an epimorphism (onto homomorphism) mapping $(S, *) \to (T, \circ)$. Prove each of the following:

(a) If $*$ is associative in S, then $\circ$ is associative in T.

(b) If $*$ is commutative in S, then $\circ$ is commutative in T.

(c) If e is the identity element for S, then $f(e)$ is the identity element for T.

(d) If $\bar{a}$ is the inverse of element a in S, then $f(\bar{a})$ is the inverse of $f(a)$ in T.

6. Let $C = \{a + bi | a, b \in R \text{ and } i^2 = -1\}$ be the system of complex numbers. Recall the definition of complex number *multiplication*: for any two complex numbers $a + bi$, $c + di$,

$$(a + bi) \cdot (c + di) = (ac - bd) + (ad + bc)i$$

Using this definition,

(a) Find the identity element for multiplication and show it works.

(b) Let $a + bi$ be any complex number except $0 + 0i$, and find its multiplicative inverse. Show it works.

7. Let S be any set, $\mathscr{P}(S)$ its power set (set of all subsets of S). Discuss the following operations on $\mathscr{P}(S)$ with respect to associativity, commutativity, existence of identity element, and existence of inverses.

(a) $A * B = A \cup B$ (union of sets)

(b) $A * B = A \cap B$ (intersection of sets)

(c) $A * B = (A \cup B) - (A \cap B)$ (symmetric difference of sets)

8. Let $A = \{1, 2, 3\}$ and $S_3 = \{\text{all permutations of } A\}$. Find the inverse of each permutation in S_3 with respect to permutation multiplication. Does this

generalize to all finite sets of permutations? That is, if A has n elements, does every permutation in S_n have an inverse? Why or why not?

9. Consider the system $F = \{ f | f: R \to R \}$ of functions mapping reals to reals.

   (a) With respect to the operation of *addition* of functions,

      (i) Is $+$ associative?

      (ii) Is $+$ commutative?

      (iii) What is the identity element relative to $+$?

      (iv) Given a function f, what is its additive inverse?

   (b) With respect to the operation of *composition* of functions,

      (i) Is $\circ$ associative?

      (ii) Is $\circ$ commutative?

      (iii) What is the identity element relative to $\circ$?

      (iv) Which elements in F have $\circ$-inverses, and what are the inverses?

10. Construct addition and multiplication tables for the residue-class (or modular) arithmetics: W_3, W_4, W_5, W_6, W_7, and W_9. In each case assume both operations are associative (they are, and the proof rests upon associativity of standard $+$ and $\cdot$). What about commutativity and identity elements? Does every element in each system have a $+$-inverse? Does every "nonzero" element in each system have a multiplicative-inverse? Any conjectures?

6.2   *Examples of Systems*

This section is devoted entirely to examples of algebraic systems $(S, *)$ consisting of a nonempty set S involving a single binary operation $*$ in each case. Some of the systems will be familiar to you, and some will not. Additionally, the *associative*, *commutative*, *identity element*, and *inverse* properties may or may not be present, and the assigned problems will deal with checking these properties in each instance. In the next section we shall classify algebraic systems having one operation by their structure.

Example 1   **The Classical Number Systems**

(a) N = natural numbers = $\{1, 2, 3, 4, \dots\}$
   Operation: addition

(b) N = natural numbers = $\{1, 2, 3, 4, \dots\}$
   Operation: multiplication

(c) W = whole numbers = $\{0, 1, 2, 3, \ldots\}$
Operation: addition

(d) W = whole numbers = $\{0, 1, 2, 3, \ldots\}$
Operation: multiplication

(e) I = integers = $\{\ldots -3, -2, -1, 0, 1, 2, 3, \ldots\}$
Operation: addition

(f) I = integers = $\{\ldots -3, -2, -1, 0, 1, 2, 3, \ldots\}$
Operation: multiplication

(g) Q = rational numbers = $\left\{\dfrac{a}{b} \,\middle|\, a, b \in I \text{ and } b \neq 0\right\}$
Operation: addition

(h) $Q - \{0\} = \left(\begin{array}{c}\text{rational numbers} \\ \text{excluding } 0\end{array}\right) = \left\{\dfrac{a}{b} \,\middle|\, a, b \in I \text{ and } a, b \neq 0\right\}$
Operation: multiplication

(i) R = real numbers
Operation: addition

(j) $R - \{0\}$ = real numbers excluding 0
Operation: multiplication

Example 2   *The Complex Number System*

(a) C = complex numbers = $\{a + bi \mid a, b \in R \text{ and } i^2 = -1\}$
Operation: complex addition, defined by

$$(a + bi) + (c + di) = (a + c) + (b + d)i$$

(b) $C - \{0 + 0i\}$ = complex numbers (excluding $0 + 0i$) = $\{a + bi \mid a, b \in R$ and $i^2 = -1, a \neq 0 \text{ or } b \neq 0\}$
Operation: complex multiplication, defined by

$$(a + bi) \cdot (c + di) = (ac - bd) + (ad + bc)i$$

Example 3   *The Residue Class, or Modular Arithmetics*

(a) $W_n = \{0, 1, 2, 3, \ldots n - 1\}$
Operation: $\oplus$ addition modulo n (add two elements as whole numbers and take remainder of sum upon division by n)

(b) $W_n - \{0\} = \{1, 2, 3, \ldots n - 1\}$
Operation: $\odot$ multiplication modulo n (multiply two elements as whole numbers and take remainder of product upon division by n)

Example 4   *Systems Involving Collections of Sets*

Let X be a set, $\mathscr{P}(X)$ its power set,

(a) $\mathscr{P}(X) = \{$all subsets of $X\}$
Operation: set union $(\cup)$

(b) $\mathscr{P}(X) = \{$all subsets of $X\}$
Operation: set intersection $(\cap)$

(c) $\mathscr{P}(X) = \{$all subsets of $X\}$
Operation: symmetric difference of sets $(\triangle)$

$$A \triangle B = (A \cup B) - (A \cap B)$$

Example 5   *Systems Involving Collections of Functions*

(a) Let X be a subset of R, $S = \{f | f\colon X \to X, f \text{ a function}\}$
Operation: function addition

$$\forall_{f,g \in S} \; \forall_{x \in X} \; [(f + g)(x) = f(x) + g(x)]$$

(b) Let X be a subset of R (not containing 0), $S = \{f | f\colon X \to X, f \text{ a function}\}$
Operation: function multiplication

$$\forall_{f,g \in S} \; \forall_{x \in X} \; [(f \cdot g)(x) = f(x) \cdot g(x)]$$

(c) Let X be any set, $S = \{f | f\colon X \to X, f \text{ a function}\}$
Operation: function composition

$$\forall_{f,g \in S} \; \forall_{x \in X} \; [(f \circ g)(x) = f(g(x))]$$

(d) Let X be a set, $S = \{f | f\colon X \xrightarrow[\text{onto}]{1\text{-}1} X\}$
Operation: function composition

(e) Let $R =$ real numbers, $D = \{$all differentiable $R \to R$ functions$\}$
Operation: function addition

(f) Let $R =$ real numbers, $D = \{$all differentiable $R \to R$ functions$\}$
Operation: function composition

Example 6   *Systems of Integral Multiples*

Let $kI = \{kn | k \text{ is some fixed positive integer}; n \text{ is any integer}\}$ (for example, $2I$ would be the set of all *even* integers; 2 is fixed, and n ranges through all integers)

(a) Set: kI for some fixed positive integer k, $I = \{$all integers$\}$
Operation: standard addition

(b) Set: kI for some fixed positive integer k, $I = \{$all integers$\}$
Operation: standard multiplication

Example 7   *Permutations of a Finite Set*

(a) Let $S = \{1, 2\}$; $S_2 = \left\{ \begin{pmatrix} 1 & 2 \\ 1 & 2 \end{pmatrix}, \begin{pmatrix} 1 & 2 \\ 2 & 1 \end{pmatrix} \right\}$ (all permutations of S)

Operation (on S_2): permutation multiplication (or function composition)

(b) Let $S = \{1, 2, 3\}$; $S_3 = \left\{ \begin{pmatrix} 1 & 2 & 3 \\ 1 & 2 & 3 \end{pmatrix}, \begin{pmatrix} 1 & 2 & 3 \\ 2 & 1 & 3 \end{pmatrix}, \begin{pmatrix} 1 & 2 & 3 \\ 1 & 3 & 2 \end{pmatrix}, \right.$

$\left. \begin{pmatrix} 1 & 2 & 3 \\ 3 & 2 & 1 \end{pmatrix}, \begin{pmatrix} 1 & 2 & 3 \\ 2 & 3 & 1 \end{pmatrix}, \begin{pmatrix} 1 & 2 & 3 \\ 3 & 1 & 2 \end{pmatrix} \right\}$

Operation (on S_3): permutation multiplication (function composition)

(c) Let $S = \{1, 2, 3, 4\}$; $S_4 = \{$all 24 permutations of $S\}$
Operation (on S_4): permutation multiplication (function composition)

(d) In general, let $S = \{1, 2, 3, \ldots n\}$, n a positive integer ($n > 1$); $S_n = $ all permutations
of S
Operation (on S_n): permutation multiplication (function composition)

Example 8   *Fourth Roots of Unity*

Let $C = \{1, -1, i, -i\}$
Operation: complex multiplication

Example 9   *Dihedral Systems (Planar)*

This is a collection of systems that are derived from certain rigid motions (rotations, reflections) of plane figures. In each case, we start with a plane regular polygon, label its vertices and axes of symmetry, then form the set of all counterclockwise rotations about its center of symmetry which permute vertices together with all reflections about axes of symmetry that permute vertices. The operation is function composition. Simply put, "transformation T_1 followed by transformation T_2 yields the same result (on the vertices arrangement) as transformation___."

(a) Model: Equilateral triangle
Set: $\{R_0, R_{120}, R_{240}, F_1, F_2, F_3\}$
Operation: composition

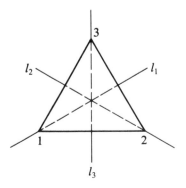

(b) Model: Square
Set: $\{R_0, R_{90}, R_{180}, R_{270}, F_1, F_2, F_3, F_4\}$
Operation: composition

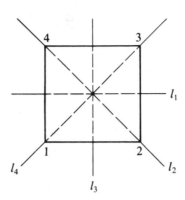

(c) Model: Rectangle (leads to a subsystem of square)
Set: $\{R_0, R_{180}, F_1, F_2\}$
Operation: composition

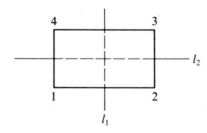

Example 10   *Gaussian Integers*

(a) Let $G = \{a + bi \,|\, a, b$ are *integers*, and $i^2 = -1\}$
Operation: complex number addition (see Example 2)

(b) Let $G = \{a + bi \,|\, a, b$ are *integers*, and $i^2 = -1\}$
Operation: complex number multiplication (see Example 2)

Example 11   *Adjunction Systems to Rational Numbers*

(a) Let Q = rational numbers
$Q\sqrt{2} = \{a + b\sqrt{2} \,|\, a, b \in Q\}$
Operation: addition of reals

(b) $Q\sqrt{2} = \{a + b\sqrt{2} \mid a, b \in Q\}$
Operation: multiplication of reals

These can be generalized to $Q(r)$, where r is any irrational number.

Example 12   **Systems of Vectors, or Ordered n-tuples**

(a) Let R = reals; $R^2 = \{(a, b) \mid a, b \in R\}$
Operation: componentwise addition

$$(a, b) + (c, d) = (a + c, b + d)$$

(b) Let R = reals; $R^2 = \{(a, b) \mid a, b \in R\}$
Operation: "special multiplication"

$$(a, b) \cdot (c, d) = (ac + bd, ad + bc)$$

(c) Let R = reals; $R^n = \{(a_1, a_2, \ldots a_n) \mid a_i \in R, i = 1, 2, \ldots n\}$
Operation: componentwise addition

$$(a_1, a_2, \ldots a_n) + (b_1, b_2, \ldots b_n) = (a_1 + b_1, a_2 + b_2, \ldots a_n + b_n)$$

Example 13   **Systems of Matrices**

(a) [2 × 2]   Let $M = \left\{\begin{pmatrix} a & b \\ c & d \end{pmatrix} \middle| a, b, c, d \in I\right\}$

Operation: matrix addition (componentwise)

$$\begin{pmatrix} a & b \\ c & d \end{pmatrix} + \begin{pmatrix} e & f \\ g & h \end{pmatrix} = \begin{pmatrix} a + e & b + f \\ c + g & d + h \end{pmatrix}$$

(b) [$n \times n$]   Let $M = \left\{\begin{pmatrix} a_{11} & a_{12} & \cdots & a_{1n} \\ a_{21} & a_{22} & \cdots & a_{2n} \\ \vdots & \vdots & & \vdots \\ a_{n1} & a_{n2} & \cdots & a_{nn} \end{pmatrix} \middle| a_{ij} \in I \left\{\begin{matrix} i = 1, 2, \ldots n \\ j = 1, 2, \ldots n \end{matrix}\right.\right\}$

Operation: matrix addition (componentwise)

(c) [2 × 2]   Let $M = \left\{\begin{pmatrix} a & b \\ c & d \end{pmatrix} \middle| a, b, c, d \in I\right\}$

Operation: matrix multiplication

$$\begin{pmatrix} a & b \\ c & d \end{pmatrix} \cdot \begin{pmatrix} e & f \\ g & h \end{pmatrix} = \begin{pmatrix} ae + bg & af + bh \\ ce + dg & cf + dh \end{pmatrix}$$

(d) [a special 2 × 2 system]

$$S = \left\{\begin{pmatrix} 1 & 0 \\ 0 & 1 \end{pmatrix}, \begin{pmatrix} 0 & 1 \\ -1 & 0 \end{pmatrix}, \begin{pmatrix} -1 & 0 \\ 0 & -1 \end{pmatrix}, \begin{pmatrix} 0 & -1 \\ 1 & 0 \end{pmatrix}\right\}$$

Operation: matrix multiplication (see (c) above)

(e) [another 2 × 2 system]

$$S = \left\{ \begin{pmatrix} 0 & 0 \\ 0 & 0 \end{pmatrix}, \begin{pmatrix} 1 & 0 \\ 0 & 0 \end{pmatrix}, \begin{pmatrix} 0 & 1 \\ 0 & 0 \end{pmatrix}, \begin{pmatrix} 1 & 1 \\ 0 & 0 \end{pmatrix} \right\}$$

Operation: matrix addition with mod 2 entrywise addition

(f) [still another 2 × 2 system]

$$S = \left\{ \begin{pmatrix} a & b \\ 0 & 0 \end{pmatrix} \middle| a, b \in I \right\}$$

Operation: matrix addition

(g) $S = \left\{ \begin{pmatrix} a & b \\ 0 & 0 \end{pmatrix} \middle| a, b \in I \right\}$

Operation: matrix multiplication

(h) $S = \left\{ \begin{pmatrix} a & 0 \\ b & 0 \end{pmatrix} \middle| a, b \in I \right\}$

Operation: matrix addition

(i) $S = \left\{ \begin{pmatrix} a & 0 \\ b & 0 \end{pmatrix} \middle| a, b \in I \right\}$

Operation: matrix multiplication

Example 14   *Systems of Polynomials of One Variable*

(a) Let $P = \left\{ \sum_{i=0}^{n} a_i x^i \middle| a_i \in R \text{ and } n \text{ is a nonnegative integer} \right\}$

Operation: polynomial addition (add coefficients of corresponding terms)

$$(n \geq m) \sum_{i=0}^{n} a_i x^i + \sum_{i=0}^{m} b_i x^i = \sum_{i=0}^{n} (a_i + b_i) x^i$$

(b) Let $P = \left\{ \sum_{i=0}^{n} a_i x^i \middle| a_i \in R \text{ and } n \text{ is a nonnegative integer} \right\}$

Operation: polynomial multiplication

$$\left(\sum_{i=0}^{n} a_i x^i \right) \cdot \left(\sum_{j=0}^{m} b_j x^j \right) = \sum_{k=0}^{n+m} \left(\sum_{i+j=k} a_i b_j \right) x^k$$

Note: $\sum_{i+j=k} a_i b_j = a_0 b_k + a_1 b_{k-1} + \cdots + a_k b_0.$

Example 15   ***Direct Sum Ring***

Let $2I = \{$even integers$\} = \{2n|n \in I\}$
$\quad 3I = \{$integral multiples of 3$\} = \{3k|k \in I\}$
Consider $(2I) \times (3I) = \{(a, b)|a \in 2I, b \in 3I\}$

(a) Set: $(2I) \times (3I)$
Operation: componentwise addition

$$(a, b) + (c, d) = (a + c, b + d)$$

(b) Set: $(2I) \times (3I)$
Operation: componentwise multiplication

$$(a, b) \cdot (c, d) = (ac, bd)$$

Example 16 $W = $ whole numbers $= \{0, 1, 2, 3, \dots\}$
Operation: $a * b = \begin{cases} a \cdot b & \text{if } a = b \\ a + b & \text{if } a \neq b \end{cases}$

Example 17 $W = $ whole numbers $= \{0, 1, 2, 3, \dots\}$
Operation: $a * b = \begin{cases} 0 & \text{if } a = b \\ a + b & \text{if } a \neq b \end{cases}$

Example 18 $W = $ whole numbers $= \{0, 1, 2, 3, \dots\}$
Operation: $a * b = \begin{cases} a + b & \text{if either is 0} \\ 2a + b & \text{otherwise} \end{cases}$

Example 19 $S = \{2^n|n \in I\}$
Operation: multiplication

Example 20 $N = $ natural numbers $= \{1, 2, 3, \dots\}$
Operation: $a * b = a^b$

Example 21 $R = $ real numbers
Operation: averaging (arithmetic mean)

$$a * b = \frac{a + b}{2}$$

Example 22 R = real numbers
Operation: $x * y = \min\{x, y\}$

(if $x = y$, then let $x * y = x$)

Example 23   **Finite Abstract Systems with Two Elements**

(a) $S = \{a, b\}$
Operation $*$ defined by table

$*$	a	b
a	a	b
b	b	a

(b) $S = \{a, b\}$
Operation $\triangle$ defined by table

$\triangle$	a	b
a	a	a
b	a	b

Example 24   **Finite Abstract Systems with Three Elements**

(a) $S = \{a, b, c\}$
Operation $*$ defined by table

$*$	a	b	c
a	a	b	c
b	b	c	a
c	c	a	b

(b) $S = \{a, b, c\}$
Operation $\triangle$ defined by table

$\triangle$	a	b	c
a	a	b	c
b	b	a	c
c	c	b	a

Example 25   **Finite Abstract Systems with Four Elements**

(a) $S = \{e, a, b, c\}$
Operation $*$ defined by table

$*$	e	a	b	c
e	e	a	b	c
a	a	b	c	e
b	b	c	e	a
c	c	e	a	b

(b) $V = \{e, a, b, c\}$
Operation $\circ$ defined by table
(This is the famous Klein-4
group, or "Vierergruppe.")

$\circ$	e	a	b	c
e	e	a	b	c
a	a	e	c	b
b	b	c	e	a
c	c	b	a	e

(c) $S = \{1, a, a^2, a^3\}$
Operation $\triangle$ defined by table
(This is called a cyclic group
with generator a.)

$\triangle$	1	a	a^2	a^3
1	1	a	a^2	a^3
a	a	a^2	a^3	1
a^2	a^2	a^3	1	a
a^3	a^3	1	a	a^2

(d) $S = \{a, b, c, d\}$
Operation $*$ defined by table

$*$	a	b	c	d
a	a	b	c	d
b	b	a	d	c
c	c	d	a	a
d	d	c	b	b

Example 26   **Finite Abstract Systems with Five Elements**

(a) $S = \{e, a, b, c, d\}$
Operation $*$ defined by table

$*$	e	a	b	c	d
e	e	a	b	c	d
a	a	b	c	d	e
b	b	c	d	e	a
c	c	d	e	a	b
d	d	e	a	b	c

(b) $S = \{x, y, z, w, t\}$
Operation $\triangle$ defined by table

$\triangle$	x	y	z	w	t
x	t	y	z	x	w
y	w	z	t	y	x
z	y	x	w	z	t
w	x	y	z	w	t
t	z	w	x	t	y

Example 27   **Vector Cross-Product**

Let $S = \{\langle a_1, a_2, a_3 \rangle | a_1, a_2, a_3 \in R\}$
(S is the set of all vectors in three-dimensional space)
Operation: vector cross-product

$$\langle a_1, a_2, a_3 \rangle x \langle b_1, b_2, b_3 \rangle = \langle a_2 b_3 - a_3 b_2, a_3 b_1 - a_1 b_3, a_1 b_2 - a_2 b_1 \rangle \cdot$$

$$\left(\text{or} \ \begin{vmatrix} i & j & k \\ a_1 & a_2 & a_3 \\ b_1 & b_2 & b_3 \end{vmatrix} \right)$$

Example 28 $P = \{$all points in the plane$\}$

Operation: $p_1 * p_2 = \begin{cases} \text{midpoint of segment } \overline{p_1 p_2} & \text{if } p_1 \neq p_2 \\ p_1 & \text{if } p_1 = p_2 \end{cases}$

Example 29 $R^2 = \{(a, b) | a, b \in R\}$
Operation: $(a, b) * (c, d) = (a, d)$

Example 30   **Real Quaternions**

Let $H = \{a + bi + cj + dk \,|\, a, b, c, d \in R\}$
where i, j, and k are elements of a system under "multiplication,"

Mnemonic

which acts as follows:

$$i^2 = j^2 = k^2 = -1$$
$$ij = k \qquad ji = -k$$
$$jk = i \qquad kj = -i$$
$$ki = j \qquad ik = -j$$

(a) Set: H, as above
    Operation: addition componentwise
$$(a + bi + cj + dk) + (e + fi + gj + hk)$$
$$= (a + e) + (b + f)i + (c + g)j + (d + h)k$$

(b) Set: H, as above
    Operation: quaternion multiplication
$$(a + bi + cj + dk) \cdot (e + fi + gj + hk)$$
$$= (ae - bf - cg - dh)$$
$$+ (af + be + ch - dg)i$$
$$+ (ag + ce + df - bh)j$$
$$+ (ah + de + bg - cf)k$$

Assigned Problems

1. Study all the examples presented in this section. Decide in your own mind which of the properties of operations—associative, commutative, identity element, inverse elements—hold in each instance. For perhaps many of the systems, this will be known; for others, it may not be easy to determine. You need not "prove" your way through all of the properties for each of the systems; obviously, that would be extremely time-consuming and tedious. However, you should be able to verify the existence or nonexistence of a given

property for a given system. Be able to make judgments about these properties in "different-looking" systems, such as those involving sets, functions, vectors, and matrices (Examples 4, 5, 12, and 13). Also, Examples 16, 22, 28, and 29 provide good practice in unfamiliar (contrived) situations, while Examples 23–26 all deal with tables. Example 30 on quaternions is an interesting extension of the complex number ideas to a "hypercomplex" system, but checking it out thoroughly is something you might reserve for a long, cold winter's night.

2. Some of the "different-looking" systems are isomorphic. For example, $(2I, +)$ is isomorphic to $(3I, +)$ (also true for multiplication); also, the dihedral systems act "like" permutation systems, $Q\sqrt{2}$ behaves like $Q\sqrt{3}$, polynomials of degree n can be regarded as ordered n-tuples, and some of the finite abstract systems (Examples 23–26) and modular arithmetics (Example 3) are structurally the same. Look over the list of examples and see how many isomorphisms you can spot.

6.3   *Classification of Systems*

The power of the abstract approach to mathematics becomes clearer when we classify systems according to their algebraic structure, prove theorems about the more general abstract systems, and recognize how the results of these theorems apply simultaneously to large collections of particular systems. We shall use the ideas of *binary operation, associativity, commutativity, identity element,* and *inverse of an element* to classify algebraic systems. As before, we shall restrict our discussion to systems involving only a single binary operation. The most important and richest structure of all such systems is embodied in the concept of *group*, which we make our starting point of classification.

Definition of Group

A *group* is an algebraic system $(G, *)$ consisting of a nonempty set G and a single binary operation $*$ on G having the following properties:

1. $*$ is *associative* in G:
$$[\forall_{a, b, c \in G} \; (a * b) * c = a * (b * c)]$$

2. There exists an *identity element e* in G:
$$[\exists_{e \in G} \; \forall_{x \in G} \; a * e = e * a = a]$$

3. Every element in G has an *inverse* (in G) with respect to $*$:
$$[\forall_{a \in G} \; \exists_{\bar{a} \in G} \; a * \bar{a} = \bar{a} * a = e]$$

Note that the group operation ∗ *need not* be *commutative*. If the operation ∗ *is* commutative, we call $(G, *)$ an *Abelian* (or *commutative*) group. The term *Abelian* derives from the Norwegian algebraist Nils Henrik Abel, who at the age of 19 (and thereafter) made significant contributions to the theory of algebra.

Following are some examples of groups.

Example 1   *Twenty Examples of Groups*

(a) $(I, +)$ Integers under addition

(b) $(Q, +)$ Rationals under addition

(c) $(Q - \{0\}, \cdot)$ Rationals excluding 0 under multiplication

(d) $(R, +)$ Reals under addition

(e) $(R - \{0\}, \cdot)$ Reals excluding 0 under multiplication

(f) $(C, +)$ Complex numbers under addition

(g) $(C - \{0 + 0i\}, \cdot)$ Complex numbers excluding $0 + 0i$ under multiplication

(h) $(kI, +)$ Integer-multiples (of k) under addition (any $k \in I^+$)

(i) $(W_n, \oplus)$ All residue-class arithmetics under addition (mod n)

(j) $(W_p, \odot)$ Prime residue-class arithmetics under multiplication (mod prime p)

(k) $(\{1, -1, i, -i\}, \cdot)$ Fourth roots of unity under multiplication

(l) $(S_n, \circ)$ Sets of permutations on n elements under permutation multiplication

(m) (Gaussian integers, $+$) Gaussian integers under complex addition

(n) $(M, +)$ All $n \times n$ matrices with integer entries under matrix addition

(o) $(D, \circ)$ Dihedral systems under composition

(p) $(P, +)$ Real polynomials (one variable) under addition

(q) $(R^n, +)$ Ordered n-tuples of reals (vectors) under componentwise addition

(r) $(V_4, *)$ Klein 4-group (abstract)

∗	e	a	b	c
e	e	a	b	c
a	a	e	c	b
b	b	c	e	a
c	c	b	a	e

(s) $(C_4, *)$ Cyclic 4-group (abstract)

∗	1	a	a^2	a^3
1	1	a	a^2	a^3
a	a	a^2	a^3	1
a^2	a^2	a^3	1	a
a^3	a^3	1	a	a^2

(t) $(F, \circ)$ all one-to-one, onto functions from a set X to itself, under function composition

The twenty groups listed above in Example 1 represent a small fraction of groups known in mathematics. However, these are perhaps better known groups, and they are given here to illustrate the wide range of systems that exhibit basically the same kind of structure—that of being a *group*. Most of the systems listed are Abelian groups; however, since function composition is a noncommutative operation, some that involve that operation are non-Abelian groups, such as Examples (l), (o), and (t).

If we gradually relax the conditions for defining a group among single binary operation systems, we produce general systems that are related to groups but that are less rich in structure. For example, omitting condition (3)—existence of inverse for every element in the system—produces a system called a *monoid*.

Definition of Monoid

> A *monoid* is an algebraic system $(S, *)$ consisting of a nonempty set S and a single binary operation $*$ on S having the following properties:
>
> 1. $*$ is *associative* in S.
>
> 2. There exists an *identity element e* in S.

Certainly every *group* is a *monoid*, but *not* every *monoid* is a *group*. The following examples illustrate this.

Example 2   *A Commutative Monoid*

Consider the system $(R, \cdot)$, the real numbers under multiplication. Multiplication is a *binary operation* in the system of real numbers. Every time you multiply two real numbers, the product is a unique real number. Multiplication of real numbers is *associative*, and there is an *identity element* for multiplication, namely 1, since

$$\forall_{x \in R} \ x \cdot 1 = 1 \cdot x = x$$

But *not every real number has an inverse with respect to multiplication*; in fact, 0 does not. If it did, there would have to be a real number a satisfying the equation

$$0 \cdot a = 1$$

and the product of 0 with any real number is 0, not 1. Therefore, $(R, \cdot)$ is a monoid and not a group. Since multiplication is commutative, we call $(R, \cdot)$ a *commutative monoid*.

By similar reasoning, $(Q, \cdot)$ and $(C, \cdot)$ are also commutative monoids since their "zero" elements have no inverses under the operation of multiplication. This is why we exclude "0" or "$0 + 0i$" in order to form commutative groups in these systems.

Example 3   *A Noncommutative Monoid*

Let X be any set and let

$$S = \{ f | f: X \rightarrow X \}$$

be the set of all functions mapping X to itself. Consider the system $(S, \circ)$ whose operation is function composition. Function composition is a *binary operation* on S since every composition of two functions mapping $X \rightarrow X$ is again a unique function mapping $X \rightarrow X$. We have proved a theorem that says function composition is *associative* (see Section 5.3). Moreover, there is an *identity element* for this system, namely the identity function i_X that maps every element in X to itself:

$$\forall_{x \in X} \; i_X(x) = x$$

This is indeed an identity element for $(S, \circ)$ since

$$\forall_{f \in S} \; \forall_{x \in X} \; (f \circ i_X)(x) = f(i_X(x))$$

$$= f(x)$$

also

$$(i_X \circ f)(x) = i_X(f(x))$$

$$= f(x)$$

Therefore, $f \circ i_X = i_X \circ f = f$ for every $f \in S$.

Now, *not* every element (function) in S has an inverse with respect to function composition. In fact, only those functions that are both one-to-one and onto have inverses under composition; all other functions do not. Therefore $(S, \circ)$ is a monoid; it is not a group; and since function composition is not commutative, $(S, \circ)$ is a *noncommutative monoid*.

Next, suppose we take the monoid classification and further relax the condition that there need be an *identity element* in the system. This produces a more general system (less rich in structure) called a *semigroup*.

Definition of Semigroup

> A *semigroup* is an algebraic system $(S, *)$ consisting of a nonempty set S and a single binary operation $*$ on S that is *associative*.

Thus, to determine that a given system is a *semigroup*, all we need to do is verify that we have a *binary operation* (closure, uniqueness) and that it is *associative*. Therefore, every group is a monoid and every monoid is a semigroup. The classifications we are developing (by relaxing conditions) are getting larger and more inclusive each time, with each new general system including those that preceded it. But the inclusion does not work in both directions; that is, there are monoids that are not groups (Examples 2 and 3) and there are semigroups that are not monoids, as the next examples demonstrate.

Example 4   A Commutative Semigroup

Consider the system $(N, +)$ where N is the set of all natural numbers

$$N = \{1, 2, 3, 4, \ldots\}$$

Addition is a *binary operation* on N (the sum of any two natural numbers is a unique natural number), and addition is *associative* in the system $(N, +)$. (We take this as an assumption in this book; it can be proved in any formal development or construction of the system of natural numbers from the Peano postulates.)

However, this system *lacks an additive identity element*. There is no natural number e such that for every natural number n

$$n + e = n$$

Since addition is a commutative operation on N, then we have $(N, +)$ as an example of a *commutative semigroup* that is *not* a monoid.

Example 5   A Noncommutative Semigroup

Let $R^2 = R \times R = \{(a, b) | a, b \in R\}$ ($R = \{$real numbers$\}$). Define operation $*$ on R^2 by:

$$\forall_{(a, b), (c, d) \in R^2} \; [(a, b) * (c, d) = (a, d)]$$

That is, the outcome of performing this operation on two ordered pairs is an ordered pair having its first component the same as the first component of the first pair and its second component the same as the second component of the second pair. It is, therefore, a *binary operation* on R^2, since each *pair* of elements in R^2 (ordered pairs) is assigned to a unique (determined by first and last components) element (ordered pair) in R^2. That operation $*$ is associative must be proved.

$$[(a, b) * (c, d)] * (e, f) = (a, d) * (e, f)$$
$$= (a, f)$$

On the other hand,

$$(a, b) * [(c, d) * (e, f)] = (a, b) * (c, f)$$
$$= (a, f)$$

It is easy to see that the "survivors" in each case after performing the operation any finite number of times will always be the "leftmost" component and the "rightmost" component. Therefore, $*$ is an *associative* operation.

The operation is *not* commutative, since

$$(2, 3) * (4, 5) = (2, 5)$$

but

$$(4,5) * (2,3) = (4,3)$$

and

$$(2,5) \neq (4,3)$$

Moreover, there is *no identity element* in this system with respect to $*$. This is clear since the operation "breaks apart" the ordered pair, using only its first (or last) component. For any element (e_1, e_2) to be an identity element, it would require every ordered pair in R^2 to have e_2 as its second component or e_1 as its first component in order to reproduce the given ordered pair in every instance. This is not the case, since the ordered pairs in R^2 vary.

Therefore, the system $(R^2, *)$ where $*$ is defined above is an example of a *noncommutative semigroup* that is *not* a monoid.

Suppose we take the final step and eliminate the associativity condition, so that all we have is a system $(S, *)$ consisting of a nonempty set S and a *binary operation* $*$ on S. This maneuver leads us to one of two possible general systems —*groupoid* or *loop*.

Definition of Loop

A *loop* is an algebraic system $(S, *)$ consisting of a nonempty set S and a binary operation $*$ on S. An additional characteristic of a loop is that the *cancellation laws* (both right and left) hold. That is,

$$\forall_{a,b,c \in S} \text{ (if } a * b = a * c, \text{ then } b = c)$$
$$\text{and}$$
$$\forall_{a,b,c \in S} \text{ (if } a * c = b * c, \text{ then } a = b)$$

Whether or not these cancellation laws hold becomes more apparent when the operation is defined by a table. To require both cancellation laws to hold is equivalent to requiring each element in the system to appear *at most* once in every row (left cancellation) and every column (right cancellation) in the table. If the set S is *finite*, this condition is equivalent to requiring each element in the system to appear *exactly once* (*once* and only once) in every row and column of the table. Following are two examples of *loops*, one having an infinite set and the other having a finite set.

Example 6   ***An Infinite Loop***

Take the system of integers under subtraction $(I, -)$. Subtraction is a binary operation on I since the difference of any two integers is a unique integer. However, subtraction is nonassociative, since

$$(5 - 3) - 2 \neq 5 - (3 - 2)$$

Also subtraction is noncommutative and there exists no (two-sided) identity element for subtraction. The right and left cancellation laws *do* hold, however, for subtraction; that is,

$$\forall_{a, b, c \in I} \left[(\text{if } a - b = a - c, \text{ then } b = c) \text{ and } (\text{if } a - c = b - c, \text{ then } a = b) \right]$$

Therefore, $(I, -)$ is an example of an infinite (noncommutative) loop.

Example 7   A Finite Loop

Let $S = \{a, b, c, d, e\}$, and define a binary operation $*$ on S by the table below.

$*$	a	b	c	d	e
a	b	c	d	e	a
b	e	d	a	c	b
c	d	a	e	b	c
d	c	e	b	a	d
e	a	b	c	d	e

The operation $*$ is a binary operation on S since the only outcomes that appear in the cells of the table are elements of S (closure) and no cell contains more than one element (uniqueness). The operation is nonassociative, since

$$(a * b) * c \neq a * (b * c)$$

Since there is not perfect symmetry with respect to the main diagonal (e.g., $a * b \neq b * a$), $*$ is noncommutative. Curiously, there is an identity element e, and every element has $*$-inverse (some authors require existence of identity element and inverses in their definition of loop). $(S, *)$ as defined above is an abstract finite loop.

The distinguishing feature between the structure characterized as a *loop* and that characterized as a *groupoid* is the right/left cancellation laws. A loop is a groupoid in which the cancellation laws hold, but not every groupoid is a loop. As a final example, we devise two systems having the most general structure—with a nonempty set and a binary operation but no other properties—the groupoid.

Example 8   A Finite Groupoid

Let $S = \{a, b, c\}$ and define $*$ by the table below:

$*$	a	b	c
a	b	a	a
b	c	a	b
c	b	c	a

It is left to you to check that $*$ is a binary operation on S, but $*$ is *nonassociative*, *noncommutative*, there is no *identity element* (therefore *inverses* make no sense), and the *cancellation laws* do not hold. $(S, *)$ is an example of a finite groupoid that is *not* a loop.

Example 9   An Infinite Groupoid

Consider the system $(N, *)$ where $N = \{1, 2, 3, \dots\}$ is the set of natural numbers, and $*$ is defined by

$$\forall_{a, b \in N} \quad a * b = \begin{cases} a \cdot b & \text{if } a = b \\ a + b & \text{if } a \neq b \end{cases}$$

The operation $*$ is a binary operation on N since products and sums of natural numbers always yield unique natural numbers. However, $*$ is *nonassociative* since

$$(5 * 3) * 2 \neq 5 * (3 * 2)$$

Operation $*$ is *commutative* since both multiplication and addition are commutative operations on N. There is *no identity element* in N for this operation (why?), and the *cancellation laws fail to hold*, since

$$3 * 3 = 3 * 6 \quad \text{but} \quad 3 \neq 6$$

Thus, $(N, *)$ is an example of an infinite commutative groupoid that is *not* a loop.

We summarize our classification of single-operation systems by the schematic in Figure 6.3, where lower systems in the lattice are included within upper ones to which they are connected.

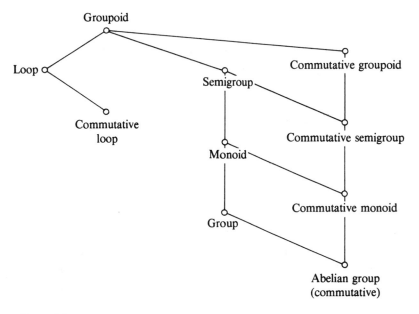

Figure 6.3

Assigned 1. Take the 30 examples that were outlined in Section 6.2 and classify each
Problems   system as closely as possible with respect to each category discussed in this
      section.

2. According to the schematic diagram relating the various systems, it appears
that every group is a loop. This means that in any group $(G, *)$, both right and
left cancellation laws hold. That is,

$$\forall_{a,\,b,\,c \in G} \; [(\text{if } a * b = a * c, \text{ then } b = c) \text{ and } (\text{if } a * c = b * c, \text{ then } a = b)]$$

Prove that this is indeed the case.

3. If $(G, *)$ and $(H, \circ)$ are groups, and if

$$h: (G, *) \to (H, \circ)$$

is a homomorphism, then h is said to be a *group homomorphism*. The set of all
elements in G that map to the identity element in H is called the *kernel of the
homomorphism h*. That is, if e_H is the identity element of H, then

$$\text{kernel } h = \{x \mid x \in G \text{ and } h(x) = e_H\}$$

Prove that the kernel of a group homomorphism is itself a group (in this case, a
subgroup of G).

4. We know that the system $(R^3, +)$ is a group, where $R^3 = \{(x, y, z) \mid x, y,$
$z \in R\}$ whose elements can be regarded as ordered triples or vectors in space,
and where $+$ is componentwise addition of ordered triples, defined by

$$(x, y, z) + (w, v, r) = (x + w, y + v, z + r)$$

Consider each of the following mappings from $(R^3, +) \to (R^3, +)$. In each
case, prove that the mapping is a homomorphism, and then determine its
kernel. Can you visualize a geometric description of the kernel in each case?

(a) $h(x, y, z) = (x, y, 0)$

(b) $h(x, y, z) = (x, y, -z)$

(c) $h(x, y, z) = (x - y, 0, 0)$

5. Let V_3 be the set of three-dimensional vectors with real components. Then

$$V_3 = \{\langle a, b, c \rangle \mid a, b, c \in R\}$$

or

$$V_3 = \{ai + bj + ck \mid a, b, c \in R, i = \langle 1, 0, 0 \rangle, j = \langle 0, 1, 0 \rangle, \text{ and } k = \langle 0, 0, 1 \rangle\}$$

Look up, in any calculus book, the *cross-product* operation on vectors in space.
It can be symbolized as though one were expanding the determinant

$$\begin{vmatrix} i & j & k \\ a & b & c \\ e & f & g \end{vmatrix}$$

where i, j, and k are symbols for the
standard basis vectors and the second and
third rows of the determinant are the components of the vectors

That is,

$$(ai + bj + ck) \, x \, (ei + fj + gk)$$

$$= (bg - fc)i - (ag - ec)j + (af - eb)k$$

What kind of system is (V_3, x)?

6.4   Proofs on Mathematical Systems

The mathematical systems we are most familiar with are quite rich in structure. These systems involve sets of natural, whole, integer, rational, real, or complex numbers and operations of addition and multiplication. Because the operations are associative and commutative, identity elements (0 and 1) exist, cancellation laws hold, and every element (in most cases) has a unique inverse, we take for granted numerous pieces of information that extend these basic properties or that can easily be derived from them. Reduction of structure to an abstract set, a single operation, and a handful of properties usually is accompanied by a feeling of "not much to work with." On the other hand, there is a positive aspect in that we begin to focus on structural properties of the given system and our own mathematical reasoning processes in an attempt to "squeeze out" a maximum of deduced information from a minimum of given information.

The examples that follow involve proofs of some very basic theorems about mathematical systems having a single binary operation. In each case, you should think about how the theorem might translate into information about familiar systems having that structure. For example, any theorem about general groups immediately holds for all the special kinds of groups involving numbers, sets, functions, vectors, matrices, permutations, polynomials, and so on. Thus, while the theorem is proved once, the interpretation of the theorem applies in many diverse situations.

Additionally, it will be helpful to be aware of the *nature* of the proof in each case. We have studied direct proof, indirect proof, conditional proof, proof by cases, existence proof, proof by counterexample, and mathematical induction. As you study the following examples, you should see how these various tools of deductive reasoning are applied to general systems and learn to apply them yourself.

Example 1   We know and have proved that the identity element of a *group* is unique (see Section 6.1, Problem 1). A *groupoid* need not have an identity element. However, if it does, is it necessarily true that the identity element is unique? We wish to prove the following statement:

If a groupoid has an identity element, then that identity element is unique.

Proof   Let $(S, *)$ be a groupoid. What do we know about groupoids? S is nonempty and $*$ is a binary operation—that's all! We wish to prove a conditional statement, so we shall start from a point of view of conditional proof. Assume $(S, *)$ has an identity element, and call it e. To prove uniqueness of e, we need to show that there cannot be two distinct identities. So, taking the stance of indirect proof, suppose e and e' are two different identity elements in the groupoid $(S, *)$. That is, $e \neq e'$ (our assumption). By the definition of identity element (e on left):

$$e * e' = e'$$

Also, by the definition of identity element (e' on right),

$$e * e' = e$$

Finally, since $*$ is a binary operation, it maps the pair (e, e') to exactly one image (result). Thus, we are forced to conclude

$$e' = e$$

But this is contrary to our assumption. Therefore, we conclude $(S, *)$ cannot have two distinct identity elements. Hence, if $(S, *)$ has an identity element, it must be unique.

Example 2   Prove that, in a group $(G, *)$,

$$\forall_{a \in G} \ \bar{\bar{a}} = a$$

(where $\bar{a}$ means the $*$-inverse of $\bar{a}$). It is conceivable that if we start with one element a of a group G and find its inverse $\bar{a}$, that element may be different from a. Then $\bar{a}$, being a group element itself, also has an inverse $\bar{\bar{a}}$ in G that may be different from $\bar{a}$. The question arises: "Is $\bar{\bar{a}}$ different from the original element a?" We try to prove that the structure of a group does *not* permit this to happen.

Proof   Let $a \in G$. We can say this since G is a nonempty set (definition of group). We focus on the nature of $\bar{\bar{a}}$. What does this really mean? $\bar{\bar{a}}$ is the inverse of $\bar{a}$. What does that mean? By definition of inverse,

$$\bar{a} * \bar{\bar{a}} = e$$

where e is the identity element in G. (We know e is an element of G since G is a group and we are using part of the definition of group.)

  Now we stop momentarily and ask ourselves, "What do we wish to prove here?" Looking back, we note that we want to show

$$\bar{\bar{a}} = a$$

We reason backwards momentarily and note that if this were true, this would mean a must play the same role as $\bar{\bar{a}}$, that of being $\bar{a}$'s inverse. That is,

$$\bar{a} * a = e$$

Now, we were thinking of a being the inverse of $\bar{a}$, acting on the right side. But $\bar{a}$ *is* the inverse of a, and we can interpret the equation as $\bar{a}$ acting on the left of a. Therefore the last equation holds. So we now have

$$\bar{a} * \bar{\bar{a}} = e$$

and

$$\bar{a} * a = e$$

This says that both $\bar{a}$ and a are inverses of the same element $\bar{a}$. Appealing to the theorem that says that inverse elements in a group are unique, we conclude $\bar{\bar{a}} = a$. (We could have set both left sides equal and used left cancellation as an alternative method.)

Example 2 proves that the inverse of the inverse of a group element is the original group element. Think of how this theorem applies to various special groups with which you are already familiar. Here are several examples that demonstrate interpretations of the theorem proved in Example 2.

1. Let X be a set, $\mathscr{P}(X)$ its power set. Then $(\mathscr{P}(X), \cap)$ is a group with identity element $\varnothing$ and the inverse of A is $\bar{A}$ (universal complement of A). The above theorem says

$$\forall_{A \in \mathscr{P}(X)} \ \bar{\bar{A}} = A$$

that is, the complement of the complement of a set is the set itself.

2. Let $(R - \{0\}, \cdot)$ be the multiplicative group of real numbers. The identity is 1 and the inverse of x is $1/x$. The above theorem says

$$\forall_{x \in R - \{0\}} \ \frac{1}{1/x} = x$$

3. Let $(I, +)$ be the additive group of integers. The identity is 0 and the inverse of n is $(-n)$. The above theorem says

$$\forall_{n \in I} \ -(-n) = n$$

4. Let X be a set and $X = \{f | f: X \xrightarrow[\text{onto}]{1\text{-}1} X\}$

Then $(S, \circ)$ is a group under function composition. Its identity is the "identity function" on X,

$$\forall_{x \in X} \ i_X(x) = x$$

The inverse of f is f^{-1} (inverse function). The above theorem asserts

$$\forall_{f \in S} \ (f^{-1})^{-1} = f$$

5. Let $(W_5, \oplus)$ be the residue class arithmetic under addition modulo 5. The identity is 0, and the inverses of each element are

$$-\bar{0} = \bar{0}$$
$$-\bar{1} = \bar{4}$$
$$-\bar{2} = \bar{3}$$
$$-\bar{3} = \bar{2}$$
$$-\bar{4} = \bar{1}$$

Note in every case that $\bar{\bar{n}} = n$. For example,

$$\bar{\bar{2}} = \bar{3} = 2$$

Thus, the group theorem of Example 2 applies immediately to all groups, and the preceding five examples illustrate some of the range of such application.

Example 3   Show that all finite groups having exactly three elements are isomorphic.

Discussion:

Since there are few elements to deal with and we are talking about groups, we shall approach this proof from a *constructive* stance. That is, we shall assume we have an abstract group $(G, *)$ having three elements and ask ourselves, "In how many different ways can its group operation table be constructed?" If there were two (or more) different ways to do this, then by matching identity elements and corresponding inverses, we would set about to establish isomorphisms. On the other hand, if there is *only one* way to construct a group operation table for a group with three elements, then this would tell us all groups having three elements must be isomorphic to the abstract group we have constructed, hence isomorphic to each other (recall that "isomorphic to" is an equivalence relation).

Proof   Let e be the identity element in $(G, *)$ and a and b be two other distinct elements. Then, our operation table must look like

$*$	e	a	b
e	e	a	b
a	a		
b	b		

The question immediately arises: "What should $a * a$ be?" Since $a * e = a$, we cannot have $a * a = a$ because by left cancellation this would force $a = e$ and the three elements are supposed to be distinct. Therefore, $a * a$ is either e or b. Suppose it is e. Then, we must have $a * b = b$, since otherwise the middle row in our table would have two matching elements, contradicting the left cancellation law that we know to hold (for groups). But if $a * b = b$, then the third column has two b's in it, arising from $e * b = b$, and again we would obtain $a = e$, a contradiction. The only way out is to define

$$a * a = b$$

This, and observing the right and left cancellation laws, forces uniquely determined outcomes in the rest of the table.

$*$	e	a	b
e	e	a	b
a	a	b	e
b	b	e	a

Therefore, *all* groups having exactly three elements must behave exactly like (be isomorphic to) our model in this table. Thus we have proved

All finite groups having exactly three elements are isomorphic.

You may have noticed that there was a disguised proof by cases in the above example. It is perhaps not readily apparent since we were only dealing with three elements. The next example involves a very explicit proof by cases and follows a line of reasoning similar to that in Example 3.

Example 4   Suppose $(G, *)$ is a group and a and b are two elements in G that do *not* commute. Under these circumstances, prove that all the elements

$$e \qquad a \qquad b \qquad a * b \qquad b * a$$

are distinct.

Discussion:

Our primary strategy will be to examine each of the $\binom{5}{2} = 10$ possible combinations of equality taking two elements at a time. In each case we will try to establish a contradiction to the assumption that a and b do *not* commute; that is,

$$a * b \neq b * a$$

To save time and space, group properties will be used but not explicitly indicated.

Proof   *Case 1:* $e = a$: Then $a * b = e * b = b = b * e = b * a$
(contradiction to $a * b \neq b * a$)

Case 2: $e = b$: Then $a * b = a * e = a = e * a = b * a$
(contradiction to $a * b \neq b * a$)

Case 3: $e = a * b$: Then $a * e = e * a$
$$= (a * b) * a$$
$$= a * (b * a)$$

By left cancellation, $e = b * a$.
(contradiction to $a * b \neq b * a$)

Case 4: $e = b * a$: Then $e * a = a * e$
$$= a * (b * a)$$
$$= (a * b) * a$$

By right cancellation, $e = a * b$.
(contradiction to $a * b \neq b * a$)

Case 5: $a = b$: Then $a * b = a * a = b * a$
(contradiction to $a * b \neq b * a$)

Case 6: $a = a * b$: Then $a * e = a * b$
By left cancellation, $e = b$.
(reduces to case 2)

Case 7: $a = b * a$: Then $e * a = b * a$
By right cancellation, $e = b$.
(reduces to case 2)

Case 8: $b = a * b$: Then $e * b = a * b$
By right cancellation, $e = a$.
(reduces to case 1)

Case 9: $b = b * a$: Then $b * e = b * a$
By left cancellation, $e = a$.
(reduces to case 1)

Case 10: $a * b = b * a$: contradicts the hypothesis.

Therefore, in every case indicating equality among pairs of these five elements, we reach a contradiction to the original hypothesis that a and b do *not* commute. We conclude that the five elements

$$e \quad a \quad b \quad a * b \quad b * a$$

are all distinct.

The above proof leads to an important theorem in the theory of finite groups, namely:

> *Any noncommutative finite group must have at least six elements.*

This says that all finite groups having 1, 2, 3, 4, or 5 elements must be commutative, or Abelian, groups. We have not quite proved the latter theorem, but the result of Example 4 is a significant, decisive step. If $(G, *)$ is any noncommutative group, it must have at least two noncommuting elements a and b. According to the proof in Example 4, the elements e, a, b, $a * b$, $b * a$ are all different. Thus $(G, *)$ has at least five distinct elements, and we need to find a sixth. To do this we might consider $a * a$ or $a * b * a$, and show that at least one of these is different from each of the original five. The proof (involving cases) is left to you as a problem.

Example 5   Let a and b be two elements of a group $(G, *)$ that commute: $a * b = b * a$.

Prove that $(a * b)^n = a^n * b^n \quad \forall_{n \in I^+}$.

Discussion:

It is interesting to note that this is a property we take for granted in (for example) the real number system under multiplication

$$(a \cdot b)^n = a^n \cdot b^n$$

We must, however, carefully interpret exactly what the nth (positive integral) power of a group element means.

$$\forall_{a \in G} \; a^n = a * a * \cdots * a \; (n \text{ times})$$

In other words, a^n means the result of operating on a, n times *via the group operation* $*$.
We proceed with a proof by mathematical induction on n.

Proof   (i) If $n = 1$, we have $(a * b)^1 = a^1 * b^1$, which is trivially true.

(ii) Assume n is a positive integer, and

$$(a * b)^n = a^n * b^n$$

(iii) We try to deduce, in view of the inductive hypothesis (ii), that

$$(a * b)^{n+1} = (a^{n+1}) * (b^{n+1})$$

Thus,

$$(a * b)^{n+1} = (a * b)^n * (a * b)$$
$$= (a^n * b^n) * (a * b)$$

These first two steps rested on the definition of $(n + 1)$st power of a group element $(a * b)$ and the inductive hypothesis. Now we are operating in a group, so we know the associative property holds. Also we know a and b commute. Using these two facts, we can write the last expression as

$$a^n * (b^n * a) * b$$

and progressively "split off" one b at a time from b^n, associate it with a, commute, and reassociate, as follows:

$$b^n * a = b^{n-1} * (b * a)$$
$$= b^{n-1} * (a * b)$$
$$= (b^{n-1} * a) * b$$

In this way we can "move" a through the b^n term to finally obtain

$$a^n * (a * b^n) * b$$

and after two applications of the associative property we have

$$(a^n * a) * (b^n * b)$$

or

$$(a^{n+1}) * (b^{n+1})$$

which was desired.

Example 6   Prove or disprove: The union of two groups (having the same binary operation and identity) is a group.

Discussion:

To show that a system is a group, we need to show that the set is nonempty, it has a binary operation defined on it, and that the binary operation is associative, there exists an identity element, and every element has an inverse. Since we have here two groups having the same binary operation and identity elements, we know the union is nonempty, the operation is associative (wherever defined), the union has an identity element, and every element in the union has an inverse (the inverse from the group it originally came from). At a glance, it would appear that the union of two groups is a group, but we must look deeper. Is it true that if $*$ is a binary operation on two sets G and H, then it is also a binary operation on $G \cup H$? That is, if we take two elements in $G \cup H$ and operate on them, will the result always be a unique element of $G \cup H$? The answer, unfortunately, is no. Consider, as a counterexample, the two groups $(3I, +)$ and $(5I, +)$.

Both are groups with respect to the same binary operation $+$ (addition of integers), and both have the same identity element, 0. However $3 \in 3I$ and $5 \in 5I$, but $3 + 5 \notin (3I \cup 5I)$. Therefore, addition is a binary operation on $3I$ and $5I$ individually, but it is *not* a binary operation on their union $3I \cup 5I$. Hence, the union of two groups is not necessarily a group.

Assigned Problems

1. Complete the proof of the theorem: *Any noncommutative group must have at least six elements.* Study Example 4 and the comments following it.

2. (a) Let $(G, *)$ be a group. Prove
$$\forall_{a,\, b \in G} \; \overline{(a * b)} = \overline{b} * \overline{a}$$
(where $\overline{(a * b)}$, $\overline{a}$, $\overline{b}$ are the $*$-inverses of $a * b$, a, and b, respectively).

(b) Interpret the theorem you proved in (a) for some known groups such as $(I, +)$, $(R - \{0\}, \cdot)$, systems of invertible functions under composition, etc.

3. Give an example of an algebraic system $(S, *)$ containing an element a such that
$$(a * a) * a \neq a * (a * a)$$

4. We know from Example 6 that the *union* of two groups $(G, *)$, $(H, *)$ having the same binary operation $*$ and identity element e is *not* necessarily a group. Is the *intersection* of two such groups a group? Prove or disprove.

5. Let $(C, +)$ be the group of complex numbers under complex addition. Let φ be a mapping that associates every complex number $(a + bi)$ with its complex conjugate $(a - bi)$. That is
$$\varphi(a + bi) = a - bi$$
Prove that φ is an *automorphism* of $(C, +)$ with itself.

6. See Example 5. What if the group in question were the integers under addition $(I, +)$? This is a commutative group (*all* pairs of elements com-

mute). Does the theorem we proved really say

$$\forall_{a,\,b \in I}\;\forall_{n \in I^+}\;(a + b)^n = a^n + b^n$$

Give an *accurate* interpretation of our theorem in the $(I, +)$ situation.

7. Let $(S, *)$ be a commutative semigroup. Show that

$$\forall_{a,\,b,\,c \in (S,\,*)}\;[a * (b * c) = c * (b * a)]$$

8. Suppose $(G, *)$ is a group. Prove that the equations

   (a) $a * x = b$ (b) $y * a = b$

   have unique solutions in G for x and y. This requires an *existence* and *uniqueness* proof. That is, you must produce a solution for each equation, and (in each case) show it is unique.

9. Prove, if in a group $(G, *)$, $a^2 = e$ for every $a \in G$, then $(G, *)$ is an Abelian group.

10. (a) Let $R^2 = R \times R$, and $*$ be a binary operation on R defined by

    $$\forall_{(a,\,b),(c,\,d) \in R^2}\;[(a, b) * (c, d) = (a + c, b + d + 2bd)]$$

    Prove $(R^2, *)$ is a commutative monoid. Why isn't it a group?

    (b) Consider the following four functions on $R - \{0\}$:

    $$f(x) = x$$
    $$g(x) = \frac{1}{x}$$
    $$h(x) = -x$$
    $$l(x) = -\frac{1}{x}$$

    Prove that $\{f, g, h, l\}$ constitutes a group with respect to function composition.

Answers and Hints to
Selected Assigned Problems

Chapter 1

Problems 1.1, p. 6.

2. (a) $(P \wedge Q) \to R$
   (c) $(P \wedge \sim Q) \to \sim R$
   (h) $R \wedge (P \vee Q)$
   (j) $\sim (P \wedge Q)$
   (l) $(P \leftrightarrow Q) \wedge \sim (\sim Q \to R)$

3. (b) $(P \vee Q) \wedge \sim (P \wedge Q)$
   (e) $(P \wedge Q) \vee (R \wedge S)$
   (h) $\sim P \to (\sim Q \wedge R)$
   (l) $(P \wedge Q) \vee R$
   (m) $P \wedge (Q \to \sim R)$

4. (a) (3), (8)
   (c) (1)
   (d) (1), (2), (3), (5), (7), (8), (9)

5. (a) $(P \vee Q) \wedge R$
   (c) $(\sim P) \wedge Q$
   (j) $P \wedge (Q \to R)$

6. (b) $\overset{②\quad①\quad⑤④\quad③\quad⑧\quad⑦⑥}{[\sim (R \vee S) \wedge \sim (P \vee Q)] \to (P \to \sim Q)};$ conditional

   (e) $\overset{③②\quad①\quad⑤\quad④\quad⑨⑧\quad⑦⑥}{\{[P \vee \sim (R \wedge S)] \to (Q \vee T)\} \vee [\sim (P \wedge \sim S)]};$ disjunction

---

Problems 1.2, p. 15.

1. (a), (b), (c), (d), (e), (f), and (h) are true; (g) and (i) are false.

2. (a), (b), (c), (f), (g), (h), and (i) are true; (d) and (e) are false.

3. Observe the table under the main connective.

5.

P	Q	R	$[(P$	$\overset{①}{\wedge} Q)$	$\overset{③②}{\vee \sim R]}$	$\overset{⑥}{\to}$	$(R$	$\overset{⑤④}{\vee \sim Q)}$
T	T	T	T	T	F	T	T	F
T	T	F	T	T	T	F	F	F
T	F	T	F	F	F	T	T	T
T	F	F	F	T	T	T	T	T
F	T	T	F	F	F	T	T	F
F	T	F	F	T	T	F	F	F
F	F	T	F	F	F	T	T	T
F	F	F	F	T	T	T	T	T

Note the mixed truth values under the main connective (6). The truth value of this compound sentence depends on the particular truth values assigned to its component parts (P, Q, and R).

6. (b), (c), (d), and (e). (Main truth tables exhibit all "T.")

8. (a) and (b) are true; (c), (d), and (e) are false. For (b), $(F \wedge F) \to F$ is a true conditional because its antecedent is false.

Problems 1.3, p. 21.

1. (b) and (d) are tautologies;
   (a) and (e) are contradictions;
   (c) is indeterminate.

2. (a) $P \wedge \sim Q$
   (e) $A \wedge (\sim B \vee \sim C)$
   (i) $[(A \rightarrow B) \wedge A] \wedge \sim B$

   (c) $A \wedge \sim B$
   (g) $(A \vee B) \wedge \sim C$

4. Yes. $P \rightarrow (Q \vee R)$ is logically equivalent to $(P \wedge \sim Q) \rightarrow R$. Use tautologies T9, T20, and T25.

6. (a) $(x \leq y)$ and $(f(x) \nleq f(y))$.   (c) f is integrable, but f is not continuous and f is not differentiable.
   (e) p is not prime and p^2 is not rational.
   (g) Irrational numbers cannot be represented as quotients of integers.

7. Which tautology says that any biconditional sentence is logically equivalent to a conjunction of two conditionals?

9. To obtain a Scheffer-stroke representation of $\sim P$, remember that $\sim P$ is equivalent to $\sim P \vee \sim P$; then use the stroke definition. To do likewise for $P \vee Q$, write each of P, Q as a double negation; then replace the inner negation of each by its Scheffer-stroke representation, and apply the stroke definition again. Apply this kind of reasoning to the others, but be careful.

Problems 1.4, p. 27.

2. (a)   *conditional:*   If $x = y$, then $x^3 = y^3$.   (True)
         converse: If $x^3 = y^3$, then $x = y$.   (True)
         inverse: If $x \neq y$, then $x^3 \neq y^3$.   (True)
   contrapositive: If $x^3 \neq y^3$, then $x \neq y$.   (True)

   (d)   *conditional:*   If $x = y$, then $[\![x]\!] = [\![y]\!]$.   (True)
         converse: If $[\![x]\!] = [\![y]\!]$, then $x = y$.   (False; take $x = 2.09$, $y = 2.56$.)
         inverse: If $x \neq y$, then $[\![x]\!] \neq [\![y]\!]$.   (False; equivalent to converse)
   contrapositive: If $[\![x]\!] \neq [\![y]\!]$, then $x \neq y$.   (True)

   (h)   *conditional:*   If $a - b = c - d$, then $a = c$ and $b = d$ (False; take $a = 5$, $b = 2$, $c = 4$, and $d = 1$)
         converse: If $a = c$ and $b = d$, then $a - b = c - d$.   (True)
         inverse: If $a - b \neq c - d$, then $a \neq c$ or $b \neq d$.   (True)
   contrapositive: If $a \neq c$ or $b \neq d$, then $a - b \neq c - d$.   (False)

   (k)   *conditional:*   If the series $\sum_{n=1}^{\infty} a_n$ converges, then $\lim_{n \to \infty} a_n = 0$.   (True; this is the nth-term test.)

         converse: If $\lim_{n \to \infty} a_n = 0$, then the series $\sum_{n=1}^{\infty} a_n$ converges. $\left(\text{False; consider the harmonic series } \sum_{n=1}^{\infty} \frac{1}{n}.\right)$

   inverse is false; *contrapositive* is true.

3. Suppose you started by assuming n is *not* prime. This means either $n = 1$ or n is composite (product of two integers, neither of which is 1 or n). Could you then prove $2^n - 1$ is *not* prime? What logic are you applying here?

5. Prove (a) first. Note that any odd integer can be represented by $2k + 1$, where k is some integer. How is (b) related to (a)?

Problems 1.5, p. 39.

1. In colloquial English:
   (b) says, "2 is both even and prime."

(c) says, "Any number (integer assumed here) that 2 divides is an even number."

(g) says, "Every prime number divides some even number."

(i) is a difficult-looking way to say, "Exactly one number (integer) is both prime and even."

2. (a) $\forall_x[J(x) \to L(x)]$ (d) $\exists_x[J(x) \wedge O(x) \wedge V(x)]$; (Why are parentheses unnecessary here?)

(e) $J(j) \wedge [\sim O(j)] \wedge [\sim V(j)]$ (i) $\forall_x[(C(x) \wedge O(x)) \to L(x)]$

(k) $\forall_x[W(x) \to \sim (P(x) \wedge H(x))]$

(n) $\exists_x [L(x) \wedge J(j) \wedge A(x, j)] \wedge \exists_y [S(y) \wedge J(j) \wedge A(y, j)]$; or its equivalent:

  $\exists_x\exists_y[L(x) \wedge S(y) \wedge J(j) \wedge A(x, j) \wedge A(y, j)]$

(o) $\exists_x[L(x) \wedge \forall_y(A(x, y) \to J(y))]$ (s) $\forall_x[J(x) \to \forall_y(A(y, x) \to J(y))]$

(t) $\forall_x[J(x) \to \forall_y(A(x, y) \to J(y))]$

3. (d) $\exists_x\forall_y\{[P(x) \wedge \sim Q(y)] \vee [\forall_z(\sim R(x, y, z))]\}$

4. $\lim\limits_{x \to a} f(x) \neq L$ means $(\exists_{\varepsilon > 0})(\forall_{\delta > 0})(\exists_{x \in \text{ dom } f})[0 < |x - a| < \delta$ and $|f(x) - L| \not< \varepsilon]$.

This means that you can find a positive (real) number ε such that no matter what positive (real) number δ you consider, there will be some point x in the domain of the function such that x is within δ units of a (the approach point), but $f(x)$ is *not* within ε units of L.

7. No; no; yes; yes.

The function $f(x) = 0$, $\forall_{x \in R}$, is both odd and even. (Are there any others?)

The function $g(x) = x^3 + x^2 - 1$ is neither odd nor even.

8. (c) A function is not one-to-one if and only if $(\exists_{x \in \text{ dom } f})(\exists_{y \in \text{ dom } f})$ $f(x) = f(y)$ but $x \neq y$.

9. Think of horizontal asymptotes. If you negate the given statement, then what must the graph of f (not) do?

Chapter 2

Problems 2.1, p. 52.

2. (a) Work backwards to create a plan. We want B; we have $\sim M \to B$; therefore, we want $\sim M$. Which premise(s) involve M? We have $M \to L$; use its contrapositive to establish that we now want $\sim L$. $\sim L$ follows from K, which can be deduced from E, which, in turn, comes from $\sim G$. But $\sim G$ is a free-standing premise, so we've built a bridge backwards to known information.

3. $(\{[(P \to W) \wedge (J \to E)] \wedge (P \vee J)\} \wedge \sim W) \to E$. Check out the law of the dilemma and disjunctive syllogism.

4. (c) Use a truth table. An alternate procedure would be to take cases on M. Suppose M is true. Then, what if M is false? Is the conclusion deducible from the premises, or is it not?

6. Let $n, n + 1, n + 2$, and $n + 3$ be the four consecutive integers. Investigate 3 or 4 special cases ($n = 1, 2, 3, 4$) and try to identify a pattern. Consider the expression $n^2 + 3n + 1 = n(n + 3) + 1$.

8. Apply the method of reversing the sum (Example 6).

9. When you are finished, you will have proved: If a rectangle having variable length l and width w has constant perimeter ($2l + 2w$), then its area (lw) is maximum if $l = w$. That is, among rectangles with constant perimeter, the square has maximum area.

Problems 2.2, p. 58.

1. (a) Add D to the existing premises, and try to deduce C. By conditional proof, the statement $D \rightarrow C$ follows from the original premises alone.

4. Have you ever seen this problem before?

5. All are justifiable.
    (a) Two conditional proofs and adjunction   (b) Exportation tautology
    (c) Exportation tautology with $P \rightarrow Q$ treated as a simple sentence.
    (d) $[(P \wedge \sim Q) \rightarrow R] \leftrightarrow [\sim (P \wedge \sim Q) \vee R]$ conditional disjunction (T9)
    $\quad\quad\quad \leftrightarrow [(\sim P \vee Q) \vee R]$ De Morgan (T19)
    $\quad\quad\quad \leftrightarrow [\sim P \vee (Q \vee R)]$ associativity (T25)
    $\quad\quad\quad \leftrightarrow [P \rightarrow (Q \vee R)]$ conditional disjunction (T9)

6. Argue geometrically, assuming x is in the first quadrant.

7. From calculus, what do you know about the square-root function? An alternate method would be to prove the contrapositive, using what you know about squaring both sides of an inequality involving nonnegative terms.

10. If r and s are roots of a quadratic equation, then
    $$(x - r)(x - s) = 0$$
    must be the equation. Now, equate corresponding coefficients to obtain
    $$-p = r + s \quad \text{and} \quad q = rs$$
    Finally, compute $-p^3 + 3pq$.

Problems 2.3, p. 66.

1. (d) Use $\sim (\sim E \vee M)$ as the added premise. This is equivalent to $E \wedge \sim M$, from which E, $\sim M$ can be deduced separately. Along with the given premises, a contradiction (such as $M \wedge \sim M$) is easily obtained.

2. Assume the negation: $(\exists_{x, y \in R})$ $[(xy = 0)$ and $(x \neq 0)$ and $(y \neq 0)]$. Here, R represents the set of real numbers (see p. 112). Select either of x or y, and multiply both sides of $xy = 0$ by its multiplicative inverse (which exists for nonzero numbers). Obtain the result that the other number is 0, which contradicts the assumption that it is not.

4. Check the definitions of intersection of two sets and empty set in Section 3.1, if you are unsure about them. The statement to prove has the structure $P \rightarrow (Q \rightarrow R)$, which is equivalent to $(P \wedge Q) \rightarrow R$. Start by assuming the negation $(P \wedge Q) \wedge \sim R$. That is, assume A and B are sets in some universe and $A \cap B \neq \varnothing$, but $A = \varnothing$. Then we have $\varnothing \cap B \neq \varnothing$. But the intersection of the empty set with any set is $\varnothing$. Thus, we have a contradiction, and our original assumption must be false. Therefore, the original statement is true.

5. Yes. Assume it is not true, that is, assume $\exists_{x > 0}\left(x + \dfrac{1}{x} < 2\right)$. This leads to $x^2 - 2x + 1 < 0$, which contradicts a well-known theorem that $a^2 \geq 0$ for all real numbers a.

6. Assume there exists a nonzero real number x that has two distinct (different) multiplicative inverses; call them y_1 and y_2.

10. Assume there *is* a tangent line like that. Call the point of tangency (a, b). Compute the slope of the tangent line in two different ways, and use the fact that since (a, b) is on the parabola, $b = a^2$. What happens with the resulting quadratic equation? How do you interpret your finding?

11. A table would help organize your thinking here. For various stages (0, 1, 2, 3, . . .), observe which toys (by number) would have to be in which sacks *(A, B, C)*.

Problems 2.4, p. 73.

1. (b) Assume B (as added premise). Obtain C.
   Next, assume $\sim B$ (as added premise). Obtain C again.
   Conclude that C follows from the original premises alone, appealing to proof by cases.
   (*Note:* You could also start by assuming A; then $\sim A$; and so on.)

2. In each part, consider the definition of absolute value. It depends on two cases — whether a number is greater than or equal to 0, or less than 0.
   (a) 2 cases: (i) $x \geq 0$ (ii) $x < 0$
   (b) 4 cases: (i) $x \geq 0$ and $y \geq 0$
             (ii) $x \geq 0$ and $y < 0$
             (iii) $x < 0$ and $y \geq 0$ (Symmetric case to (ii). Why?)
             (iv) $x < 0$ and $y < 0$

5. Take cases on m, which is either even or odd.

7. $\sin (a + b) = \cos \left[\dfrac{\pi}{2} - (a + b) \right]$

   $= \cos \left[\left(\dfrac{\pi}{2} - a \right) - b \right]$

   $= \cos \left(\dfrac{\pi}{2} - a \right) \cos b + \sin \left(\dfrac{\pi}{2} - a \right) \sin b$

   $= \sin a \cos b + \cos a \sin b$

8. (a) Consider the isosceles right triangle shown in Figure A1.

   $A = \dfrac{1}{2} a^2$ by standard area formula

   $A^2 = \left(\dfrac{2a + c}{2} \right) \left(\dfrac{2a + c}{2} - a \right) \left(\dfrac{2a + c}{2} - a \right) \left(\dfrac{2a + c}{2} - c \right)$

   $= \dfrac{1}{4} a^4$ by Heron's formula (and Pythagorean theorem).

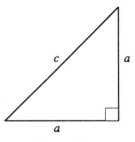

Figure A1

9. Let n, $n + 1$, and $n + 2$ be three consecutive integers. Take cases on n. Either $n = 3k$, or $n = 3k + 1$, or $n = 3k + 2$ for some integer k. If $n = 3k$, then n is a multiple of 3 and we're done. If $n = 3k + 1$, then $n + 2$ is a multiple of 3; if $n = 3k + 2$, then $n + 1$ is a multiple of 3. In every possible case, we obtain the result that one of the integers n, $n + 1$, or $n + 2$ is a multiple of 3.

10. Call the inner circle C_2 and the outer one C_1. Think of C_2 as being a tiny circle (radius near 0). Now imagine the radius of C_2 increasing slowly. Can it attain a value of 4? a value of more than 4?

Problems 2.5, p. 80.

2. Let $p(y) = y^{70} + 2y - 1$. Then $p(0) = -1$ and $p(1) = 2$. Since p is a polynomial function, it is continuous everywhere (in particular, on [0, 1]). Thus, by the Intermediate-Value Theorem (of calculus), $p(x) = 0$ for at least one $x \in [0, 1]$. We conclude that the polynomial defining p has a real zero in the interval [0, 1].

3. No. Assume there were real numbers a and b such that

$$\frac{1}{a+b} = \frac{1}{a} + \frac{1}{b}$$

This leads to the quadratic equation $a^2 + ab + b^2 = 0$, which has no real solution for a or b. Why not?

4. For visual thinking, sketch graphs of the functions on each side of each inequality (each pair on the same graph), and compare their intersections and relative positions.

   (a) True for any number in $(-\infty, 1]$ (b) True for any number in $(-\infty, -1) \cup (0, 1]$

   (c) True for any $x \geq \dfrac{\pi^2}{4}$ (d) True if $0 < a < 1$, and $x \geq \log_a x$; never true if $a > 1$

   (e) True for any number in $(0, .5)$. Actually, the x-coordinate of the intersection of the two graphs is approximately $.56714 \ldots$

   (f) True for all negative numbers

6. Yes. $1 - i$ and $-1 + i$ are the two square roots of $0 - 2i$ in the system of complex numbers.

7. Reason backwards. You want $|(5x + 4) - 14| < \varepsilon$. This would be true if $|5x - 10| < \varepsilon$. That would follow if we knew $5|x - 2| < \varepsilon$. Similarly, we could obtain the last inequality if we knew $|x - 2| < \dfrac{\varepsilon}{5}$. Therefore, choose $\delta \leq \dfrac{\varepsilon}{5}$. Then if $|x - 2| < \delta$ (by assumption) and $\delta \leq \dfrac{\varepsilon}{5}$, we would have $|x - 2| < \dfrac{\varepsilon}{5}$, and all these steps would reverse.

8. 46 works; so does 1234. What if you relax the condition that the digits be strictly increasing?

10. Lots of functions work. For example,

$$f(x) = \frac{1}{x - 2} \quad \text{and} \quad g(x) = \frac{-1}{x - 2}, \qquad \text{or} \qquad f(x) = \frac{x}{x - 2} \quad \text{and} \quad g(x) = \frac{-2}{x - 2}$$

Problems 2.6, p. 88.

2. $D_n = \dfrac{n(n - 3)}{2}$

   In this mathematical-induction proof, the key question to ask is "How many *new* diagonals are created when we add one new vertex to the convex polygon?" Note that from the $n + 1^{\text{st}}$ vertex, there would be $n - 2$ new diagonals, and what *was* a side connecting the two adjacent vertices now becomes a new diagonal.

3. Use analogy to problem 2, where people are points (vertices) and handshakes are segments joining vertices (diagonals or sides). Problem 2 really contains a geometric model of the handshaking problem.

4. (i) $\displaystyle\sum_{i=1}^{1} \frac{1}{i^2} = \frac{1}{1^2} = 1$ and $\displaystyle\frac{2 \cdot 1}{1+1} = 1$. Thus, the statement is true for $n = 1$. ($1 \le 1$)

(ii) Assume $\displaystyle\sum_{i=1}^{k} \frac{1}{i^2} \le \frac{2(k+1)}{k+2}$ for some integer $k \ge 1$.

(iii) Consider $\displaystyle\sum_{i=1}^{k+1} \frac{1}{i^2} = \sum_{i=1}^{k} \frac{1}{i^2} + \left(\frac{1}{(k+1)^2}\right)$

Thus, the inductive step requires that we show $\displaystyle\frac{2k}{k+1} + \frac{1}{(k+1)^2} \le \frac{2(k+1)}{k+2}$. A little algebra finishes the job.

7. The case for $n = 2$ sets is given in the problem. Assume the statement is true for k sets ($k \ge 2$) and consider the complement of a generalized union of $k + 1$ sets. Keeping the complement bar on, break up the union into a union of k sets in union with the $k + 1^{st}$ set. Now apply the De Morgan law for two sets (one being the union of k sets; the other being the $k + 1^{st}$ set). Apply the inductive hypothesis to the complement of the union of k sets, and regroup the resulting intersection.

8. In the inductive step, split up the sum of $k + 1$ terms into the sum of the first k, plus the $k + 1^{st}$ term. Then, use algebra to complete the step.

9. Be careful here. Use the fact that $(k + 1)! = (k + 1)k!$

10. Note that the $k + 1^{st}$ derivative of a differentiable function is the kth derivative of the first derivative.

Problems 2.7, p. 94.

1. (a) True for all $n \ge 10$

(b) (i) The statement is true for $n = 10$, since $1000 < 1024$.

(ii) Assume $k^3 < 2^k$ for some integer $k \ge 10$

(iii) We need to show: $(k + 1)^3 < 2^{k+1}$

To do this, note that $2^{k+1} = 2 \cdot 2^k = 2^k + 2^k$. Also, expand $(k + 1)^3 = k^3 + 3k^2 + 3k + 1$.

Now, for $k \ge 10$, $3k < 3k^2$ and $1 < 3k^2$, so that $3k^2 + 3k + 1 < 3k^2 + 3k^2 + 3k^2 = 9k^2$ and $9k^2 < k^3$.

Thus, we have $k^3 + 3k^2 + 3k + 1 < 2k^3 < 2 \cdot 2^k = 2^{k+1}$.

2. (b) No. The statement $P_n : n^2 + 5n + 1$ is even, is false for all natural numbers n. Part (a) demonstrates that the inductive step holds, but the assertion was never verified for any basis step (for any specific natural number). Note also (from logic) that a false antecedent yields a true conditional, regardless of the consequent.

5. Study the proof in Example 5. Use strong induction.

7. (a) For $n = -1$, $(-1)^3 - (-1) = 0$, and 6 divides 0. In the inductive step, write

$$\begin{aligned}
(k - 1)^3 - (k - 1) &= (k - 1)[(k - 1)^2 - 1] \\
&= (k - 1)[k^2 - 2k] \\
&= k^3 - 3k^2 + 2k \\
&= (k^3 - k) + (-3k^2 + 3k) \qquad \text{(Note the rearrangement.)} \\
&= (k^3 - k) + (-3k)(k + 1)
\end{aligned}$$

Both of the summands are divisible by 6; hence the sum is divisible by 6.

(b) Use an analogous method to that applied in part (a).

(c) Conclusion: 6 divides $n^3 - n$, for all integers n.   (d) No. Replace "n" by "$-n$."

(e) No. We could have used an argument about divisibility on three consecutive integers, since
$$n^3 - n = (n-1)n(n+1)$$

8. (a) $f_{10} = 55,$ $f_{13} = 233,$ $f_{n+3} - f_{n+1} = f_{n+2}$

(b) Prove by induction on $n \geq 1$. Use the fact that the definition of Fibonacci numbers implies
$$f_{k+1} + f_{k+2} = f_{k+3}$$

Problems 2.8, p. 104.

1. Not true. Let $f(x) = x^3 - 3x^2 + x + 4$ and compute $f(-1) = -1,$ $f(0) = 4$. What do you conclude from this about the given equation?

3. (a) Counterexample: $a = 1$ and $b = 1$ (b) Counterexample: $a = 2$ and $b = 4$

Modification: $\forall_{a,\, b > 0} \left(\dfrac{1}{a+b} < \dfrac{1}{a} + \dfrac{1}{b} \right)$ Modification: $\dfrac{a+b}{a} = \dfrac{1+b}{1}$ if and only if $a = 1$

4. (a) This says a sum of squares of numbers is the square of the sum of the numbers. It is not true in general. As a counterexample, let $n = 2$ and $x_i = i$. Then
$$\sum_{i=1}^{2} i^2 = 5, \quad \text{but} \quad \left(\sum_{i=1}^{2} i \right)^2 = 9$$

5. The pattern implied here is not an overgeneralization; it holds for all positive integers n. Note that
$$n^3 \div (n+1) = n(n-1) + \left(\frac{n}{n+1} \right)$$

7. (a) False. Take $n = 2$; the left side yields $\dfrac{17}{15}$, the right side yields $\dfrac{3}{5}$.

(b) True. This can be proved by mathematical induction on $n \geq 1$.

8. (a) $x = \dfrac{1}{2}, y = 2.$ (b) Any $x < -1$ will do.

(c) Any two real numbers such that one is positive and the other is negative.

(d) Use $0 < a < 1$ and $x > 0.$ (e) Choose x close to 0 and positive.

(f) Any two real numbers such that one is positive and the other is negative.

(g) Let $f(x) = x$ and $x = 2.$ (h) What if $x = \dfrac{9\pi}{2}$?

Chapter 3

Problems 3.1, p. 117.

1. Use an indirect proof. Assume there exists a set A such that $\varnothing \not\subseteq A$.

3. (a) $\{2, 5, 7\}$ (b) $\{1, 2, 4, 5, 6, 7, 8, 9, 10\}$ (c) $\{2, 5, 6, 7, 9\}$ (d) $\{2, 5, 7\}$

(e) $\{2, 3\}$ (f) $\{2\}$ (g) $\{3, 6, 9\}$ (h) $\{3, 6, 9\}$

4. (b) Think of row 5 (starting at row 0) of Pascal's triangle.

5. (a) $C = A$ (A function is constant $\leftrightarrow$ it has derivative 0 everywhere.)

    (b) $P \subseteq C$ (Every polynomial function is continuous; but not conversely.)

7. (d) $\varnothing$ $(A * A = (A - A) \cup (A - A) = \varnothing \cup \varnothing = \varnothing)$ (e) A $(A * \varnothing = (A - \varnothing) \cup (\varnothing - A) = A \cup \varnothing = A)$

    (f) $\bar{A}$ $(A * U = (A - U) \cup (U - A) = \varnothing \cup \bar{A} = \bar{A})$

8. (a) If $A \subseteq B$, then $(A \cup C) \subseteq (B \cup C)$.

    (e) If $a + b = 0$, then $a = 0$ or $b = 0$.   (*Note*: This is not true for all integers.)

10. Yes, the student is correct. Use an indirect proof. Assume there are sets A and B such that $A \cap B = A \cup B$, but $A \nsubseteq B$. Then there must be an element x such that $x \in A$ but $x \notin B$. If $x \in A$, then certainly $x \in (A \cup B)$. But then $x \in (A \cap B)$. Therefore, $x \in B$, a contradiction.

Problems 3.2, p. 129.

2. (a) $k!$ different ways. Prove by mathematical induction on k.   (b) Permutations

3. Use mathematical induction on $n \geq 1$ to prove: $\forall_{n \in N}$ (if A is a finite set and if B is a set with exactly n elements, then $A \cup B$ is finite).

4. Use mathematical induction on $n \geq 0$ to prove: $\forall_{n \geq 0}$ (if A has exactly n elements, then every subset of A is finite).

6. (a) $n(A \cup B \cup C) = n(A) + n(B) + n(C) - n(A \cap B) - n(A \cap C) - n(B \cap C) + n(A \cap B \cap C)$

    (b) Use the notation $\displaystyle\sum_{\substack{i < j}}^{k} n(A_i \cap A_j)$ to avoid counting duplications. (Here $i = 1, 2, \ldots k$ and $j = 1, 2, \ldots k$, but only for i less than j.)

7. $\displaystyle\bigcup_{r \in \Delta} B_r = [0, \infty)$ $\displaystyle\bigcap_{r \in \Delta} B_r = \{0\}$

8. Test some special cases for $n = 1, 2, 3, \ldots$ Obtain graphs that include the boundaries and interiors of even powered functions. (For $n = 1$, the graph of $y \geq x^2$ consists of the boundary and interior of the parabola $y = x^2$.) Imagine what happens as n gets very large, and consider the union of all these sets of points.

9. (c) $2^{(2^k)}$ 10. (b) $\{\varnothing, \{(1, x)\}, \{(2, x)\}, \{(1, x), (2, x)\}\}$ (d)   (i) $\{((1, x), \Delta), ((2, x), \Delta)\}$

                                                           (ii) $\{(1, (x, \Delta)), (2, (x, \Delta))\}$

                                                            (iii)   No. Recall the definitions of ordered-pair equality (in this problem) and set equality (see p. 107)

Problems 3.3, p. 136.

1. Two possible solutions are depicted in Figure A2.

    (a) Use the linear function f that contains the points (a, c) and (b, d).

$$\forall_{x \in [a, b]} \; f(x) = \left(\frac{d - c}{b - a}\right)(x - a) + c$$

Figure A2(a)

(b) Project one segment onto the other.

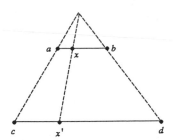

Figure A2(b)

4. If (x, y) is any point on the unit circle, draw a radius from $(0, 0)$ to (x, y) and extend it so as to intersect the unit square at (x', y'). Let $f(x, y) = (x', y')$; then f is a $1 - 1$ onto function.

5. The association of (x, y) with z is a function since decimal representation of real numbers is unique. Moreover, given any such decimal z, we can easily and uniquely recover both x and y from alternating digits of z; hence the function is $1 - 1$ from $[0, 1] \times [0, 1]$ into $[0, 1]$. To go the other way, use the identity function to map $[0, 1]$ to itself, which is equivalent to the subset $\{(x, 0) | 0 \leq x \leq 1\}$ of the unit square. Now apply the Schröder-Bernstein theorem.

6. Use indirect proof and the fact that $(A - B) \cup B = A \cup B$.

Problems 3.4, p. 143.

1. (a) 16   (b) 9   (c) $n(A \cup B) = 24, 25$; $n(A \cap B) = 23, 22$ 3. (a) 45   (b) 49

5. A Venn diagram might be as shown in Figure A3.   7. (a) $\displaystyle\prod_{i=1}^{t} (a_i + 1)$.

(a) 10   (b) 3   (c) 2

Increase each exponent by 1, and take the product of all these factors.

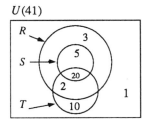

Figure A3

8. This is a long, tedious problem, but it is good practice in divisibility characteristics and counting. First, arrange a table with headings A (divisible by 2), B (divisible by 3), C (divisible by 5), and D (divisible by 7). Then form a lexicographical ordering of "truth values," where "T" means a number is in the set, and "F" means a number is not in the set. For example, the row $T F F T$ would refer to numbers that are divisible by 2, not by 3, not by 5, and also divisible by 7. That is, count all multiples of 14 that had not been counted previously (this would exclude 42, 70, and 84). The associated Venn diagram would be as shown in Figure A4.

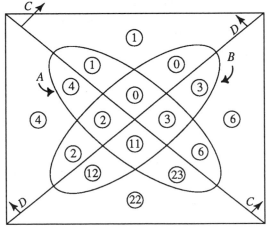

Figure A4

9. (a) No. It should be 7324.

   (b) Count each 2-way intersection, excluding the 3-way intersection. Obtain 2037 households.

   (c) 3742 households.

Problems 3.5, p. 151.

1. An example for S32 is given in Figure A5.

2. S5:　$A \cap \left(\bigcup\limits_{i=1}^{n} B_i \right) = \bigcup\limits_{i=1}^{n} \left(A \cap B_i \right)$, where

   A is a fixed set.

   S11:　$\overline{\left(\bigcup\limits_{i=1}^{n} A_i \right)} = \bigcap\limits_{i=1}^{n} \overline{A_i}$

   S33:　$A - \bigcap\limits_{i=1}^{n} B_i = \bigcup\limits_{i=1}^{n} \left(A - B_i \right)$, where A is a fixed set.

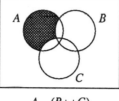

$A - (B \cup C)$

(a)

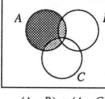

$(A - B) \cap (A - C)$

(b)

Figure A5

3. (a) $(R \cap S) \cup (R \cap T)$ (c) $(Z \cap X) \cup (Y \cap X)$ (f) $T \cup (R \cap S)$ (h) $(B \cap C) \cup D$

4. (a) U (c) $\bar{B}$ (f) $\varnothing$ (h) $A \cup \bar{B}$

5. (b) $A \cup T$ (d) $\varnothing$ (h) B (Use the fact that $A \cap B \subseteq B$.)

   (k) D (o) $\bar{C} \cap D$ (Use S30.)   (p) U (Use S31.)

6. Since $A \subseteq B$, then $\bar{A} \cup B = U$ (by S29). Now simplify $\overline{(\bar{A} \cup B)}$.

Problems 3.6, p. 159.

3. (a) $(A \subseteq B) \leftrightarrow \forall_x$ (if $x \in A$, then $x \in B$)   (definition of subset)   (b) The contraposition tautology (T6)

            $\leftrightarrow \forall_x$ (if $x \notin B$, then $x \notin A$)   (logical contrapositive)

            $\leftrightarrow \forall_x$ (if $x \in \bar{B}$, then $x \in \bar{A}$)   (definition of complement)

            $\leftrightarrow \bar{B} \subseteq \bar{A}$ (definition of subset)

5. The statement is true. *Proof*:

$$
\begin{aligned}
A - (A - B) &= A \cap \overline{(A - B)} & \text{(S15)} \\
&= A \cap \overline{(A \cap \bar{B})} & \text{(S15)} \\
&= A \cap (\bar{A} \cup \bar{\bar{B}}) & \text{(S12)} \\
&= A \cap (\bar{A} \cup B) & \text{(S22)} \\
&= (A \cap \bar{A}) \cup (A \cap B) & \text{(S5)} \\
&= \varnothing \cup (A \cap B) & \text{(S8)} \\
&= A \cap B & \text{(S19)}
\end{aligned}
$$

Now, use symmetry — interchange the roles of A and B in the above derivation — and we obtain $B - (B - A) = B \cap A$. Since $A \cap B = B \cap A$ (S2), we have

$$A - (A - B) = B - (B - A)$$

6. Use mathematical induction on $n \geq 1$.

7. Use the strategy of working independently on each side of the equation. Try to reduce each side to the same set.

9. (a) True $(M * M = (M - M) \cup (M - M) = \varnothing \cup \varnothing = \varnothing)$
   (b) False   (Let $M = \{1, 2, 4\}$; $L = \{1, 2\}$; $N = \{1\}$. Then $M \cup (L - N) = \{1, 2, 4\}$ and $(M \cup L) - (M \cup N) = \varnothing$.
   (c) True   (Use the strategy given in Problem 7. The challenge here is to keep the sets and operations straight.)

Chapter 4

Problems 4.1, p. 173.

1. (1) None   (3) s only   (5) r, t, a (7) r, s, t, a (9) r, s, t (10) r, s (14) r, s, t

3. The graph is a closed disk with its center at $(0, 0)$ and radius $\sqrt{2}$. The domain and range of R are both
$$\left[-\sqrt{2}, \sqrt{2}\right]$$

5. Prove by double inclusion. Start with a suitable notation for the "point" selected in $(A \cup B) \times C$, namely (x, y), where $x \in A \cup B$ and $y \in C$.

6. Any binary relation on A is a subset of $A \times A$. Therefore, the number of binary relations on A is the same as the number of subsets of $A \times A$, which is 2^{100}, or approximately 1.268×10^{30}.

7. (a) $2^{(k^2)}$, by the same reasoning as in Problem 6.
   (b) $A \times A$ has k^2 ordered pairs as its elements. Arrange these ordered pairs in k rows, where the ith row would be $(a_i, a_1), (a_i, a_2), \ldots (a_i, a_k)$. Now, to obtain a binary relation having all of A as its domain, we need to select at least one ordered pair (possibly more) from each row. The number of ways we can do this (for each row) is the same as the number of subsets of each row (except for $\varnothing$), which is $2^k - 1$. Since selection from each row is an independent event, the total number of ways to select at least one pair from each row is $(2^k - 1)^k$.
   (c) Use reasoning similar to part (b), except that now we must have all ordered pairs of the form (a_i, a_i) in the relation. So we must count all possible subsets of each row (including $\varnothing$) from the remaining ordered pairs in the row.

9. Shrink down the sets. What if $A = D = \varnothing$?

10. (a) r, s, t (b) The domain and the range are each the set of all real numbers.
    (c) The union of the lines $y = x$, $y = -x$

Problems 4.2, p. 185.

1. (a) If $P \rightarrow Q$ and $Q \rightarrow R$ are both true, can $P \rightarrow R$ be false?   (c) s only

2. (a) r, t (b) None   (e) s only   (j) r, s, t (k) r only
   (m) Think of the collection of subsets of the set as you answer this question.   (n) r, t

4. 15. In general, the number B_n of different equivalence relations on an n-element set is given by the recursive formula:

$$B_0 = 1, \ B_1 = 1, \ B_2 = 2, \ B_3 = 5, \ B_{n+1} = \sum_{i=0}^{n} \binom{n}{i} B_i.$$

(See *American Mathematical Monthly*, May 1964; article by Gian Carlo-Rota.)

5. There is a flaw in logic here. Watch the quantification on the reflexive property ($\forall_{a \in A} \ aRa$), and the condition on the symmetric property ($\forall_{a, b \in A}$ *if* $aRb, \ldots$).

7. (b) Ten equivalence classes — one for each units digit 0, 1, 2, . . . 9

(d) Infinitely many equivalence classes, each of the form $\{x, -x\}$ (e) One equivalence class, the set A itself

10. The equivalence classes are concentric circles, centered at the origin, with radii varying from $r = 0$ (a point circle), and increasing through all real numbers. The set of these equivalence classes fills the plane.

Problems 4.3, p. 198.

2. (a) (1) If $E_1 = \{(1, 1), (2, 2), (3, 3), (4, 4), (5, 5), (1, 3), (3, 1), (4, 5), (5, 4)\}$, then $\mathcal{P}_1 = \{\{1, 3\}, \{2\}, \{4, 5\}\}$.
   (2) If $E_2 = \{(1, 1), (2, 2), (3, 3), (4, 4), (5, 5), (2, 3), (2, 4), (3, 2), (4, 2), (3, 4), (4, 3)\}$, then $\mathcal{P}_2$
       $= \{\{1\}, \{2, 3, 4\}, \{5\}\}$.
   (b) As one example, if $\mathcal{P}_3 = \{\{1\}, \{2\}, \{3, 5\}, \{4\}\}$, then $E_3 = \{(1, 1), (2, 2), (3, 3), (4, 4), (5, 5), (3, 5), (5, 3)\}$.

3. (b) Equivalence classes are all singleton sets of the form $\{x\}$, one for each real number x. The relation is really that of equality.
   (c) Two equivalence classes — {all true statements},   {all false statements}
   (d) Three equivalence classes — $\{1, 2, \ldots 9\}$, $\{10, 11, \ldots 99\}$, $\{100\}$

5. The number of binary relations on a 3-element set is 2^9, or 512. Of these, only 5 are equivalence relations (check this out). Therefore, the probability is $\dfrac{5}{512} \approx .01$.

6. We can regard the members of each equivalence class as (equivalent) *fractions*; the classes themselves define *rational numbers*. Thus, the class

$$\{\ldots, (-8, -12), (-6, -9), (-4, -6), (-2, -3), (2, 3), (4, 6), (6, 9), (8, 12), \ldots \}$$

contains the "fraction" $(2, 3)$, but the entire class might be regarded as the rational number $\dfrac{2}{3}$. Note that the fractions $(2, 3)$ and $(4, 6)$ are not equal (recall ordered-pair equality) — they are equivalent; the rational numbers $\dfrac{2}{3}, \dfrac{4}{6}$ are equal (different names for the same class).

8. To show the cells are mutually disjoint, suppose $(A_i \times B_j) \cap (A_k \times B_l) \neq \varnothing$.
   This means there is some element $(x, \dot{y})$ in the intersection set.
   Thus $(x, y) \in A_i \times B_j$ and $(x, y) \in A_k \times B_l$.
   But then $x \in A_i$ and $y \in B_j$, as well as $x \in A_k$ and $y \in B_l$.
   Since $x \in A_i \cap A_k$ and $\mathcal{P}_1$ is a partition of A, then $i = k$ and $A_i = A_k$.
   Similarly, since $y \in B_j \cap B_l$, and $\mathcal{P}_2$ is a partition of B, then $j = l$ and $B_j = B_l$.
   Therefore, $A_i \times B_j = A_k \times B_l$.

9. (a) Each $L_r \neq \varnothing$ since $(0, r) \in L_r$. Also, to show $R^2 \subseteq \bigcup_{r \in R} L_r$, take any point $(x, y) \in R^2$. Then $(x, y) \in L_{y-x}$. Therefore, $(x, y) \in \bigcup_{r \in R} L_r$.

   (b) The collection of lines parallel to $y = x$ with varying y-intercepts.

   (c) Define E on R^2 by:
$$(x, y) \, E \, (z, w) \leftrightarrow (y - x = w - z)$$

Problems 4.4, p. 209.

1. (a) Partial order   (b) Neither   (Two different segments having the same length are not comparable.)
   (c) Total order   (d) Partial order   (For example, 2 and 3 are not comparable.)
   (e) Partial order   (Trichotomy fails because both $r \subseteq r$ and $r = r$.)

4. Use indirect proof. For example, suppose there are two (distinct) least upper bounds for M with respect to the partial order relation R. Call them L_1 and L_2. Then $L_1 \, R \, L_2$ since L_2 is an upper bound and L_1 is a least upper bound. But also, $L_2 \, R \, L_1$ since L_1 is an upper bound and L_2 is a least upper bound. By antisymmetry, we have $L_1 = L_2$, which contradicts the assumption of distinctness. An analogous proof can be made for the uniqueness of greatest lower bound, assuming its existence.

5. (a) Factor each number in S into its prime factorization. Any lower bound on S relative to the "divides" relation must divide every element in S_i, that is, it must be a common divisor of all elements in S. Some lower bounds are 1, 2, 4, 14, and 28.
   (b) Upper bounds for S with respect to the "divides" relation are common multiples of all the elements in S. The least common multiple (least upper bound here) is 120,120. Other upper bounds would be integral multiples of 120,120.

6. (b) Find any pair of elements that are not comparable.
   (c) Take special cases for $x = 0$ or $y = 0$ (or both). If neither is 0, use lcm and gcd of x, y.
   (d) Note that $1 \cdot x = x$.   (e) Note that $x \cdot 0 = 0$

7. Yes   8. (a) $M \cup N$ (b) $M \cap N$

10. Let $f(x) = 1 + \dfrac{1}{x}$, and ask what happens to f as $x \to 0^+$ or $x \to \infty$. Sketch a graph.

Problems 4.5, p. 222.

1. (a) 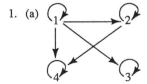   (d)    2. (a) r, t, a
   (c) r, s, t
   (d) s only

3. (b) Let $(x, y) \in R \cap R^{-1}$. Then $(x, y) \in R$ and $(x, y) \in R^{-1}$.
   Since $(x, y) \in R$, then $(y, x) \in R^{-1}$, by the definition of R^{-1}.
   Since $(x, y) \in R^{-1}$, then $(y, x) \in (R^{-1})^{-1}$. But $(R^{-1})^{-1} = R$ (Why?)
   We have, $(y, x) \in R^{-1}$ and $(y, x) \in R$, so $(y, x) \in R \cap R^{-1}$.
   Therefore, $R \cap R^{-1}$ is a symmetric relation.

4. (b)   (e)

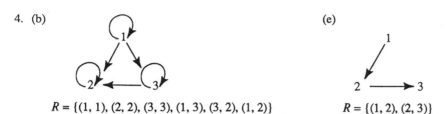

$R = \{(1, 1), (2, 2), (3, 3), (1, 3), (3, 2), (1, 2)\}$ $R = \{(1, 2), (2, 3)\}$

6. (a) G has order 10 and size 16. The degrees of vertices $a, b, c, \ldots j$ are 2, 3, 3, 4, 4, 2, 4, 6, 2, 2, respectively.

   (b) No. We desire an Euler circuit, but not all vertices have even degree.

   (c) Yes. The path $(b, a, h, i, j, h, b, c, e, h, g, e, d, g, f, d, c)$ works. Note Theorem 2 and the fact that b and c are the only two vertices of odd degree in the graph G.

   (d) No. Look at the subgraph (h, i, j, h). If we enter from h, we traverse h twice; if we start at i or j, we must exit via h, and we cannot complete the circuit without traversing h again.

8. One Euler circuit would be $(a, b, d, i, c, d, e, f, j, e, i, h, c, b, f, g, j, h, a)$.

9. (a) In a complete graph of n vertices, each vertex must be connected to every other vertex. Therefore, any vertex has $(n - 1)$ adjacent vertices. Since there are n vertices, each having $(n - 1)$ adjacent vertices, the sum of the degrees of all vertices is $n(n - 1)$.

10. Yes. It is true for all convex polyhedra. Imagine any convex polyhedron sitting on one of its faces and resting upon a plane. Now suppose the edges become nonrigid (turn to rubber, if you will). What happens to the polyhedron?

Problems 4.6, p. 232.

1. Let $x, y \in R$ and $z > 0$.
   $$\begin{aligned}
   (x < y) &\leftrightarrow y - x > 0 && \text{by } O_4 \\
   &\leftrightarrow (y - x)z > 0 && \text{by } O_3 \text{ and Axiom 3} \\
   &\leftrightarrow yz - xz > 0 && \text{algebra (distributive property)} \\
   &\leftrightarrow xz < yz && \text{by } O_4
   \end{aligned}$$

2. ($\leftarrow$) If $x > 0$ and $y < 0$, then $-y > 0$ (by O_5), and $x(-y) = -(xy) > 0$ (by O_{9a})
   Therefore, $xy < 0$ (by O_5)
   The argument is symmetric in x and y (they can be interchanged.) So, if $x < 0$ and $y > 0$, we still obtain $xy < 0$.

   ($\rightarrow$) Assume $xy < 0$ and argue by cases on x, y. Consider $x = 0$ or $y = 0$, both positive, both negative, and in each case produce a contradiction, leaving only the desired inequalities as remaining possibilities.

5. (a) If $x \neq 0$, then by trichotomy, $x > 0$ or $x < 0$. In either case O_{9a} guarantees that $x^2 > 0$. A logical equivalent to this is $\forall_{x \in R} (x = 0 \ \text{ or } \ x^2 > 0)$.

   (c) Use indirect proof. Suppose $\exists_{x \in R} (x > 0 \text{ and } x + \dfrac{1}{x} < 2)$. By various properties (O_8, O_7) and algebra, obtain the inequality $(x - 1)^2 < 0$, a contradiction to problem 5a (if $x \neq 1$) or trichotomy (if $x = 1$).

6. If $a = 0$ and $b = 0$, take $r = 0$; if $a > 0$ and $b > 0$, take $r = \dfrac{a}{b}$.

8. Use mathematical induction on $n \geq 1$. For the inductive step, use the facts that
$$\frac{1}{2n} < \frac{1}{n+1}, \quad \text{for } n > 1$$
Therefore,
$$\frac{1}{n+1} + \frac{1}{n+2} + \ldots + \frac{1}{n+n} > n \cdot \frac{1}{2n} = \frac{1}{2} \quad \text{and}$$
$$\frac{1}{2n+1} + \frac{1}{2n+2} + \ldots + \frac{1}{2n+2n} > 2n \cdot \frac{1}{4n} = \frac{1}{2}$$
Then choose $m = 4n$ and show that $\displaystyle\sum_{i=1}^{m} \frac{1}{i} > k + \frac{1}{2} + \frac{1}{2}$ whenever $\displaystyle\sum_{i=1}^{m} \frac{1}{i} > k$.

10. The inequality is not true if one of a or b is positive and the other negative. If we restrict a and b to be both positive or both negative, the statement is true.

Problems 4.7, p. 238.

1. This is just a special case of D_6, where $x = y = 1$.

2. (a) $a|c$ (b) $c|b$

   (c) $c|a$ and $c|b$. (Therefore, $c|ma + nb$ for any $m, n \in I$ and $c|k(ab)$ for any $k \in I$.)

   (d) $a|b$ and $c|b$. (If gcd $(a, c) = 1$, we could conclude $ac|b$ by D_9.)

3. Factor $n^5 - n$ into $(n - 1)n(n + 1)(n^2 + 1)$. Then argue the case for divisibility by 2 and 3 by noting the presence of three consecutive integers in the factorization. For divisibility by 5, use a case argument on n. Either $n = 5k$, $n = 5k + 1, \ldots n = 5k + 4$. In each case, show that one of the factors above must be a multiple of 5. It follows from D_9, that if 2, 3, and 5 divide $n^5 - n$, then so does 30.

4. 7. You can prime factor each integer and look for the largest product of shared primes, or look up the Euclidean algorithm and proceed by pairs of integers.

6. (a) $a|b$ implies $b = ka$ for some $k \in I$. Therefore, $b^3 = (ka)^3 = k^3 a^3$. Hence $a^3|b^3$.

7. In the inductive step, rewrite the expression for $k + 1$ in an algebraically equivalent form that contains the expression in the inductive hypothesis.

9. This is analogous in reasoning to the solution for Problem 3. Factor $n^9 - n$, watch for consecutive factors, and argue by parity (evenness versus oddness).

Chapter 5

Problems 5.1, p. 254.

1. (a) Function; the domain is the set of all reals except 2; the range is the set of all reals except 4.

   (e) Nonfunction; $f\left(\sqrt{2}\right)$ is both 0 and 1.

2. (b) Domain = $[-1, 1]$; range = $[0, \pi]$.

   (d) The domain is the set of all reals except 2 and 3; the range is $(-\infty, -4] \cup (0, \infty)$.

4. True, because some function is applied (think of function as operator) to both sides of each equation. Note that in (b), x and y must be positive reals.

5. (a) Nonfunction; $(1, 1)$ and $(1, 2)$ destroy the uniqueness property.

   (b) Function; constant function of value 2 over a finite domain.

   (c) Nonfunction; consider $(4, 2)$ and $(4, -2)$.

   (d) Function; any triangle in the plane has a unique perimeter.

   (e) Nonfunction; many different triangles can have the same perimeter.

8. Because domain $f = A$, it follows that every one of the k elements in A must be used at least once in every mapping. Also, for every element in A, there are exactly m possible images under any mapping. These choices are independent, since all we require is that all elements in A are paired, but not necessarily all elements in B are paired. Since there are k such elements, each having m independent choices for an image, the number of possible mappings is m^k. (One could also argue that each of the m elements in B has exactly k possible preimages, and all preimages must be used for each mapping.)

10. Index the elements in A and B, and form $A \times B$ as a set of ordered pairs displayed in rows of the form (x_i, y_1), (x_i, y_2), . . . (x_i, y_m). Then ask yourself the question, "In how many ways can we select *at most one* (possibly none) of the pairs from each row?"

Problems 5.2, p. 264.

1. Think visually. Draw a graph that meets the conditions, then define the function.

2. (a) $1 - 1$ and onto   (b) Neither $1 - 1$ nor onto   (c) $1 - 1$ and onto

   (d) Neither $1 - 1$ nor onto   (e) Neither $1 - 1$ nor onto

3. (b) Try a cubic polynomial function that has a local maximum and a local minimum. How would you construct such a function by working backwards?

5. Remember that a permutation is just a $1 - 1$, onto function of a set onto itself.

6. (a) There are real numbers (negatives) with no preimage.

7. No. f is neither $1 - 1$ nor onto, since given any possible remainder upon division by 7 $(0, 1, 2, \ldots 6)$, many different integers can have the same remainder. Also, a remainder cannot be a negative integer.

9. (b) No. Distances cannot be negative real numbers.

   (c) No. Many different points (think of circles centered at $(0, 0)$) can have the same distance from the origin.

Problems 5.3, p. 276.

1. The initial assumption (by indirect proof) here would be: Suppose there are functions $f: B \rightarrow A$ and $g: A \rightarrow B$, where $g \circ f$ is the identity function on B, and f is not $1 - 1$ or g is not onto. Take the latter two statements as cases, and show that each leads to some contradiction.

2.

$\cdot$	f_1	f_2	f_3	f_4	f_5	f_6
f_1	f_1	f_2	f_3	f_4	f_5	f_6
f_2	f_2	f_1	f_5	f_6	f_3	f_4
f_3	f_3	f_6	f_1	f_5	f_4	f_2
f_4	f_4	f_5	f_6	f_1	f_2	f_3
f_5	f_5	f_4	f_2	f_3	f_6	f_1
f_6	f_6	f_3	f_4	f_2	f_1	f_5

4. Take cases on $g(1)$ under the restrictions of the problem. You should produce only 3 (of the 27 possible) mappings.

6. Let $B = \{$all reals$\}$ with $A = C = [-1, 1]$. Then let f be the sine function and g be the arcsine function.

8. No. The theorem requires that dom $(f \circ g)$ be the entire set $R - \{-1, 1\}$. Here, the domain is a singleton set.

10. (a) h is a composition of the reciprocal, natural logarithmic, and sine functions, acting in that order.

$$h'(x) = \cos\left(\ln\left(\frac{1}{x}\right)\right) \cdot \frac{1}{\left(\frac{1}{x}\right)} \cdot \left(-\frac{1}{x^2}\right), \quad \text{or} \quad -\cos\left(\ln\left(\frac{1}{x}\right)\right) \cdot \left(\frac{1}{x}\right), \quad \text{or} \quad -\cos(\ln x) \cdot \left(\frac{1}{x}\right)$$

(b) As one example, $(f \circ g \circ h)(x) = \tan\left(x^{3/2} + 2x^{1/2}\right)$. To be a meaningful composition, x must be positive, and $\left(x^{3/2} + 2x^{1/2}\right)$ cannot be an odd integral multiple of $\frac{\pi}{2}$. The derivative of the composition is

$$\sec^2\left(x^{3/2} + 2x^{1/2}\right) \cdot \left(\frac{3}{2}x^{1/2} + x^{-1/2}\right)$$

Problems 5.4, p. 287.

1. (a) Show f is $1 - 1$ and onto.

   (b) g is not $1 - 1$ since $g(1) = g(-1)$.
   Therefore, g^{-1} fails to exist (as a function).

2. $h^{-1}(x) = \sqrt[5]{\frac{x-7}{2}} + 1$

4. The case for $n = 2$ functions is a consequence of the theorem on the inverse of a composition. Assume the statement is true for $n = k$ functions $(k \geq 2)$. That is,

$$(f_1 \circ f_2 \circ \ldots \circ f_k)^{-1} = f_k^{-1} \circ f_{k-1}^{-1} \circ \ldots \circ f_1^{-1}$$

Now, consider the case for $n = k + 1$ (functions):

$$\begin{aligned}
(f_1 \circ f_2 \circ \ldots \circ f_k \circ f_{k+1})^{-1} &= [(f_1 \circ f_2 \circ \ldots \circ f_k) \circ f_{k+1}]^{-1} && \text{function composition is associative} \\
&= f_{k+1}^{-1} \circ (f_1 \circ f_2 \circ \ldots \circ f_k)^{-1} && \text{case for } n = 2 \\
&= f_{k+1}^{-1} \circ (f_k^{-1} \circ f_{k-1}^{-1} \circ \ldots \circ f_1^{-1}) && \text{inductive hypothesis} \\
&= f_{k+1}^{-1} \circ f_k^{-1} \circ \ldots \circ f_1^{-1} && \text{function composition is associative}
\end{aligned}$$

5. (b) Let $A = [-1, 1]$, $B = \{\text{all reals}\}$, $g(x) = \sin^{-1}x$, and $f(x) = \sin x$.

Then $\forall_{x \in [-1, 1]} \sin(\sin^{-1}x) = x$.

But it is not true that $\forall_{x \in R} \sin^{-1}(\sin x) = x$.

Take $x = \dfrac{\pi}{2} + 0.1$. $\sin^{-1}\left(\sin\left(\dfrac{\pi}{2} + 0.1\right)\right) = \dfrac{\pi}{2} - 0.1$, not $\dfrac{\pi}{2} + 0.1$.

9. Take f', the first derivative of f. If f' is always positive on domain f, then f is increasing and f^{-1} exists. If f' is always negative on domain f, then f is decreasing and f^{-1} exists. If f' takes on both positive and negative values on domain f, then f^{-1} fails to exist. Here

$$f'(x) = 3x^2 + 6x = 3x(x + 2)$$

f' is positive on $(-\infty, -2) \cup (0, +\infty)$ and negative on $(-2, 0)$.

10. A function is a binary relation from one set to another (meeting certain conditions). Thus, a function is a set of ordered pairs. If we reverse the order of each pair, we obtain the inverse relation. The geometric effect of interchanging every x- and y-coordinate is a reflection in the line $y = x$. To see this, consider the line segment connecting (x, y) with (y, x). It has slope -1, hence it is perpendicular to the line $y = x$. Moreover, the midpoint of the segment joining (x, y) and (y, x) is $\left(\dfrac{x + y}{2}, \dfrac{y + x}{2}\right)$, and this is also a point on the line $y = x$. Thus, the line $y = x$ is the perpendicular bisector of all segments joining various points (x, y) with others (y, x), and the graph of f is symmetric to the graph of f^{-1} with respect to this line.

Problems 5.5, p. 296.

1. Every binary operation on S is a function from $S \times S$ to S. There are $(k^2)^k = k^{2k}$ functions possible from $S \times S$ to S. Why?

2. All are binary operations.

3. All are binary operations.

4. Ask yourself questions such as: "Is the outcome always a point?" "Is it a *unique* point, or could there be more than one such point?"

5. (c) Do you recall the Cantor diagonalization function which was used to prove that the set of rational numbers is countable?

7. The subsets in parts (a), (b), (c), (e), and (f) are closed sets relative to function composition. In (d) and (g) the subsets are not closed. For (d), consider the fact that $f(0)$ could be any number (in particular, nonzero); for (g) consider $f(x) = \sin 2\pi x$ and $g(x) = x - \dfrac{1}{2}$. Both integrals yield 0 on $[0, 1]$, but the integral of $g \circ f$ yields $-\dfrac{1}{2}$ on $[0, 1]$.

9. (c) All are powers of i and also powers of $-i$.

10. All are binary operations on the set of all $R \to R$ functions.

Problems 5.6, p. 308.

1. (a), (b), (c): monomorphisms;
   (f): epimorphism;
   (h), (i), (j): isomorphisms;
   (k): homomorphism.
   (d), (e), (g), (l): not morphism functions.

   For example:

   (a)
   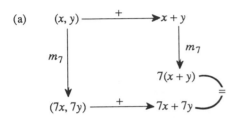

   (g)
   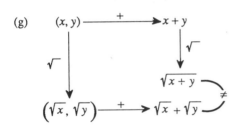

3. We want $\forall_{x,\,y \in R}\ f(x+y) = f(x) \oplus f(y)$, where $f(x) = x+1$. To obtain $x + y + 1$ on both sides, the operation $\oplus$ must subtract 1 from the sum of any two numbers. That is, define $\oplus$ by,

$$\forall_{x,\,y \in R}\ x \oplus y = x + y - 1$$

   This makes f an isomorphism.

4. This has to be checked for each of the 16 products of elements in this system.

5. Rely on the $1 - 1$, onto, and morphism properties of the individual functions to demonstrate the respective properties of the composition.

7. What would be a very "natural" way to define a function linking sets with elements of the form $a + b\sqrt{2}$ and $a + b\sqrt{3}$, so that the function turns out to be an isomorphism?

9. Consider the diagram

   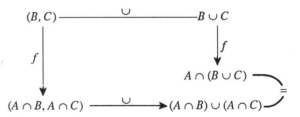

Chapter 6

Problems 6.1, p. 321.

1. What method of proof is often used in uniqueness arguments?

2. This is analogous to Problem 1.

3. (a) You can simplify this operation by considering two cases: $x \geq y$ and $x < y$.
   (b) 0 is the *-identity element.   (c) * is commutative.

5. (b) Let $t_1, t_2 \in T$. Then since f is onto (epimorphism), $\exists_{s_1,\,s_2 \in S}$ such that $f(s_1) = t_1$ and $f(s_2) = t_2$.

   Now $t_1 \circ t_2 = f(s_1) \circ f(s_2)$
   $\qquad\qquad = f(s_1 * s_2) \qquad$ morphism property
   $\qquad\qquad = f(s_2 * s_1) \qquad$ * is commutative
   $\qquad\qquad = f(s_2) \circ f(s_1) \qquad$ morphism property
   $\qquad\qquad = t_2 \circ t_1$
   $\therefore \circ$ is commutative in T.

6. (a) The identity element for multiplication of complex numbers is $1 + 0i$. To produce it, let $a + bi$ be any complex number with not both a and b equaling 0, and suppose $c + di$ is the identity element. Then $(a + bi) \cdot (c + di) = a + bi$. Find c and d by equating corresponding coefficients.

   (b) If $a + bi$ is any complex number (except $0 + 0i$), then the multiplicative inverse of $a + bi$ is

   $$\left(\frac{a}{a^2 + b^2} \right) - \left(\frac{b}{a^2 + b^2} \right) i$$

   Use a method similar to that in Part (a).

7. See the theorems in Section 3.4.

8. See Section 5.3, Problem 2. Also, since every permutation is a $1 - 1$, onto function, every permutation on any finite set has an inverse.

10. Compare, particularly, the modular arithmetics that have prime, versus composite, moduli.

Problems 6.3, p. 342.

2. For left cancellation, let a, b, c, $\in G$ and let $a * b = a * c$. We show $b = c$. Since G is a group, then a has a $*$-inverse; call it $\bar{a}$. Now operate on the left sides of both members of the equation with $\bar{a}$ (we can do this since $*$ is a binary operation); we obtain

$$\bar{a} * (a * b) = \bar{a} * (a * c)$$

$\therefore$ $\quad (\bar{a} * a) * b = (\bar{a} * a) * c$ $\quad$ $*$ is associative in G

$\therefore$ $\quad\quad\quad e * b = e * c$ $\quad$ $\bar{a}$ is $*$-inverse of a

$\therefore$ $\quad\quad\quad\quad\quad b = c$ $\quad$ e is $*$-identity in G

So we have proved, $\forall_{a, b, c \in G}$ if $a * b = a * c$, then $b = c$.

3. To show closure, let x, $y \in ker\ h$, and consider $x * y$.

$$h(x * y) = h(x) \circ h(y) \quad\quad \text{morphism property}$$

$$= e_H \circ e_H \quad\quad\quad x, y \in ker\ h$$

$$= e_H$$

$\therefore x * y \in ker\ h$,   if $x, y \in ker\ h$.

Associativity is hereditary from G since $ker\ h \subseteq G$, and $e_G \in ker\ h$ by Problem 5c of Section 6.1 Finally, we show that if $x \in ker\ h$, then $\bar{x} \in ker\ h$, where $\bar{x}$ is the $*$-inverse of x. First,

$$h(x) \circ h(\bar{x}) = h(x * \bar{x}) \quad\quad \text{morphism property}$$

$$= h(e_G) \quad\quad\quad \bar{x} \text{ is } *\text{-inverse of } x$$

$$= e_H$$

Next, we know $h(x) = e_H$ since $x \in ker\ h$. Therefore,

$$e_H \circ h(\bar{x}) = e_H$$

which says

$$h(\bar{x}) = e_H$$

$\therefore$ $\quad\quad \bar{x} \in ker\ h$

We have thus proved that $ker\ h$ is a group with respect to the operation of G.

4. (a) h is the projection to the xy-plane.

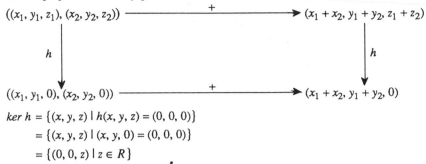

$$ker\ h = \{(x, y, z) \mid h(x, y, z) = (0, 0, 0)\}$$
$$= \{(x, y, z) \mid (x, y, 0) = (0, 0, 0)\}$$
$$= \{(0, 0, z) \mid z \in R\}$$

This is the z-axis. That is, the set of points in R^3 that project to the origin under the xy-plane projection is the set of points on the z-axis.

5. $(V_3, \times)$ is a groupoid.

Problems 6.4, p. 350.

1. Examine Example 4. The set $\{e, a, b, a * b, b * a\}$ has 5 distinct members. Show that either $a * a$ or $a * b * a$ is distinct from each of these 5 by taking cases on whether $a * a \neq e$ or $a * a = e$.

2 (a) Compute $(a * b) * (\bar{b} * \bar{a})$ using various group properties. Remember that, in a group, inverse elements are unique.

3. An abstract example would be $S = \{e, a, b, c, d\}$, with operation $*$ defined by the table

$*$	e	a	b	c	d
e	e	a	b	c	d
a	a	b	c	d	e
b	b	e	d	a	c
c	c	d	a	e	b
d	d	c	e	b	a

The element b is characterized by $(b * b) * b \neq b * (b * b)$.

6. In $(I, +)$, the property in Example 5 translates to:

$$\forall_{a, b \in I} \quad \underbrace{(a + b) + (a + b) + \ldots + (a + b)}_{n\ \text{times}} = \underbrace{(a + a + \ldots + a)}_{n\ \text{times}} + \underbrace{(b + b + \ldots + b)}_{n\ \text{times}}$$

or simply, $n(a + b) = na + nb$,   for every $n \in I^+$

7. This is straightforward from the associative, binary-operation, and commutative properties.

9. If $a^2 = e$ for every $a \in G$, then every element in the group is its own inverse.

10. (a) This is not a group because elements lack inverses.

(b) Make a composition table for $\{f, g, h, l\}$ and observe its characteristics.

Index

Abel, Nils Henrik, 335
Abelian group, 335
Absolute (universal) complement, 114
Abstract systems, 331, 332, 346, 347
Added premise, 55, 56
Addition
  of complex numbers, 324, 327
  of consecutive natural numbers, 52, 85
  of functions, 298, 323
  of matrices, 328
  of natural numbers, 27
  of polynomials, 329
  of residue classes, 296
Additive identity element, 33, 64, 317
Additive inverse, 33, 320, 322
Adjacent vertices in graphs, 212
Adjunction, 47
Aleph-null, 132, 135, 136
Algebraic systems, 311
  Abelian groups as, 335
  by adjunction to rationals, 327
  classical number, 112, 323, 335
  collections of sets as, 325
  of complex numbers, 304, 321, 322, 324, 326, 327
  dihedral, 326, 335
  direct sum rings as, 330
  finite abstract, 331, 332, 335, 340, 346
  fourth roots of unity as, 326, 335
  of functions, 325
  Gaussian integers as, 327, 335
  groupoids, 340, 341

groups, 334
  of integral multiples, 325, 335
  loops as, 339, 340
  of matrices, 328
  modular arithmetics as, 324
  monoids as, 336, 337
  of numbers, 323–325, 330, 335
  with operation defined by table, 316, 319, 331, 332, 335, 340, 346
  of ordered *n*-tuples, 328, 335
  of permutations, 276, 325, 335
  of polynomials, 329, 335
  of quaternions, 333
  of residue classes, 324, 335
  semigroups as, 337, 338
  with vector cross-product operation, 332
  of vectors, 328, 335
Algebra of sets, 147
Algorithm, 102
Analogy, 114
  between graphs and binary relations, 213
  between order and divisibility, 234
  between sets, sentences, and arithmetic, 116
Antecedent, 3, 24
Antisymmetric relation, 172, 213, 222, 223
Argument, 45
Associative binary operation, 312
Associative property
  of binary operation, 312
  of conjunction, 19
  of disjunction, 19

of set intersection, 147
of set union, 147
Assumption, 44
Atomic sentence, 2
Automorphism, 302, 306, 307, 350
Axiom
  of choice, 128, 135
  of extent, 107
  of specification, 108
  of Zermelo, 128
Axioms of Inequality, 225

Biconditional, 3
  replacement, 47, 48
  truth table for, 13
  truth value of, 13
Bijection, 261
Binary operation, 114, 115, 289, 311
  addition as, 291
  multiplication as, 291
  See also Operation
Binary relation, 163–210, 213–215
  antisymmetric, 172, 213
  congruence, 178, 186, 190, 194
  directed graph of, 213
  divisibility, 200, 206, 207, 233
  domain of, 168
  equivalence, 174, 214, 215
  examples of, 166–171, 213–215
  from A to B, 168, 241
  graph of, 169, 213-215
  greatest integer, 179, 195, 196
  identity, 170
  inequality, 200, 225, 226
  inverse of, 168, 214
  linear order, 202